AF356153

Gem Characterisation

Gem Characterisation

Editors

Stefanos Karampelas
Emmanuel Fritsch

Basel • Beijing • Wuhan • Barcelona • Belgrade • Novi Sad • Cluj • Manchester

Editors
Stefanos Karampelas
Laboratoire Français de
Gemmologie (LFG)
Paris
France

Emmanuel Fritsch
Institut des Matériaux de
Nantes Jean Rouxel (I.M.N.),
Centre National de la
Recherche Scientifique
Nantes Université
Nantes
France

Editorial Office
MDPI
St. Alban-Anlage 66
4052 Basel, Switzerland

This is a reprint of articles from the Special Issue published online in the open access journal *Minerals* (ISSN 2075-163X) (available at: https://www.mdpi.com/journal/minerals/special_issues/GEMC).

For citation purposes, cite each article independently as indicated on the article page online and as indicated below:

Lastname, A.A.; Lastname, B.B. Article Title. *Journal Name* **Year**, *Volume Number*, Page Range.

ISBN 978-3-7258-1333-9 (Hbk)
ISBN 978-3-7258-1334-6 (PDF)
doi.org/10.3390/books978-3-7258-1334-6

Cover image courtesy of Ugo Hennebois, LFG

Contents

About the Editors

Stefanos Karampelas

Dr. Stefanos Karampelas is the Chief Gemmologist at the Laboratoire Français de Gemmologie (LFG), Paris, and also teaches for the Advanced Gemmology Diploma at the University of Nantes, France. He holds a BSc in Geology from Aristotle University of Thessaloniki (Greece), an MSc in Geosciences and an Advanced Gemmology Diploma, both from the University of Nantes (France), and he completed his PhD in Materials Physics at the University of Nantes (France) and in Mineralogy at Aristotle University of Thessaloniki (Greece). For the last 15 years, he has worked as a researcher in different gemmological laboratories around the world. His research interests include advanced non-destructive techniques which he applies to all types of gem materials, and he has authored over 100 articles and conference abstracts, two book chapters, and a book. He is the Secretary of the Gem Materials Commission of the International Mineralogical Association (IMA), a member of the GeoRaman International Advisory Committee (GRISAC), a delegate to the International Gemmological Conference (IGC), a member of the Editorial Board of *Gems and Gemology*, and an Associate Editor of *the Journal of Gemmology and of Minerals*.

Emmanuel Fritsch

Dr. Emmanuel Fritsch is a professor of physics at the University of Nantes in France. He holds a geological engineering degree from the ENSG, Nancy, France, and a Ph.D. from the Sorbonne in Paris. He worked for nearly ten years at the Gemological Institute of America (GIA) in the USA, where he became a Graduate Gemologist (GG), and was promoted to the manager of GIA Research from 1992 to 1995. His research is currently conducted at the Institut des Matériaux Jean Rouxel (IMN-CNRS) in Nantes, France. Dr. Fritsch's interests span a wide scope, from laboratory techniques applied to gemology, to color and luminescence in gems (especially diamonds), geographical origin, treated and synthetic gems, opals, and pearls. His self-assigned mission is to make gemology more scientific. He has authored over 350 articles, 18 book chapters, a book on the gems from Mogok in 2018, and delivered over 350 conferences all over the world, both scientific and popular. He is responsible for the DUG (Diplôme d'Université de Gemmologie) diploma. He has received many awards, including the 2013 Bonnano Award for Excellence in Gemology, and is one of the three honorary fellows of the Gemmological Association of Great Britain.

Preface

The purpose of this Special Issue is to present the advances in gem characterisation using principally non-destructive means. The proper identification of gems is an ever-increasing challenge, especially with harder-to-detect treatments and new demands in terms of geographical origin determination. To grow, gemmology requires valuable inputs from other fields of science, as well as more fundamental studies concerning gems.

This reprint includes 14 articles, published by around 50 different researchers from about 20 different institutions situated in different continents. This further confirms that a growing number of scientists are working on various aspects of gemmology around the globe. Hopefully, this series of articles will contribute to a better understanding of the characteristics of some gems.

Stefanos Karampelas and Emmanuel Fritsch
Editors

Editorial

Editorial for the Special Issue "Gem Characterisation"

Stefanos Karampelas [1,*] **and Emmanuel Fritsch** [2]

1   Laboratoire Français de Gemmologie (LFG), 30 Rue de la Victoire, 75009 Paris, France
2   IMN, Institut des Matériaux de Nantes Jean Rouxel, Centre National de la Recherche Scientifique, Nantes Université, 44000 Nantes, France; emmanuel.fritsch@cnrs-imn.fr
*   Correspondence: s.karampelas@lfg.paris

Citation: Karampelas, S.; Fritsch, E. Editorial for the Special Issue "Gem Characterisation". *Minerals* **2024**, *14*, 350. https://doi.org/10.3390/min14040350

Received: 5 March 2024
Revised: 16 March 2024
Accepted: 26 March 2024
Published: 28 March 2024

Gem characterisation is an ever-increasing challenge, especially with hard-to-detect treatments and new demands regarding origin determination. For this purpose, only non- (or micro-) destructive methods can be used [1–3]. This Special Issue includes 14 articles published by around 50 different researchers from about 20 different institutions situated in different continents.

Luminescence has largely been used for the characterisation of minerals and gems during the last decade [4–6]. In the present issue, a review on fluorescence and phosphorescence spectroscopies and their application in gem characterisation is presented by Zhang and Shen [7]. This article briefly summarises luminescence spectroscopy and illustrates the experiments that the authors performed on diamonds, fluorite, jadeite jade, hauyne and amber. Vigier et al. [8] review blue shortwave-excited luminescence (BSL) in natural minerals as well as synthetic materials. These authors also describe the BSL of several minerals and gems such as beryl (morganite), dumortierite, hydrozincite, pezzotaite, tourmaline (elbaite), some silicates glasses, and synthetic opals. They conclude that the BSL of these minerals is caused by titanate groups (TiO_6). A luminescence study using a 405 nm laser on Cr-containing samples such as alexandrite and spinel is published in an article authored by Xu et al. [9]. It is demonstrated that the photoluminescence lifetime displayed notable differences between natural, heated, and synthetic versions of these materials.

Treatments to improve the colour, appearance and/or durability of gems have been used for several years; over recent decades, these have increased and become more sophisticated [10–13]. In the present issue, four studies on the treatment of gems are published [14–17]. Zhou et al. [14] present the results of the method they developed for the high-temperature copper diffusion process for the surface recolouring of faceted labradorites. In parallel, they describe their gemmological and spectroscopic characteristics. The heat treatment of pink zoisite is presented by Schwarzinger [15]. This paper describes the heat treatment of zoisites under pure oxygen which allowed the manganese to remain oxidized, while the brownish yellow colour component was still successfully removed. Detection of this treatment is not easy as the temperature is relatively low and induces little change in internal features, but a combination of UV-Vis-NIR spectroscopy and trace element chemistry provided by LA-ICP-MS might give evidence of such treatment. Two papers on the detection of the low-temperature heat treatment (i.e., below 1200 °C) of corundum by studying inclusions are also published in this Special Issue [16,17]. Krzemnicki et al. [16] present a study on rubies and sapphires containing diaspore and goethite inclusions. Based on their experiments and in agreement with the literature, the dehydration of diaspore in corundum occurs between 525 and 550 °C, whereas goethite transforms to hematite between 300 and 325 °C. As both diaspore and goethite might be present as inclusions in gem-quality corundum, these dehydration reactions and phase transformations can be considered important criteria to separate unheated from heated stones. Karampelas et al. [17] focus on the analysis of zircon inclusions found in pink to purple sapphires from Ilakaka (Madagascar). It is found that by using Raman spectroscopy, the full width at half maximum (FWHM) of the main Raman band of zircon inclusions in the unheated samples from Ilakaka (Madagascar) is slightly larger than in heated samples.

The origin determination of gems using non-destructive methods is of increasing importance in gemmology [18–21]. Three studies on emeralds from different mining areas are published in this Special Issue [22–24]. Chen et al. [22] present results on emeralds from the Panjshir Valley in Afghanistan. In terms of inclusions, these might be confused with those from Colombia, some from China as well as from a small mine in Zambia (Musakashi). It is suggested that these emeralds might be separated by using chemical plots. A study on rough emerald single crystals and rough emeralds in the host rock from the ruins of Alexandria and from Mount Zabargad in Egypt, preserved in the collection of the museum of the Ecole des Mines (Mines Paris—PSL) since the late 19th or early 20th century, is presented by Nikopoulou et al. [23]. Tube-like inclusions as well as mineral inclusions of quartz, calcite, dolomite, albite and phlogopite, among others, are observed. Moreover, high concentrations of alkali elements but low amounts of caesium (below 200 ppm) are measured, and it is further confirmed that iron together with chromium contribute to their colour. Gao et al. [24] publish an update regarding emeralds from Kagem mine in Zambia. This is nowadays one of the most important mines in terms of economic value and market share in the world. The most common inclusions in Kagem emeralds are two-phase inclusions of prism shape as well as mineral inclusions which typically include actinolite, graphite, magnetite, and dolomite. These emeralds present a high amount of alkali elements as well as a relatively high amount of caesium (average amount above 500 ppm). This article adds information to the extensive studies already published on samples from the same area [25,26]. In another article relevant to the origin determination of gems, the results on purple-violet spinels from Tanzania and Myanmar are presented [27]. The samples found in these two countries differ in terms of inclusions as well as some spectroscopic characteristics.

Quartz and other silica minerals are some of the most abundant minerals in the Earth's crust [28,29]. These minerals have been used as gems since the antiquity. In this Special Issue, three studies on quartz and chalcedony are published [30–32]. Caucia et al. [30] present the characteristics of grey to black quartz from the Burano Formation situated in the province of Reggio Emilia in Italy. They demonstrate, among other things, that the colour of these samples is linked to inclusions of disordered graphite. The characteristics and the possible geological origin of agates from Mesoproterozoic volcanics situated in Pasha-Ladoga Basin (NW Russia) are presented in one of the published articles in this Special Issue [31]. It is revealed that two different phases of agate formation took place, which were most likely controlled by two different fluids and/or their mixture. Finally, Svetova et al. [32] present a study of amethyst-bearing agate from the Tevunskoye deposit situated in the Northern Kamchatka (Russia). Agate mineralization is found to occur in lavas and tuffs, as amygdales, geodes, lenses and veins present similarities with those from Ijevan (Armenia) and Ametista do Sul (Brazil). Amethysts in these agates formed at relatively low temperatures (100 °C) from low-salinity fluids under an oxidizing environment.

Overall, this present Special Issue confirms that a growing number of scientists are working on various aspects of gemmology around the globe. Hopefully, this series of articles will contribute to a better understanding of the characteristics of some gems.

Acknowledgments: The Guest Editors would like to sincerely thank all authors, reviewers, and the editorial staff of *Minerals* for their excellent efforts to successfully publish this Special Issue.

Conflicts of Interest: The authors declare no conflicts of interest.

References

1. Gramaccioli, C.M. Application of mineralogical techniques to gemmology. *Eur. J. Mineral.* **1991**, *3*, 703–706. [CrossRef]
2. Fritsch, E.; Rondeau, B. Gemology: The developing science of gems. *Elements* **2009**, *5*, 147–152. [CrossRef]
3. Devouard, B.; Notari, F. Gemstones: From the naked eye to laboratory techniques. *Elements* **2009**, *5*, 163–168. [CrossRef]
4. Waychunas, G.A. Luminescence Spectroscopy. *Rev. Mineral. Geochem.* **2014**, *78*, 175–217. [CrossRef]

5. Gaft, M.; Reisfeld, R.; Panczer, G. *Modern Luminescence Spectroscopy of Minerals and Materials*, 2nd ed.; Springer: Cham, Switzerland, 2015.
6. Hainscwang, T.; Karampelas, S.; Fritsch, E.; Notari, F. Luminescence spectroscopy and microscopy applied to study gem materials: A case study of C centre containing diamonds. *Mineral. Petrol.* **2013**, *107*, 393–413. [CrossRef]
7. Zhang, Z.; Shen, A. Fluorescence and Phosphorescence Spectroscopies and Their Applications in Gem Characterization. *Minerals* **2023**, *13*, 626. [CrossRef]
8. Vigier, M.; Fritsch, E.; Cavignac, T.; Latouche, C.; Jobic, S. Shortwave UV Blue Luminescence of Some Minerals and Gems Due to Titanate Groups. *Minerals* **2023**, *13*, 104. [CrossRef]
9. Xu, W.; Tsai, T.-H.; Palke, A. Study of 405 nm Laser-Induced Time-Resolved Photoluminescence Spectroscopy on Spinel and Alexandrite. *Minerals* **2023**, *13*, 419. [CrossRef]
10. Kammerling, K.C.; Koivula, J.I.; Kane, R.E. Gemstone enhancement and its detection in the 1980s. *Gems Gemol.* **1990**, *26*, 32–49. [CrossRef]
11. McClure, S.F.; Smith, C.P. Gemstone enhancement and detection in the 1990s. *Gems Gemol.* **2000**, *36*, 336–359. [CrossRef]
12. McClure, S.F.; Kane, R.E.; Sturman, N. Gemstone enhancement and its detection in the 2000s. *Gems Gemol.* **2010**, *36*, 218–240. [CrossRef]
13. Shigley, J.E.; McClure, S.F. Laboratory-treated gemstones. *Elements* **2009**, *5*, 175–178. [CrossRef]
14. Zhou, Q.; Wang, C.; Shen, A.-H. Application of High-Temperature Copper Diffusion in Surface Recoloring of Faceted Labradorites. *Minerals* **2022**, *12*, 920. [CrossRef]
15. Schwarzinger, C. The Heat Treatment of Pink Zoisite. *Minerals* **2022**, *12*, 1472. [CrossRef]
16. Krzemnicki, M.S.; Lefèvre, P.; Zhou, W.; Braun, J.; Spiekermann, G. Dehydration of Diaspore and Goethite during Low-Temperature Heating as Criterion to Separate Unheated from Heated Rubies and Sapphires. *Minerals* **2023**, *13*, 1557. [CrossRef]
17. Karampelas, S.; Hennebois, U.; Mevellec, J.-Y.; Pardieu, V.; Delaunay, A.; Fritsch, E. Pink to Purple Sapphires from Ilakaka, Madagascar: Insights to Separate Unheated from Heated Samples. *Minerals* **2023**, *13*, 704. [CrossRef]
18. Hänni, H.A. Origin determination for gemstones: Possibilities, restrictions and reliability. *J. Gemmol.* **1994**, *24*, 139–148. [CrossRef]
19. Rossman, G.R. The geochemistry of gems and its relevance to gemology: Different traces, different prices. *Elements* **2009**, *5*, 159–162. [CrossRef]
20. Giuliani, G.; Groat, L.; Ohnenstetter, D.; Fallick, A.E.; Feneyrol, J. The geology of gems and their geographic origin. In *Geology of Gem Deposits*; Raeside, E.R., Ed.; Mineralogical Association of Canada: Tucson, AZ, USA, 2014; Volume 44, pp. 113–134.
21. Groat, L.A.; Giuliani, G.; Stone-Sundberg, J.; Sun, Z.; Renfro, N.D.; Palke, A. A review of analytical methods used in geographic origin determination of gemstones. *Gems Gemol.* **2019**, *55*, 512–535. [CrossRef]
22. Chen, Q.; Bao, P.; Li, Y.; Shen, A.H.; Gao, R.; Bai, Y.; Gong, X.; Liu, X. A Research of Emeralds from Panjshir Valley, Afghanistan. *Minerals* **2023**, *13*, 63. [CrossRef]
23. Nikopoulou, M.; Karampelas, S.; Gaillou, E.; Hennebois, U.; Maouche, F.; Herreweghe, A.; Papadopoulou, L.; Melfos, V.; Kantiranis, N.; Nectoux, D.; et al. Non-Destructive Study of Egyptian Emeralds Preserved in the Collection of the Museum of the Ecole des Mines. *Minerals* **2023**, *13*, 158. [CrossRef]
24. Gao, R.; Chen, Q.; Li, Y.; Huang, H. Update on Emeralds from Kagem Mine, Kafubu Area, Zambia. *Minerals* **2023**, *13*, 1260. [CrossRef]
25. Zwaan, J.C.; Seifert, A.; Vrána, S.; Laurs, B.; Anckar, B.; Simmons, W.; Falster, A.; Lustenhouwer, W.; Koivula, J.; Garcia-Guillerminet, H. Emeralds from the Kafubu Area, Zambia. *Gems Gemol.* **2005**, *41*, 116–148. [CrossRef]
26. Seifert, A.; Žáček, V.; Vrána, S.; Pecina, V.; Zacharias, J.; Zwaan, J.C. Emerald mineralization in the Kafubu area, Zambia. *Bull. Geosci.* **2004**, *79*, 1–40.
27. Wu, J.; Sun, X.; Ma, H.; Ning, P.; Tang, N.; Ding, T.; Li, H.; Zhang, T.; Ma, Y. Purple-Violet Gem Spinel from Tanzania and Myanmar: Inclusion, Spectroscopy, Chemistry, and Color. *Minerals* **2023**, *13*, 226. [CrossRef]
28. Götze, J. Chemistry, textures and physical properties of quartz—Geological interpretation and technical applications. *Mineral. Mag.* **2009**, *73*, 645–671. [CrossRef]
29. Götze, J. Editorial for Special Issue "Mineralogy of Quartz and Silica Minerals". *Minerals* **2018**, *8*, 467. [CrossRef]
30. Caucia, F.; Scacchetti, M.; Marinoni, L.; Gilio, M. Black Quartz from the Burano Formation (Val Secchia, Italy): An Unusual Gem. *Minerals* **2022**, *12*, 1449. [CrossRef]
31. Svetova, E.N.; Svetov, S.A. Agates from Mesoproterozoic Volcanics (Pasha–Ladoga Basin, NW Russia): Characteristics and Proposed Origin. *Minerals* **2023**, *13*, 62. [CrossRef]
32. Svetova, E.N.; Palyanova, G.A.; Borovikov, A.A.; Posokhov, V.F.; Moroz, T.N. Mineralogy of Agates with Amethyst from the Tevinskoye Deposit (Northern Kamchatka, Russia). *Minerals* **2023**, *13*, 1051. [CrossRef]

 *minerals*

Review

Fluorescence and Phosphorescence Spectroscopies and Their Applications in Gem Characterization

Zhiqing Zhang [1] and Andy Shen [2,*]

[1]   Hubei Land Resources Vocational College, Wuhan 430090, China; zhangzq2249@163.com
[2]   Gemmological Institute, China University of Geosciences, Wuhan 430074, China
*   Correspondence: shenxt@cug.edu.cn

Abstract: Fluorescence and phosphorescence are listed as mineral optical–physical properties in classical gemology textbooks. The trace elements which exist in gems, certain defects in the crystal lattice, and some luminous molecules contribute to luminescence phenomena in gem materials, including fluorescence and phosphorescence. A systematic luminescence study using an excitation-emission matrix (EEM) not only provides detailed information about the emission and excitation peaks, but also indicates the presence of specific trace elements, lattice defects, or luminous substances in gem materials. This provides reliable evidence for the characterization of gems. In this review paper, we briefly summarize luminescence spectroscopy and illustrate its applications in gem materials in our laboratory, including diamonds, fluorite, jadeite jade, hauyne, and amber. Meanwhile, this project is in process and needs more samples from reliable sources to confirm the described data.

Keywords: gemstones; luminescence; fluorescence spectroscopy; application

1. Introduction

Luminescence is the ability to emit light from any substance. In minerals and organic materials, it has been extensively studied for centuries [1–5]. Based on the nature of the excitation sources, luminescence can be grouped into several categories: photoluminescence (photons from UV-Vis-NIR sources, X-ray, and gamma rays; PL) and cathodoluminescence (electrons; CL), etc. As for the nature of the excited state, luminescence is subdivided to the fluorescence and phosphorescence (or persistent luminescence). The fluorescence results from the excited singlet states with spin-allowed transition, lasting near 10 ns following cessation of the excitation light. While the phosphorescence is emission of light from triplet excited states with spin-forbidden transition, it can last continuous photon emission milliseconds to seconds and even minutes, [2,6]. For example, the rare earth elements, Eu^{2+} and Dy^{3+}, cause the yellow-green color persistent luminescence up to 2000 min in mineral $SrAl_2O_4$ [7,8]. Luminescence varies in different substances. The most versatile continuous-light Xe lamps or pulsed lasers could excite some materials to emit not only longer-wavelength photons, but also anti-stokes PL at a shorter wavelength, albeit weaker (and of barely any interest in gem testing). Additionally, the tenebrescent minerals, such as sodalite, known for their reversible photochromism, could change color when exposed to sunlight, but have their color restored by heating [9]. This effect can be repeated indefinitely. Owing to the diagnostic reactions to light in substance, PL has become an essential physico-chemical property in mineralogy, gemology, biology, and many other research fields [1,2,10–12], and it especially becomes a useful tool for characterizing trace elements and imperfect crystal lattices or structural defects in minerals [13–15].

Gemology is the study of all materials that are used as gems. Luminescence analysis is useful in this field [13,16]. Typically, the ultraviolet (UV) lamps with 365 nm (longwave; LW) and 254 nm (shortwave; SW) light are used to excite gemstones' fluorescence and phosphorescence. A rapid-development excitation-emission matrix (EEM) was developed

Citation: Zhang, Z.; Shen, A. Fluorescence and Phosphorescence Spectroscopies and Their Applications in Gem Characterization. *Minerals* **2023**, *13*, 626. https://doi.org/10.3390/min13050626

Academic Editors: Stefanos Karampelas and Emmanuel Fritsch

Received: 26 February 2023
Revised: 25 April 2023
Accepted: 27 April 2023
Published: 29 April 2023

in a modern laboratory to characterize the emission and excitation peaks. These are closely related to the specific luminescence centers and fluorescence and phosphorescence behavior of the materials [2]. For instance, in benitoite, Ti-doped synthetic sapphire, and spinel, their blue shortwave-excited luminescence is caused by titanate groups (TiO_6) [12]. For natural and synthetic scapolites, $(S_2)^-$ activators trapped in [Na_4] square cages produce the yellow-orange luminescence [17]. Phosphorescence spectra and fluorescence EEMs are usually recorded as the measured luminescence intensity versus emission and excitation wavelengths in 2D or 3D plots [1]. Relevant gemstone tests require non- or micro-destruction, visualization, and repeatability, which finally lead to the introduction of a front-surface geometry model for EEM detection of gems [18].

With the EEMs analysis gradually performed in gemology [12,17], the aim of this article is to briefly review the results of our group research on several gems, including HPHT synthetic diamonds, hauyne, fluorite, jadeite jade, and amber. Similarly, with an instrument equipped with the xenon lamp, we mainly reviewed their luminescence spectra, the EEMs patterns, and the currently available applications, such as an understanding of the optical behavior of gem crystals, distinguish natural gems from their synthetic and treated counterparts, and allow traceability of the geographic origins of natural gemstones.

2. Mechanisms and Instruments

2.1. Mechanisms

Ligand field, crystal field, molecular orbital theories, and the role of the defects (intrinsic, extrinsic, and structural defects, as well as defects derived from chemical reactions) are the primary methods used to reveal the luminescence behavior of luminescent materials. Waychunas et al. [1], Gaft et al. [2], and Lakowicz [6] have systematically reviewed these aspects. In the classic Jablonski diagram (Figure 1), absorption is the basis for the emission process, ensuring that the material gains photon energy at higher-energy electronic singlet levels (e.g., S1 and S2). Luminescence is the radiation path from an excited electron to its initial singlet state. The emission of light occurs from the first high-energy excited electronic singlet and triplet states (i.e., S1 and T1) returning to the lowest-energy ground states (S0). It formally contains two categories, fluorescence and phosphorescence, according to the nature of the excited state: singlet or triplet, respectively. Thus, the emission energy tends to be lower than the excitation energy. Mirror symmetry between absorption and fluorescence curves has also been investigated for many compounds [3].

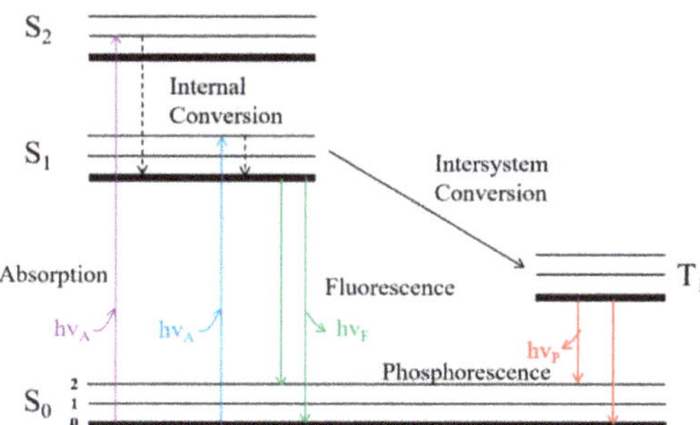

Figure 1. One form of a Jablonski diagram. Vibration levels 0, 1, and 2 show the different energies in the lowest-energy ground states (S0). Adapted from [6].

Most gems consist of inorganic minerals; particularly, valent ions such as rare earth elements, transition metal elements, and electron-withdrawing elements, which usually occupy specific crystal sites and combine with surrounding ions to form a stable ligand field or some unexpected chemical structural defects. In these situations, abundant electrons keep moving in their orbitals once excited by appropriate energies for their transition to other higher-energy orbitals, and then undergo internal and intersystem conversions and luminescence to their initial states. These processes are associated with the crystal lattice and certain elements and determine the detailed and distinct "fingerprint" luminescence information for a gemstone [15,19]. Gaft et al. [2] summed up three essential conditions for a mineral to be luminescent: (1) a suitable type of crystal lattice favorable to form emission centers; (2) sufficient content of luminescence center; and (3) a small number of quenchers. They found that, in common, the transition metal ions and REE^{3+} are related to mineral luminescence, such as the Cr^{3+} in corundum showing a *d-d* electronic transition which contributes to the red fluorescence, and the Ce^{3+} substitution in two different Ca sites in apatite leading the fluorescence at 340 nm, 360 nm, and 430 nm. However, the intrinsic Fe^{3+}, Co^{2+}, and Cu^{2+}, as well as a high concentration of Fe^{3+}, Ti^{4+}, and Ce^{3+}, might restrain the luminescence.

It should be noticed that the xenon lamps have a continuous radiation which are mostly used to induce a steady-state luminescence. However, the laser excitation sources are not only for the PL spectra, but also for calculating the decay time by the time-resolved luminescence spectra, due to the monochromatic radiation, pulsed, or modulated [2,6]. Moreover, fluorescence spectra at low temperatures show clearer details, owing to the weakness of the molecular movement. Zaitsev et al. performed the laser excitation at wavelengths 324.8, 354.8, 457.0, 473, 532, 633 and 830 nm at liquid nitrogen temperature, to discuss the 468 nm luminescence center as being closely related to vacancy clusters in brown CVD diamonds [19].

However, the chemical compositions of organic gems without crystal structures, such as amber, are more complex than those of minerals. Thus, the molecular orbital theory is more suitable for organic luminescent materials. Electronic transitions in organic molecular structures that emit photons usually involve π-bonding molecular orbitals, and thus are generally associated with aromatic structures and molecules with multiple C=C double bonds [1,3].

2.2. Instruments

Fluorescence detection is suitable for gem characterization because of its high sensitivity, rapid analysis, and lack of sample preparation. Hence, luminescence studies of gems have increased dramatically over the past few years, especially for diamonds excited by different laser wavelengths [13–15,19,20]. Classical fluorescence spectroscopy is conducted for liquids with the required measurement geometry for the excitation beam and sample, with the emission beam set to $90°$ [6]. The emitted light travels through the 1 cm^2 cuvette holding the sample before reaching the detector. The measured emission is the accumulated luminescence along the entire optical path, which is significantly affected by self-absorption and secondary fluorescence [21]. Fortunately, for gem species, these can be substantially removed through front-surface geometry determination [21–23].

Fluorescence properties consist of: (i) an emission spectrum, which records detailed information about the emission under an excitation wavelength; (ii) an excitation spectrum, showing the excitation efficiency of a particular emission wavelength; (iii) a time-resolved fluorescence spectrum, characterizing the emission intensity varying with time; and (iv) luminescence centers, related to the trace elements in minerals or lattice defects in crystals and electronic transitions in organic molecular structures [1]. The main parameters include emission peaks (λ_{em}), excitation peaks (λ_{ex}), lifetime (τ), and quantum yield (Q) [2,24]. Due to the fact that, together, they represent the fluorescence intensity, emission, and excitation wavelengths in a contour or 3D plot, EEM illustrates a complete image of the luminescence pattern of the sample. The advantage of this approach is that it directly provides the λ_{em} and λ_{ex} and decreases the load of detailed processing of statistical and chemometric analysis of raw spectra. Alternatively, the EEM plot can be sliced to provide emission or excitation spectra at a particular excitation or emission wavelength.

Therefore, in our experiments, we used a spectrofluorometer (FP-8500, JASCO, Tokyo, Japan) equipped with a 150 W xenon lamp, double monochromators, and a specially designed accessory (EFA-833). This accessory provides a preferred geometry for gem luminescence studies: a front-surface mode with an incident angle (between the excitation beam and the normal of the sample surface) of approximately 45°, schematically shown in Figure 2, and a special sample chamber for rough gems and mounted ones without any surface cover. This ensures that no sample pretreatment is required, which makes it a truly nondestructive test. In addition to transparent gems, luminescence from opaque minerals can be detected. At the same time, the sample platform was a special quartz window plate without emissions, which made it suitable for all powder and solid samples, with the only limitation being gems' size, of about 5 cm (length) × 10 cm (width) × 5 cm (height). In our laboratory, the collected EEMs were composed of a series of emission spectra (220–750 nm) irradiated at excitation wavelengths of 200–700 nm. Different excitation and emission bandwidths, scan speeds, the sensitivity level of the photomultiplier detector, response time, and the number of accumulations were set according to the specific gem species, for example, in amber and diamonds [18,25].

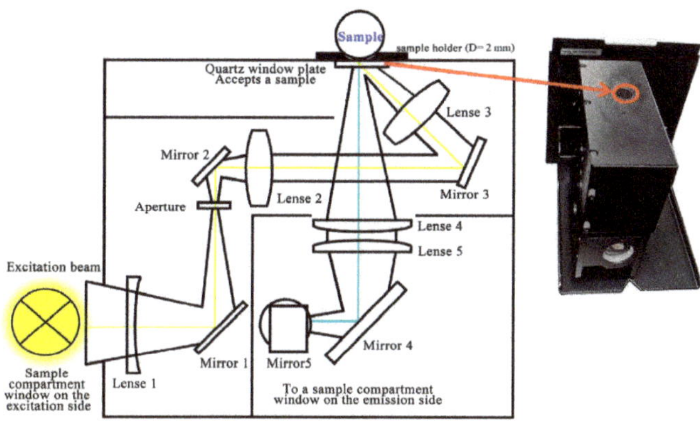

Figure 2. Schematic of the EFA-833 accessory attached in the spectrofluorometer used in this study (redrawn by Z. ZHANG, from the instrument handbook) [18].

In addition to the fluorescence EEM, this machine can record phosphorescence spectra and steady-state time-resolved fluorescence data. The latter provides raw data for calculating the decay time (or lifetime) of phosphorescence. It is suitable for decay times on the order of tens of microseconds. Therefore, the applications of time-resolved fluorescence data analysis are ongoing. We focused on analyzing the phosphorescence spectra and EEMs of fluorescence in various gems.

3. EEM Applications in Gem Characterization

3.1. Investigating the Trace Elements and Lattice Defects in Gem Materials

Until now, we found that the complete EEM fluorescence patterns and phosphorescence spectra and luminescence time-resolved data indeed extend our understanding on optical properties in gem materials, and further our investigation of the trace elements and lattice defects in diamond, fluorite and hauyne.

3.1.1. HPHT Synthetic Diamonds

Owing to the limitations of natural diamond reserves and their increasing industrial demand, millions of synthetic or lab-grown diamonds are produced using chemical vapor deposition (CVD) or high-pressure-high-temperature (HPHT) techniques. Synthetic diamonds share virtually all the chemical, optical, and physical characteristics of their natural counterparts. An ideal diamond only consists of carbon, but at least minute nitrogen-related impurities are unavoidably found in diamonds. These impurities lead to colored diamonds and cause different defect centers in diamond crystal lattices, contributing to varying luminescence [15,19,20]. For centuries, exploration and explanation of the luminescence phenomenon that occurs in diamonds have attracted the attention of mineralogists and gemologists rather physicists [26].

Luo et al. used a 365 nm UV-LED source to measure 25 pieces of diamonds (including 19 natural ones and 6 treated ones) with an N3 center (due to a vacancy surrounded by three nitrogen atoms on a {111} plane in the diamond lattice), H3 center (from a vacancy trapped at an A aggregate of nitrogen, e.g., $(N-V-N)^0$) or H4 center (related to four nitrogen atoms separated by two vacancies), a 480 nm band (because of a 480 nm absorption band), and N-V center (due to a single nitrogen atom adjacent to a vacancy) for their EEM spectra [14,15]. They found that the treated diamonds showed much more $(N-V)^-$ center related fluorescence, while naturally colored diamonds tended to show a dominant $(N-V)^0$ defect emission. As for the HPHT synthetic diamonds, Shao et al. studied a batch of type IIb samples with greenish-blue phosphorescence [25]. The EEM spectrum and emission profile (Figure 3a,b) of these diamonds clearly depicted the main emission extending from 370 to 600 nm, centered around 470 nm, and a shoulder band at 503 nm after exposure to shortwave UV light at 215–240 nm. Simultaneously, the FTIR spectra showed the existence of uncompensated boron, and Electron Paramagnetic Resonance (EPR) signals consistent with neutral isolated nitrogen centers were observed in these synthetic samples. Referring to previous work on the mechanism of the 500 nm phosphorescence band, Shao et al. proposed the donor-acceptor pair (DAP) recombination theory to explain the 470-nm band. In this model (Figure 3c), when the crystals are irradiated with UV light, excited electrons in the valence band are trapped by the ionized positive donor, and the holes remaining in the valence band capture the electrons from the ionized negative acceptor. This results in neutral donor and acceptor recombination associated with phosphorescence release [27]. Uncompensated acceptors can accelerate the recombination process by forming additional holes. Furthermore, based on the DAP theory and radiation energy of 2.64 eV (i.e., 470 nm), the theoretical energy of the bandgap is 5.47 eV, the binding energy of the acceptor is 0.37 eV, and that of the donor is 2.54 eV. In addition to considering lattice relaxation, isolated nitrogen impurities were proposed as the donors in these tested synthetic diamonds, and the phosphorescence at 470 nm and 503 nm could be explained by isolated nitrogen-boron pair recombination. The different distance distributions between the donor-acceptor pairs resulted in these two bands.

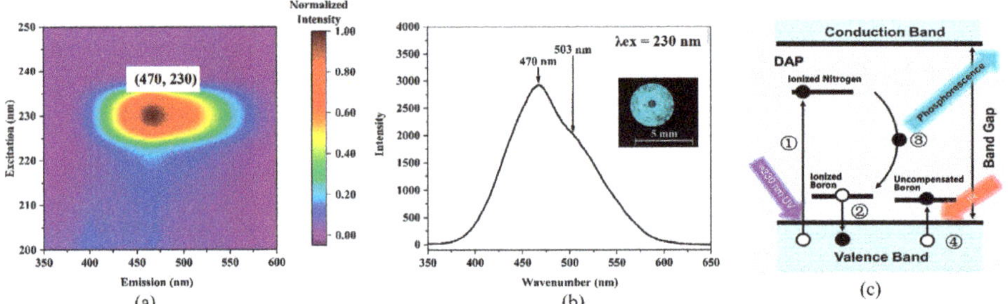

Figure 3. (**a**) Normalized EEM spectrum for type IIb HPHT synthetic diamond; (**b**) high-resolved emission curve by 230 nm excitation; (**c**) concise mechanism model of the 470 nm band in HPHT synthetic diamonds: ① When UV-light irradiated diamond, N_s^+ trapped an electron from valence band to become N_s^0; ② Then the hole left in valence band was filled with an electron from B^- to make B^0; ③ N_s^0 and B^0 recombined with phosphorescence emission. ④ Meanwhile, uncompensated boron provided extra holes in valence band which would accelerate the recombination process between N_s^0 and B^0. Adapted from Reference [25].

3.1.2. Fluorite

The Rogerley Mine in Frosterley, County Durham, England, is particularly notable for the production of many well-crystallized specimens of fluorite that exhibit strong daylight fluorescence unique to fluorite from this region [28]. Fluorite is a well-known mineral with a wide range of colors and a CaF_2 lattice. Numerous studies have recorded the luminescence from rare earth elements (REE) content [29–32] and the very common impurity in the CaF_2 crystal structure, such as Mn^{2+} placed in Ca^{2+} sites in the crystal lattice [33]; however, complete EEM fluorescence patterns are still lacking. Therefore, we had focused on the most common fluorite species, a princess-cut green fluorite from Rogerley Mine. It exhibits violet-blue fluorescence under UV irradiation at 365 nm, as shown in the inset of Figure 4.

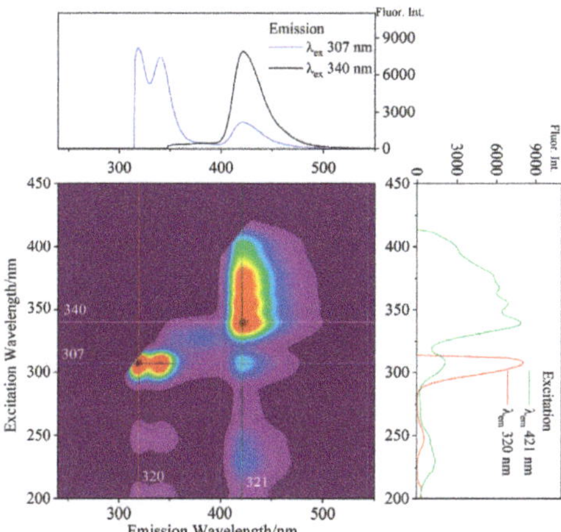

Figure 4. Detailed EEM pattern and emission and excitation curves from green fluorite with violet-blue fluorescence by LW ultraviolet light irradiation.

The EEM spectra and individual profiles (Figure 4) clearly show a mixture of emission centers caused by different ions that substitute for calcium in the crystal lattice. (i) The violet-blue part (with a center at 421 nm) of the emission spectrum could be efficiently excited by radiation at 200–260 nm and 285–415 nm, with the highest peak at 340 nm, which was accompanied by a series of low-energy excitation sub-bands. This emission corresponded to a high concentration of the rare-earth ion Eu^{2+}, and these excitation acromions could be associated with its ground multiplets [29]. (ii) The invisible ultraviolet part of the emission curve exhibited a doublet at 320 and 340 nm when excited by light at 307, 250, and 205 nm. These emission bands originate from f–d electronic transitions of Ce^{3+} [34]. Furthermore, this was supported by the excitation bands at 250 and 205 nm.

Moreover, the green color of the crystal coincided with the interstitial atom samarium, which is a very important coloring ion that substitutes for calcium in the lattice, according to previous results [35,36]. Notably, the diagnostic emissions of Sm^{2+} (708 nm) and Sm^{3+} (596 nm) (Figure 5) did not appear in our results. Nevertheless, the combination of coloration and EEM spectra could demonstrate a very useful role for the rapid characterization and detection of specific impurity elements that occupy the crystal lattice in fluorite minerals, even if chemical element analysis is not performed.

Additionally, a few details of the phosphorescence spectrum of green fluorite have been reported. Upon excitation at 308 nm, this fluorite exhibited visible phosphorescence at 382, 420, 542, 567, and 604 nm and ultraviolet emission at 320 and 340 nm, as shown in Figure 6. The lifetime of all normalized decay scatters was in the order of 604 nm $\approx$ 567 nm > 542 nm > 420 nm > 340 nm $\approx$ 320 nm, indicating the presence of four potential triplet-state luminescence centers in this crystal. However, the exact assignment of each long-lifetime emission band requires a combination of other analytical methods, such as EPR spectrometry.

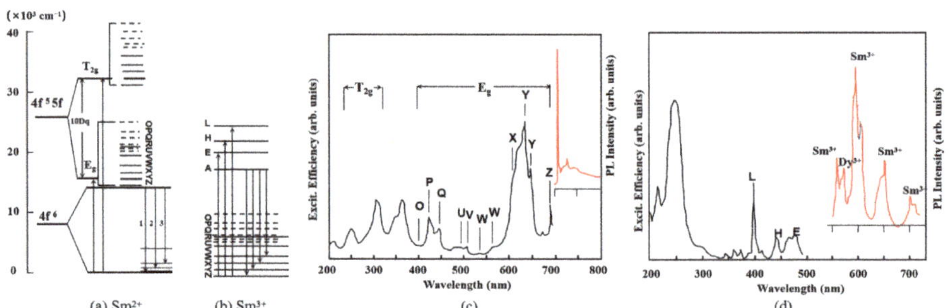

Figure 5. Schematic energy levels of (**a**) Sm^{2+} and (**b**) Sm^{3+} in the CaF_2 lattice. Arrows show the excitation and emission (arrows 1, 2, 3) transitions. (**c**) Excitation spectrum (black, λ_{em} = 730 nm), PL spectrum (red, Y-band at λ_{ex} – 634 nm) at 10 K. O–Z excitation bands correspond to the transitions in Sm^{2+} (see a). (**d**) Excitation spectrum (black, λ_{em} = 596 nm), PL spectrum (red, λ_{ex} = 401 nm) at room temperature. E, H, and L correspond to the Sm^{3+} levels (see (**b**)). Adapted from Reference [34].

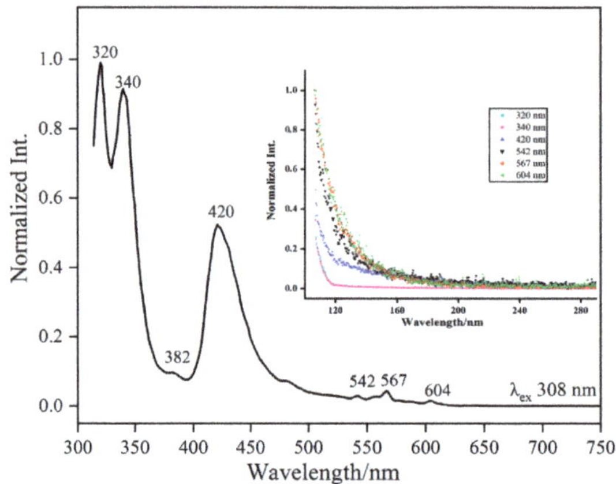

Figure 6. Normalized phosphorescence spectra by excitation at 308 nm (left) and normalized luminescence decay scatters (inset) obtained by monitoring at each phosphorescence peak by irradiation at 308 nm.

3.1.3. Hauyne

Blue hauyne $((Na,Ca)_{4-8}[AlSiO_4]_6(SO_4)_{1-2})$ is a rare gem species, and only a few studies have recorded its fluorescence spectra. Lv et al. studied 27 gem-quality huayne stones from the Eiffel District, Germany [37]. These crystals show pale grayish-blue to dark blue body color but present inert to bright orange fluorescence under 365 nm ultraviolet illumination, while all exhibit inert emission under 254 nm excitation. The absorption, 3D fluorescence, and finer emission spectra were collected using a fluorescence instrument (FP-8500, JASCO, Tokyo, Japan) and a PerkinElmer Lambda 650S UV-Vis spectrometer (Lambda 650S, PerkinElmer, PerkinElmer, MA, USA), as shown in Figures 7 and 8. The results showed two distinct emission regions: the blue region with a center at 425 nm and the yellow region with a peak at 566 nm, which was efficiently excited by 300–500 nm irradiation. This yellow-orange region emission indicates the orange fluorescence of hauyne. It was proposed that the fluorescence peak at 566 nm and its acromions at 581, 600, 621,

642, and 666 nm were related to the $(S_2)^-$ center. The absorption peak at 400 nm coincided with the most efficient excitation light at 397 nm, which supports this view.

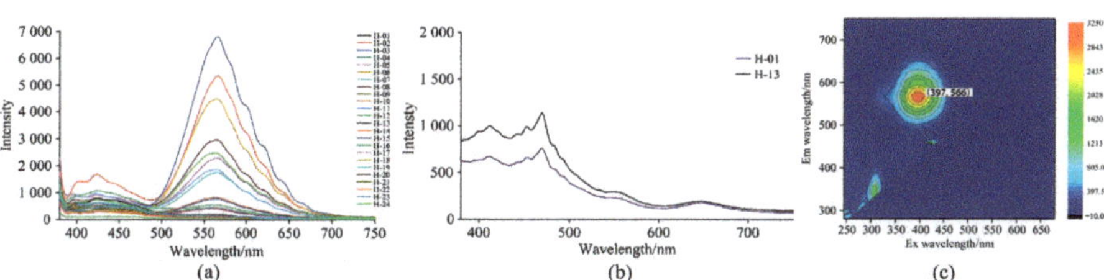

Figure 7. Fluorescence emission spectra of hauyne crystals: (**a**) λ_{ex} = 365 nm and (**b**) λ_{ex} = 254 nm. (**c**) 3D fluorescence contour of hauyne (H-03) with strong orange fluorescence. Adapted from Reference [37].

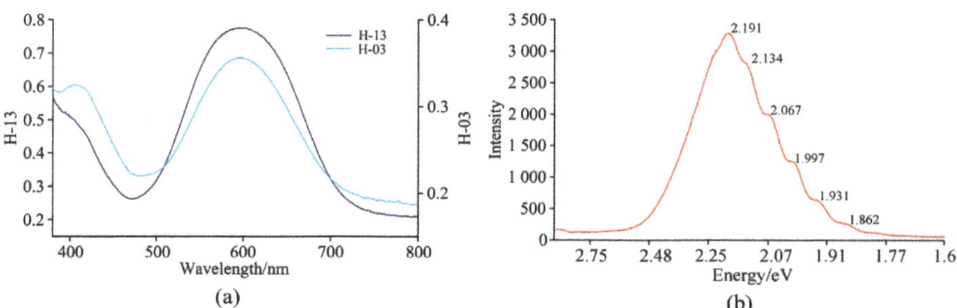

Figure 8. (**a**) Typical absorption spectra of hauyne. (**b**) Finer emission spectrum from hauyne (H-03) (λex = 397 nm). Adapted from Reference [37].

In addition, others found that the sodalite and the natural and synthetic scapolites also emit a similar pattern, mentioned in references [17,38–41]. Blemetritt et al. compared the emission spectra of the natural scapolite and the sulfur-doped synthetic scapolite, confirmed that their yellow-orange fluorescence is related to the $(S_2)^-$ in [Na$_4$] square [17]. Meanwhile, Lv et al. noticed that in the finer emission spectrum from hauyne, the fluorescence peak at 566 nm excited by 397 nm is accompanied by a series of stepped longer wavelength emissions, which have an adjacent energy differential of approximately 0.07 eV [37]. They also thought it caused from the sub-vibration level from S_2^- center excited state, as Gaft et al. proposed in the tugtupite, sodalite, and hackmanite [42].

3.2. Application for Distinguishing Natural and Treated Gems

Discerning how to rapidly and accurately distinguish natural and treated gems is a consistently popular subject in gem characterization. High-price jadeite jade, Oregon sunstone, and some of their treated products can also be recognized by the differences in their EEMs patterns.

3.2.1. Jadeite Jade vs. Impregnated Jadeite Jade

Jadeite jade, a mineral aggregate mainly composed of pyroxene-group minerals ($NaAlSi_2O_6$), is known to exhibit no fluorescence. Artificial treatments to enhance its color and appearance quality can result in visible fluorescence due to extraneous substances during the treatment processes. These processes involve removing intrinsic impurities, such as brownish iron-containing minerals, impregnating them with colorless resin for enhancing stability, or with colored resin for enhancing color. The treated jadeite jades are named B-Jade (impregnated by colorless resin), C-Jade (dyed), and B+C-Jade (impregnated by colored resin), while the untreated jadeite jade is called A-Jade in the gems market.

Ma et al. found that in addition to micro-inclusions, luster, and absorption spectra, fluorescence plays a useful role in discriminating between A-, B-, C-, and B+C-Jade [43]. The detailed EEM fluorescence patterns blaze a new trail to distinguish natural from treated specimens and determine their exact treatment process. In general, the EEM spectra (Figure 9) showed no emissions from A-jade, but different emission patterns appeared in the treated jadeite jades. B-Jades usually have a strong blue fluorescence with a triplet center at λ_{em} = 417, 441, 464 nm/λ_{ex} = 340–410 nm, owing to the impregnated fluorescent aromatic materials and epoxy resin. Similar to B-Jade, C-Jade also radiates a triplet emission but at a longer wavelength of λ_{em} = 418, 444, 470 nm/λ_{ex} = 340–410 nm, possibly owing to dyeing with colored organic material.

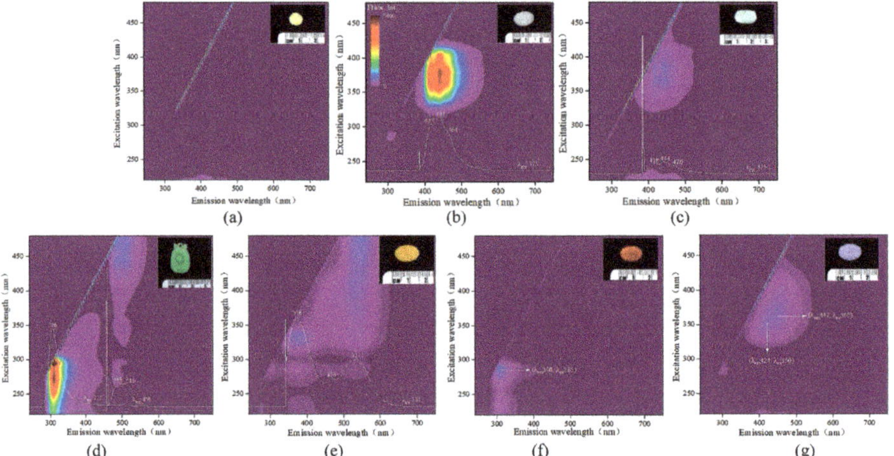

Figure 9. EEM spectra from (**a**) A-Jade, (**b**) B-Jade, (**c**) C-Jade, and (**d–g**) B+C-Jade. Adapted from Reference [43].

The different staining materials used in B+C-Jade produce a wide range of colors, including green, purple, yellow, and red. The chromium ion chelate in green-colored B+C-Jades results in an ultraviolet emission at λ_{em} = 308 nm/λ_{ex} = 220–300 nm. Meanwhile, the chelate of iron ions, a type of fluorescence quenching agent, filled in the yellow B+C-Jades and led to the weaker emissions at λ_{em} = 377 nm/λ_{ex} = 335 nm and λ_{em} = 537 nm/λ_{ex} = 335–480 nm. The high concentration of iron results in the red-colored B+C-Jade and further weakens their radiation at λ_{em} = 308 nm/λ_{ex} = 285 nm. However, the purple B+C-Jade has a whitish blue fluorescence at λ_{em} = 380–500 nm/λ_{ex} = 340–410 nm due to its Mn-bearing extraneous polymers. These EEM fluorescence spectra provide evident features for the identification of natural versus treated jadeite jades as well as different enhancement processes.

3.2.2. Oregon Sunstone vs. Copper-Diffused Plagioclase Crystals

Natural red and greenish sunstone were obtained from Oregon, USA. Zhou et al. [44–46] found that copper-diffused plagioclase crystals exhibited red, light red, and blue-green colors under different conditions. Cu diffusion in both red and light red sunstones results in orange fluorescence when illuminated by ultraviolet light at 320 nm, while the Cu-Li diffusion-treated blue-green plagioclase crystal only exhibits weak fluorescence similar to natural sunstone [45,46]. Compared with natural Oregon sunstone, which only has a weak emission at ~395 nm, the Cu-diffused sunstone with orange fluorescence has a stronger emission band at ~394 nm and an additional emission band at ~555 nm (Figure 10). Through a series of copper-related diffusion experiments, Zhou et al. found that the Cu^+–Cu^+ dimer, rather than the Cu^{2+} ions, diffused into the sunstone crystal lattice responsible for the ~555 nm emission [44,47]. The weak emission peak at ~395 nm also indicates that it was caused by Cu^+ [46]. Therefore, the additional band at ~555 nm is an indicative feature for distinguishing red copper-diffused sunstones from natural sunstones.

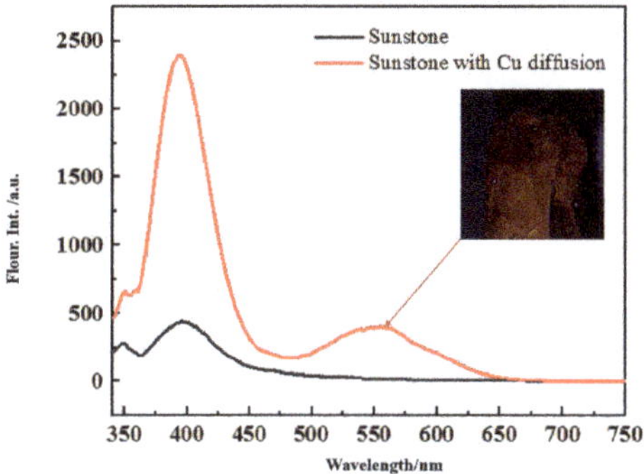

Figure 10. Fluorescence spectra of sunstone and Cu-diffused sunstone (λ_{ex} = 320 nm). (Data from Zhou et al. [44–46]).

3.3. Tracing the Specific Geographic Producing Areas of Amber

Amber belongs to maturated fossilized resin generated from resin-producing plants (including Conifer and Fabaceae families) millions of years ago that underwent rapid solidification. These solidified plants then accumulated and underwent extremely long diagenesis in the primary sediment region, or in secondary deposits by river transportation. It has been documented that amber is widely distributed across all continents of the world, except Antarctica, but commercial gem-quality amber is found in the Baltic Sea region, Dominican Republic, Myanmar, Mexico, and Fushun in China [48–50].

Fluorescence has been observed in amber for decades [51–54], and LWUV light irradiation usually excites Baltic amber to a greenish-yellow glow; Dominican and Mexican amber exhibits blue emission; and Burmese and Fushun amber exhibits violet-blue fluorescence. After long-term exposure to this condition, some Mexican, Burmese, and Fushun amber can even exhibit a noticeable yellowish phosphorescence. By introducing a front-surface fluorescence detection instrument, we first measured the 3D fluorescence of amber from the aforementioned geographic locations; details are provided in references [12,55]. For ordinary golden and brown amber species, Baltic amber has a stable broad irradiation center at λ_{em} = 435 nm/λ_{ex} = 350 nm. The emissions of Mexican and Dominican amber resembled those of the Baltic amber but were stronger. Fushun and Burmese amber both exhibited ultraviolet radiation at λ_{em} = 334, 347 nm/λ_{ex} = 240, 295 nm [55]. Meanwhile, Jiang et al. [56] and Shi et al. [57] found that EEM fluorescence patterns can not only help to identify whether the amber came from Myanmar, but also further distinguish seven kinds of special Burmese amber from some other ordinary species. What is more, when illuminated by a strong LWUV light for one minute, the order of phosphorescence intensity and lifetime of the amber was as follows: Myanmar > Fushun > Dominican Republic ≈ Mexico > Baltic Sea (Figure 11) [58]. This is the first time the phosphorescence in amber from Fushun, Dominican Republic, Mexico, and Baltic Sea regions has been noticed and measured. Additionally, FTIR provides another significant feature to distinguish Baltic amber from other locations' amber, because it shows a "Baltic shoulder", that is, an absorption mode within 1260 cm^{-1}–1150 cm^{-1} [58]. These properties of amber provide reliable information regarding the traceability of amber deposit locations.

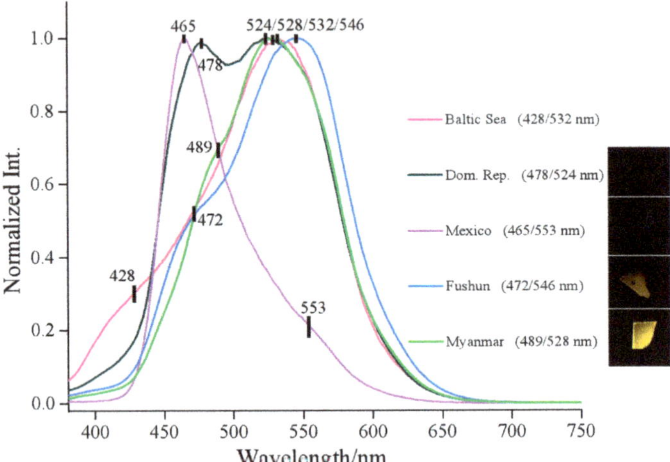

Figure 11. Normalized phosphorescence spectra of golden amber from five locations. The lifetime of the phosphorescence of Baltic amber is too short to capture its irradiation by camera; the inset presents phosphorescence images of amber from four locations. Adapted from Reference [58].

The rarest blue amber with extraordinarily visible fluorescence was mainly sourced from the Dominican Republic; however, its Mexican and Burmese counterparts were forced to imitate Dominican blue amber. Through a combination of EEM and UV-Vis spectra, we constructed a flowchart to rapidly discriminate blue amber from these three geographic origins according to their diagnostic triplet emission peaks at λ_{em} = 445, 474, 508 nm/λ_{ex} = 415, 440 nm and the invisible ultraviolet part of fluorescence at λ_{em} = 334, 347 nm/λ_{ex} = 240, 295 nm, as well as the 412 nm and 442 nm absorption bands (Figure 12). Recently, Li et al. [59] noticed that Indonesian fossil resins show an extremely similar appearance and EEM patterns to Dominican blue amber. However, the 1384 cm^{-1}, 1377 cm^{-1} and 1367 cm^{-1} in FTIR spectra from Indonesian fossil resin indicate its Dipterocarpaceae plant origin. Therefore, it is dissimilar to Dominican blue amber from the Fabaceae plant. Thus, the luminescence properties could at times trace the specific geographic producing areas of amber, but other methods are also needed.

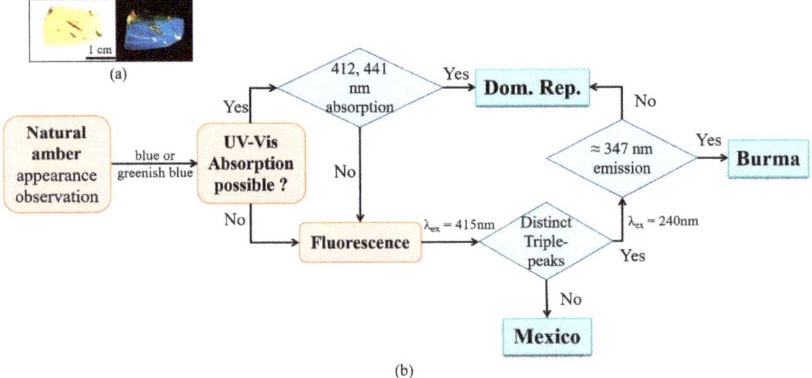

Figure 12. (**a**) Dominican blue amber. (**b**) A preliminary flowchart for blue amber to assist in determining its geographic origin, adapted from Reference [18].

3.4. Assisted Analysis on the Potential Luminous Molecules in Fossil Resins

Amber contains complex chemical components that can be used to extract and purify a single substance in a single step. The luminous matter in amber is difficult to obtain. Yellow-colored Dominican blue amber glows with intense blue fluorescence when illuminated by sunlight. We extracted the fluorescent components using ethanol (Figure 13a) and then performed EEM, gas chromatography–mass spectrometry (GC–MS), and FTIR analyses [58]. In the GCMS analysis, the retention times of the concentrated solutions of amber and perylene appeared in 14.786–15.393 min with a 252 m/z base peak. The FTIR spectra of the enriched solution from amber also showed characteristic peaks of the perylene-ethanol solution at 817 and 771 cm^{-1}, as well as at 3076, 1600, and 1580 cm^{-1} related to aromatic molecular structures with multiple C=C double bonds. The EEM results from the extracted solution are shown in Figure 13b), with typical emissions at 448, 473, and 505 nm. When the unknown fluorescent agent was enriched in this solution by blowing dry nitrogen, the EEM pattern underwent significant changes, as shown in Figure 13c. We prepared a saturated pure perylene-ethanol solution (Figure 13d), and its EEM pattern (Figure 13e) was substantially different from that of the amber-extracted solution. Nevertheless, when diluting the perylene-ethanol solution by adding pure ethanol to factors of 2.5 and 16, the EEM patterns changed dramatically, as shown in Figure 13f,g, and were very similar to those of the enriched and original extracted solutions. Therefore, we propose that perylene and its derivatives probably led to the three typical emission peaks in Dominican blue amber. However, we have not purified the fluorescent components directly, which will be the main subject of future research.

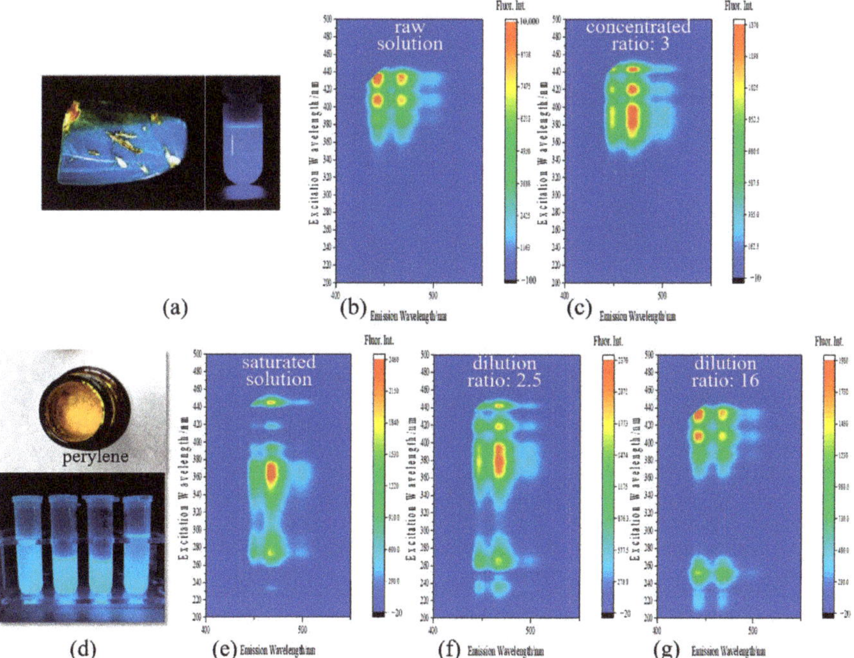

Figure 13. EEM spectra of solid samples and their solutions, adapted from Reference [58]. (**a**) the Dominican blue amber and its ethanol solution; (**b**) EEM from the solution in (**a**); (**c**) EEM of the concentrated solution from (**a**); (**d**) perylene and its ethanol solutions with different concentrations; (**e**) EEM of the saturated perylene- ethanol solution; (**f**) EEM from the diluted perylene- ethanol solution; (**g**) EEM from the perylene- ethanol solution with a dilution ratio 16.

4. Summary

With the rapid development of relevant fluorescence spectroscopy and gemological science, luminescence from gem materials has been extensively studied. Gemology is closely connected to scientific research and commercial markets. The luminescence properties of some gem materials, combined with new-tech fluorescence spectroscopy and widely used theories, act as a window for understanding the exact trace element ions or lattice defects in certain gem crystals or potential fluorescent aromatic molecules in organic materials. In addition, they provide an abundance of useful information to distinguish natural from treated jadeite jades, indicate the specific enhancement process in treated jadeite jades, and discern blue ambers of different geographic origins. Hence, we look forward to fluorescence spectroscopy being applied to other gem materials, offering more details on photoluminescence properties (EEMs, phosphorescence spectra, lifetime, and quantum yield), and applying them more extensively in mineralogical and gemological characterization, identification, and traceability processes. This is a project in process, and our research group are collecting more samples from reliable sources to confirm the data described above.

Author Contributions: Conceptualization, A.S. and Z.Z.; validation, A.S. and Z.Z.; formal analysis, Z.Z.; resources, A.S.; data curation, A.S. and Z.Z.; writing—original draft preparation, Z.Z. and A.S.; writing—review and editing, Z.Z. and A.S.; visualization, Z.Z.; supervision, A.S.; funding acquisition, A.S. and Z.Z. All authors have read and agreed to the published version of the manuscript.

Funding: This research was funded by 2023 Geological Scientific Research Project in Hubei Province, China, the grant number KJ2023-51, and supported by Hubei Geological Bureau.

Data Availability Statement: The data used to support the findings of this study are available from the first author upon request.

Acknowledgments: We acknowledge our research colleagues: Shao T., Ma P., Lv F.L. and Zhou Q.C., for their generous help on this article. Especially thanks Zhou Q.C. for providing his experiments data of sunstone and copper diffusion sunstone.

Conflicts of Interest: The authors declare no conflict of interest.

References

1. Waychunas, G.A. Luminescence Spectroscopy. *Rev. Mineral. Geochem.* **2014**, *78*, 175–217. [CrossRef]
2. Gaft, M.; Reisfeld, R.; Panczer, G. *Modern Luminescence Spectroscopy of Minerals and Materials*, 2nd ed.; Springer: Cham, Switzerland, 2015. [CrossRef]
3. Berlman, I.B. *Handbook of Fluorescence Spectra of Aromatic Molecules*, 2nd ed.; Academic Press: New York, NY, USA; London, UK, 1971.
4. Wolfbeis, O.S. The Fluorescence of Organic Natural Products. In *Molecular Luminescence Spectroscopy: Methods and Applications— Part 1*; Schulman, S.G., Ed.; Wiley-Blackwell: Hoboken, NJ, USA, 1985; pp. 167–370.
5. Romppanen, S.; Häkkänen, H.; Kaski, S. Laser-induced time-resolved luminescence in analysis of rare earth elements in apatite and calcite. *J. Lumin.* **2021**, *233*, 117929. [CrossRef]
6. Lakowicz, R.J. *Principles of Fluorescence Spectroscopy*, 3rd ed.; Springer: Berlin/Heidelberg, Germany, 2006.
7. Liu, Y.L.; Ding, H. Research developments of long lasting phosphorescent materials. *Chin. J. Inorg. Chem.* **2001**, *17*, 181–187.
8. Dorenbos, P. Mechanism of Persistent Luminescence in Eu^{2+} and Dy^{3+} Codoped Aluminate and Silicate Compounds. *J. Electrochem. Soc.* **2005**, *152*, H107–H110. [CrossRef]
9. Kondo, D.; Beaton, D. Hackmanite/Sodalite from Myanmar and Afghanistan. *Gems Gemol.* **2009**, *45*, 38–43. [CrossRef]
10. Crelling, J.C.; Bensley, D.F. Characterization of Coal Macerals by Fluorescence Microscopy. In *Chemistry and Characterization of Coal Macerals*; Winans, R.E.A., Ed.; American Chemical Society: Washington, DC, USA, 1984; pp. 33–45.
11. Liu, Y.; Zhan, G.; Liu, Z.-W.; Bian, Z.-Q.; Huang, C.-H. Room-temperature phosphorescence from purely organic materials. *Chin. Chem. Lett.* **2016**, *27*, 1231–1240. [CrossRef]
12. Vigier, M.; Fritsch, E.; Cavignac, T.; Latouche, C.; Jobic, S. Shortwave UV Blue Luminescence of Some Minerals and Gems Due to Titanate Groups. *Minerals* **2023**, *13*, 104. [CrossRef]
13. Eaton-Magaña, S. Comparison of luminescence lifetimes from natural and laboratory irradiated diamonds. *Diam. Relat. Mater.* **2015**, *58*, 94–102. [CrossRef]
14. Luo, Y.; Breeding, C.M. Fluorescence produced by optical defects in diamond: Measurement, characterization, and challenges. *Gems Gemol.* **2013**, *49*, 82–97. [CrossRef]

15. Luo, Y.; Nelson, D.; Ardon, T.; Breeding, C.M. Measurement and Characterization of the Effects of Blue Fluorescence on Diamond Appearance. *Gems Gemol.* **2021**, *57*, 102–123. [CrossRef]
16. Sehgal, A. Phosphorescence of synthetic sapphire. *Gems Gemol.* **2015**, *51*, 184–186.
17. Blumentritt, F.; Latouche, C.; Morizet, Y.; Caldes, M.T.; Jobic, S.; Fritsch, E. Unravelling the Origin of the Yellow-Orange Luminescence in Natural and Synthetic Scapolites. *J. Phys. Chem. Lett.* **2020**, *11*, 4591–4596. [CrossRef] [PubMed]
18. Zhang, Z.; Jiang, X.; Wang, Y.; Kong, F.; Shen, A.H. Fluorescence Characteristics of Blue Amber from the Dominican Republic, Mexico, and Myanmar. *Gems Gemol.* **2020**, *56*, 484–496. [CrossRef]
19. Zaitsev, A.; Kazuchits, N.; Moe, K.; Butler, J.; Korolik, O.; Rusetsky, M.; Kazuchits, V. Luminescence of brown CVD diamond: 468 nm luminescence center. *Diam. Relat. Mater.* **2021**, *113*, 108255. [CrossRef]
20. Eaton-Magana, S.; Breeding, C.M. An Introduction to Photoluminescence Spectroscopy for Diamond and Its Applications in Gemology. *Gems Gemol.* **2016**, *52*, 2–17. [CrossRef]
21. Birks, J. Fluorescence parameters and their interpretation. *J. Lumin.* **1974**, *9*, 311–314. [CrossRef]
22. Parker, C.A. *Photoluminescence of Solutions with Applications to Photochemistry and Analytical Chemistry*; Elsevier Publishing Company: New York, NY, USA, 1968; p. 544.
23. Airado-Rodríguez, D.; Durán-Merás, I.; Galeano-Díaz, T.; Wold, J.P. Front-face fluorescence spectroscopy: A new tool for control in the wine industry. *J. Food Compos. Anal.* **2011**, *24*, 257–264. [CrossRef]
24. David, M.J. *Introduction to Fluorescence*; CRC Press: Boca Raton, FL, USA, 2015; p. 308.
25. Shao, T.; Lyu, F.; Guo, X.; Zhang, J.; Zhang, H.; Hu, X.; Shen, A.H. The role of isolated nitrogen in phosphorescence of high-temperature-high-pressure synthetic type IIb diamonds. *Carbon* **2020**, *167*, 888–895. [CrossRef]
26. Thomas, H.; Stefanos, K.; Emmanuel, F.; Franck, N. Luminescence spectroscopy and microscopy applied to study gem materials: A case study of C centre containing diamonds. *Mineral. Petrol.* **2013**, *107*, 393–413.
27. Watanabe, K.; Simon, C.L.; Junichi, I.; Hisao, K.; Yoichro, S. Phosphorescence in high-pressure synthetic diamond. *Diam. Relat. Mater.* **1997**, *6*, 99–106. [CrossRef]
28. Fisher, J.; Greenbank, L. The Rogerley Mine, Weardale, County Durham, England. *Rocks Miner.* **2000**, *75*, 54–61. [CrossRef]
29. Sidike, A.; Lee, K.-H.; Kusachi, I.; Yamashita, N. Photoluminescence properties of a natural fluorite. *J. Miner. Pet. Sci.* **2000**, *95*, 228–235. [CrossRef]
30. Przibram, K. Fluorescence of Fluorite and the Bivalent Europium Ion. *Nature* **1935**, *135*, 100. [CrossRef]
31. Czaja, M.; Bodył-Gajowska, S.; Lisiecki, R.; Meijerink, A.; Mazurak, Z. The luminescence properties of rare-earth ions in natural fluorite. *Phys. Chem. Miner.* **2012**, *39*, 639–648. [CrossRef]
32. Calderon, T.; Millan, A.; Jaque, F.; Garcia Solé, J. Optical properties of Sm^{2+} and Eu^{2+} in natural fluorite crystals. *Int. J. Radiat. Appl. Instrum. Part D. Nucl. Tracks Radiat. Meas.* **1990**, *17*, 557–561. [CrossRef]
33. Topaksu, M.; Correcher, V.; Garcia-Guinea, J. Luminescence emission of natural fluorite and synthetic CaF2:Mn (TLD-400). *Radiat. Phys. Chem.* **2016**, *119*, 151–156. [CrossRef]
34. Sidike, A.; Kusachi, I.; Yamashita, N. Natural fluorite emitting yellow fluorescence under UV light. *Phys. Chem. Miner.* **2003**, *30*, 478–485. [CrossRef]
35. Bill, H.; Calas, G. Color centers, associated rare-earth ions and the origin of coloration in natural fluorites. *Phys. Chem. Miner.* **1978**, *3*, 117–131. [CrossRef]
36. Jeran, M.; Smerkolj, N.; Horváth, P. Phenomenon of Light Emission in Inorganic Materials: Fluorescence Activity of Fluorite Mineral. In Proceedings of the Socratic Lectures: 3rd International Minisymposium, Ljubljana, Slovenia, 17 April 2020; pp. 90–96.
37. Lv, F.L.; Shen, A.H. The Fluorescence Spectra of Gem-Quality Hauyne. *Spectrosc. Spectr. Anal.* **2020**, *40*, 3468–3471. (In Chinese)
38. Colinet, P.; Gheeraert, A.; Curutchet, A.; Bahers, T.L. On the Spectroscopic Modeling of Localized Defects in Sodalites by TD-DFT. *J. Phys. Chem. C* **2020**, *124*, 8949–8957. [CrossRef]
39. Stoliaroff, A.; Schira, R.; Blumentritt, F.; Fritsch, E.; Jobic, S.; Latouche, C. Point Defects Modeling Explains Multiple Sulfur Species in Sulfur-Doped $Na_4(Al_3Si_3O_{12})Cl$ Sodalite. *J. Phys. Chem. C* **2021**, *125*, 16674–16680. [CrossRef]
40. Sidike, A.; Sawuti, A.; Wang, X.M.; Zhu, H.J.; Kobayashi, S.; Kusachi, I.; Yamashita, N. Fine structure in photoluminescence spectrum of S_2^- center in sodalite. *Phys. Chem. Miner.* **2007**, *34*, 477–484. [CrossRef]
41. Blumentritt, F.; Vigier, M.; Fritsch, E. Blue Persistent Luminescence (Phosphorescence) of Sodalite. *J. Gemmol.* **2021**, *37*, 571. [CrossRef]
42. Gaft, M.; Panczer, G.; Nagli, L.; Yeates, H. Laser-induced time-resolved luminescence of tugtupite, sodalite and hackmanite. *Phys. Chem. Miner.* **2009**, *36*, 127–141. [CrossRef]
43. Ma, P.; Shen, A.H.; Shao, T.; Zhang, Z.Z.; Luo, H. Study on Three-Dimensional Fluorescence Spectrum Characteristics of Common Jadeite Jade. *Spectrosc. Spectr. Anal.* **2021**, *41*, 961–966. (In Chinese)
44. Zhou, Q.; Wang, C.; Shen, A.H. Copper Nanoparticles Embedded in Natural Plagioclase Mineral Crystals: In Situ Formation and Third-Order Nonlinearity. *J. Phys. Chem. C* **2022**, *126*, 387–395. [CrossRef]
45. Zhou, Q.C.; Wang, C.S.; Shen, A.H. High-Temperature Diffusion of Cu^{2+} Ions into Natural Plagioclase Crystals: The Role of Li_2O. *J. Phys. Chem. C* **2022**, *126*, 12244–12250. [CrossRef]
46. Zhou, Q.; Wang, C.; Shen, A.-H. Application of High-Temperature Copper Diffusion in Surface Recoloring of Faceted Labradorites. *Minerals* **2022**, *12*, 920. [CrossRef]
47. Puppalwar, S.P.; Dhoble, S.J.; Kumar, A. Improvement of photoluminescence of Cu^+ ion in Li_2SO_4. *Luminescence* **2011**, *26*, 456–461. [CrossRef]

48. Hong, Y. *Atlas of Amber Insects of China*; Hennan Scientific and Technological Publishing House: Zhengzhou, China, 2002; p. 394. (In Chinese)

49. Langenheim, J.H. *Plant Resins: Chemistry, Evolution, Ecology, and Ethnobotany*; Timber Press: Portland, OR, USA; Cambridge, UK, 2003; p. 586.

50. Seyfullah, L.J.; Beimforde, C.; Corso, J.D.; Perrichot, V.; Rikkinen, J.; Schmidt, A.R. Production and preservation of resins—Past and present. *Biol. Rev.* **2018**, *93*, 1684–1714. [CrossRef]

51. Mysiura, I.; Kalantaryan, O.; Kononenko, S.; Zhurenko, V.; Chishkala, V.; Azarenkov, M. Ukrainian amber luminescence induced by X-rays and ultraviolet radiation. *J. Lumin.* **2017**, *188*, 319–322. [CrossRef]

52. Chekryzhov, I.Y.; Nechaev, V.P.; Kononov, V.V. Blue-Fluorescing Amber from Cenozoic Lignite, Eastern Sikhote-Alin, Far East Russia: Preliminary Results. *Int. J. Coal Geol.* **2014**, *132*, 6–12. [CrossRef]

53. Bellani, V.; Giulotto, E.; Linati, L.; Sacchi, D. Origin of the blue fluorescence in Dominican amber. *J. Appl. Phys.* **2005**, *97*, 016101. [CrossRef]

54. Matuszewska, A.; Czaja, M. Aromatic compounds in molecular phase of Baltic amber—Synchronous luminescence analysis. *Talanta* **2002**, *56*, 1049–1059. [CrossRef] [PubMed]

55. Zhang, Z.Z.; Jiang, X.R.; Wang, Y.M.; Shen, A.H.; Fong, F.L. Fluorescence Spectral Characteristics of Amber from Baltic Sea Region, Dominican Republic, Mexico, Myanmar and Fushun, China. *J. Gems Gemmol.* **2020**, *22*, 1–11.

56. Jiang, X.; Zhang, Z.; Wang, Y.; Kong, F. Gemmological and Spectroscopic Characteristics of Different Varieties of Amber from the Hukawng Valley, Myanmar. *J. Gemmol.* **2020**, *37*, 144–162. [CrossRef]

57. Shi, Z.; Xin, C.; Wang, Y. Spectral Characteristics of Unique Species of Burmese Amber. *Minerals* **2023**, *13*, 151. [CrossRef]

58. Zhang, Z.Z. *Spectral Characteristics of Amber—Their Application in Provenance Determination, and Study on Fluorescent Components*; China University of Geosciences: Wuhan, China, 2021; p. 174.

59. Li, Y.Y.; Zhang, Z.Z.; Wu, X.H.; Shen, A.H. Photoluminescence in Indoesian Fossil Resins. *Spectrosc. Spectr. Anal.* **2022**, *42*, 814–820.

 *minerals*

Article

Shortwave UV Blue Luminescence of Some Minerals and Gems Due to Titanate Groups

Maxence Vigier, Emmanuel Fritsch *, Théo Cavignac, Camille Latouche and Stéphane Jobic

IMN, Institut des Matériaux de Nantes Jean Rouxel, Centre National de la Recherche Scientifique, Nantes Université, F-44000 Nantes, France
* Correspondence: emmanuel.fritsch@cnrs-imn.fr

Abstract: This article reviews blue shortwave-excited luminescence (BSL) in natural minerals and synthetic materials. It also describes in detail the emission of seven minerals and gems displaying BSL, as well as three references in which BSL is caused by titanate groups (TiO_6): benitoite, Ti-doped synthetic sapphire and spinel. Emission (under 254 nm shortwave excitation) and excitation spectra are provided, and fluorescence decay times are measured. It is proposed that BSL in beryl (morganite), dumortierite, hydrozincite, pezzotaite, tourmaline (elbaite), some silicates glasses, and synthetic opals is due to titanate groups present at a concentration of 20 ppmw Ti or above. They all share a broad emission with a maximum between 420 and 480 nm (2.95 to 2.58 eV) (thus perceived as blue), and an excitation spectrum peaking in the short-wave range, between 230 and 290 nm (5.39 to 4.27 eV). Furthermore, their luminescence decay time is about 20 microseconds (from 2 to 40). These three parameters are consistent with a titanate emission, and to our knowledge, no other activator.

Keywords: luminescence; titanate; beryl; pezzotaite; tourmaline; synthetic opal; lead glass; hydrozincite; dumortierite

Citation: Vigier, M.; Fritsch, E.; Cavignac, T.; Latouche, C.; Jobic, S. Shortwave UV Blue Luminescence of Some Minerals and Gems Due to Titanate Groups. *Minerals* **2023**, *13*, 104. https://doi.org/10.3390/min13010104

Academic Editors: Leonid Dubrovinsky and Ulf Hålenius

Received: 1 December 2022
Revised: 28 December 2022
Accepted: 6 January 2023
Published: 9 January 2023

1. Introduction

The purpose of this study is to propose an interpretation for lesser-known blue shortwave ultraviolet luminescence (BSL) observed in natural and synthetic minerals and gems. We consider a luminescence blue if the emission is perceived as blue. If one takes into account the mechanism of human color vision, this comprises emissions with maxima stretching from near-UV to about 500 nm, provided they are relatively wide (full width at half maximum –FWHM- of about 100 nm/0.65 eV). Shortwave ultra violet (SWUV or simply SW) radiation is usually defined as 254 nm (4.88 eV), as in standard UV lamps. Here, we extend that to a somewhat larger spectral range, to take into account the various values chosen by authors in the literature, ranging from about 240 to 300 nm (2.58 to 2.95 eV). However, for the experimental section, we used strictly 254 nm excitation. Only shortwave luminescence is required here, but blue longwave luminescence is not excluded as in anatase TiO_2 [1].

SWUV excited photoluminescence (SWPL) used in this work is distinct from excitation with a beam of electrons (cathodoluminescence–CL) which also produces blue luminescence in many minerals, and this may cause confusion. CL can be described as an irradiation experiment, creating defects as they are probed. Samples often heat, and sometimes change color. This does not happen with SWPL. CL uses particles with mass and charge, not SWPL. Also, because of generally high electron energy, and broad-band excitation, CL may use multiple excitation paths, with different emissions being prompted [2]. This may result in coupling with different defect states in the material. These states would not be probed with the comparatively narrow-band, lower energy SWPL. Also, CL probes strictly the surface (limited penetration of electrons) whereas SWPL tests the bulk of the material (see Figure 1).

One example is surface hydroxyl groups, that, although induced by SWUV exposure, are detected by blue Cl, not SWPL [3]. In jadeite, blue CL is attributed to Na^+ or Al^{3+} defect centers, whereas there is no blue SWPL of jadeite [4]. In summary, SWPL and CL are not identical but more complementary, emissions being induced by one (CL for example), being often not detected with the other, in particular surface- or oxygen-related defects.

Figure 1. Photo under daylight and under shortwave UV (254 nm) of the ten representative samples of BSL phenomena. (Photo: Maxence Vigier).

We aim to demonstrate that BSL is in many cases related to the presence of titanate groups, even if other, sometimes very different interpretations have been proposed. We sought BSL across a large number of species, not taking into account at first their structural or chemical characteristics. We then documented emission, excitation and time decay for those samples satisfying our criteria. We complemented this approach by an extensive bibliographical search for BSL across the chemical, mineralogical and gemmological literature.

BSL has previously been attributed to a number of activators. Our attention turned first to closed-shell compounds, that is complexes of transition elements without d electrons surrounded by oxygen atoms [5]. This family includes titanates, but also permanganates, chromates, vanadates, as well as molybdates, niobates, zirconates, tungstates and tantalates. Blasse (1980) has summarized their luminescence behavior. From this vast number of possibilities, we are interested in those absorbing around 4.8 eV (254 nm) and emitting light perceived as blue A well-studied, but rather unique example is scheelite ($CaWO_4$) where the BSL is ascribed to tungstate (WO_6) groups [6–8]. The only other closed shell compounds with similar constraints would be octahedral TiO_6 (again, [5]). This is confirmed by the study of benitoite by Gaft (2004), which proposes a detailed energy level scheme responsible for the emission of the titanate group in this titanate-containing mineral [9]. The relatively long decay time (2.6 μs) is explained by the presence of a forbidden state acting as a trap level.

BSL is also noted in rare-earth containing solids, with activators such as Ce^{3+} in apatite ($Ca_5(PO_4)_3(Cl,F,OH)$) [10]. More rarely, luminescence can also be attributed to rarer elements or ions: UO^{2+} in turquoise $(CuAl_6(PO_4)_4(OH)_8 \cdot 4H_2O)$ [11], Pb^{2+} in hydrozincite $(Zn_5(CO_3)_2(OH)_6)$ [12], Tl+ in pezzotaite $(Cs(Be_2Li)Al_2Si_6O_{18})$ [13], or sulphur-doped $(S_2)^-$ in doped sodalite $(Na_8Al_6Si_6O_{24}Cl_2)$ [14].

BSL is most frequently related to the presence of titanium, according to literature. Yet often details of the luminescence are missing (proof of Ti presence, excitation spectra or decay time measurements). In addition, the presence of isolated Ti^{3+} (*d-d* transitions) was assumed as a source of luminescence in several oxides like ZrO_2 or Al_2O_3 [15–17]. Recent theoretical modeling refutes this hypothesis in ZrO_2 and ascribes it to Ti^{4+} [18]. Minerals and gems containing Ti^{4+} cation in their nominal chemical formula are also known to present BSL phenomena. The best known example is benitoite $(BaTiSi_3O_9)$, mentioned above, [9,19] and to a lesser degree baratovite $(KCa_7(Ti,Zr)_2Li_3Si_{12}O_{36}F_2)$ and katayamalite $(KLi_3Ca_7Ti_2(SiO_3)_{12}(OH)_2)$ [16].

Other authors have mentioned F-center as luminescence activators for BSL. In oxides and silicates, these are oxygen vacancies filled with two electrons, also noted "V^x_O" in Kröger–Vink notation. Examples include spinel $(MgAl_2O_4)$ [20–22] and corundum (Al_2O_3) [23,24]. Some authors have attributed the pair formed by a F-centers and Ti^{4+} as the activator in Al_2O_3 [25], rutile (TiO_2) [26], hackmanite $(Na_8Al_6Si_6O_{24}(Cl_2,S))$ [27], and aluminosilicate glasses [28]. Each of these activators display a unique combination of emission/excitation/lifetime decay characteristics that makes it possible to pin down the origin of luminescence.

In the course of examining the luminescence of hundreds of minerals and gems, we have encountered a small number of BSL, which are considered unusual, sometimes not even described in articles, textbooks or websites. They seem to share comparable properties, mostly that the blue emission color and emission spectra are similar. Among those, BSL was observed in dumortierite $(Al_7BO_3(SiO_4)_3O_3)$, synthetic spinel $(MgAl_2O_4)$, synthetic opal and some rare tourmalines (dravite or elbaite, $NaMg_3Al_6(BO_3)_3Si_6O_{18}(OH)_4$/$Na(Li,Al)_3Al_6(OH)_4(BO_3)_3$.

Many titanium-containing minerals are known to have blue-to-blue-green shortwave-excited luminescence. Table 1 provides a summary of those minerals. The presence of titanium as chemical component, generally as a well-defined octahedral titanate group, makes these compounds the materials of choice as reference for the luminescence properties of the titanate group.

Table 1. Summary of luminescent properties of Ti-bearing natural minerals for which BSL has been documented. For all, the perceived luminescence color is blue and the purported is TiO_6.

Minerals	Emission nm (FWHM eV)	Excitation nm (FWHM eV)	Lifetime Decay	Reference
Baratovite $KCa_7(Ti,Zr)_2Li_3Si_{12}O_{36}F_2$	406 (1.00)	250 (1.63)	–	[16]
Benitoite $BaTiSi_3O_9$	420 (0.57)	240 (1.77) and 280 (1.29)	2.6 µs	[9]
Berezanskite $Ti_2\square_2KLi_3(Si_{12}O_{30})$	480	290	–	[29]
Katayamalite $KLi_3Ca_7Ti_2(SiO_3)_{12}(OH)_2$	406 (1.00)	250 (1.62)	–	[16]
Natisite $Na_2TiO[SiO_4]$	450 (0.94)	250 (1.00)	–	[29]
Penkvilksite $Na_4Ti_2Si_8O_{22} \cdot 4H_2O$	440 (0.99)	250 (1.00)	–	[29]

All of these minerals display a broad emission band centered between 406 and 450 nm (2.75 to 3.05 eV) with a 100 nm (0.65 eV) FWHM. In mineralogical studies, the excitation spectrum and lifetime decay associated with the luminescence maxima are not systematically measured (Table 1). Most BSL excitation spectra exhibits a large band about 50 nm (1 eV) FWMH in the 240–300 nm (4.13 to 5.16 eV) spectral range. The decay time is of the order of a µs. For natural minerals in Table 1, all BSL is linked to $[TiO_6]$ titanate groups.

It must be noticed that all the minerals in Table 1 do not contain iron. Other titanate containing minerals do contain iron as a major constituent and do not show BSL. Examples include non-luminescing ilmenite $FeTiO_3$, neptunite $KNa_2Li(Fe^{2+})_2Ti_2Si_8O_{24}$, warwickite $(Mg,Fe)_3Ti(O, BO_3)_2$, and titanomagnetite $Fe^{2+}(Fe^{3+}, Ti)_2O_4$. It is well-known that iron acts as a luminescence poison [27–29]. Often, the $O^{2-} \geq Fe^{3+}$ charge transfer participates in a well-known excitation-recombination process [30–33].

We also looked for evidence of BSL components in the material chemistry literature (Table 2). A small number of synthetic materials with titanate groups as a major constituent satisfy the conditions for BSL as well.

Table 2. Summary of luminescent properties of Ti-bearing synthetic compounds for which BSL has been documented. For all, the perceived luminescence color is blue.

Crystals	Emission nm	Excitation nm	Lifetime Decay	Proposed Cause	Reference
$LaMgS_{n-x}Ti_xO_8$	464–480	265		Ti^{4+}	[34]
Mg_2TiO_4	441	266			[35]
Na_2TiGeO_5	447	254		Ti^{4+}	[36]

The only FWHM available was 50 nm (0.74 eV) for the 254 nm excitation in $NaTiGeO_5$, comparable with that for natural minerals in Table 1.

After considering natural and synthetic materials containing titanate as a major component, and given the small number and relative rarity of some such compounds, we turned to materials that do not contain titanate as a component, but as an impurity (natural minerals and gems) or deliberate dopant (synthetic materials). In the synthetic compounds, titanium was mostly introduced in an effort to obtain luminescence (LED, emission and other scientific purposes). These materials are listed in Tables 3 and 4 with what is published of their luminescence properties.

Table 3. Summary of luminescent properties of natural compounds containing extrinsic Ti for which BSL has been documented. For all, the perceived luminescence color is blue.

Minerals	Emission nm (FWHM eV)	Excitation nm (FWHM eV)	Lifetime Decay	Proposed Cause	Reference
Aluminosilicate glass	490	270	0.7 and 5.6 µs	TiO_6	[28]
Bazirite $BaZrSi_3O_9$	460 (1.23)	250 (1.00)		Ti^{4+}	[37]
Diopside $CaMgSi_2O_6$	415 (1.53)	"Shortwaves"			[38]
Hydrozincite $Zn_5(CO_3)_2 (OH)_2$	430 (0.68)	240 (1.54)	0.7 µs	Pb^{2+}	[12]
Pezzotaite $CsAl_2Si_6O_{18}$	425 (0.84)	266	2–8.3 µs	Tl^+	[13]
Corundum: Sapphire Al_2O_3	425 (1.45)	250	42 µs		[39,40]
Spinel $MgAl_2O_4$	465 (0.88)	233 (1.64) and 260 (1.11)	3.5; 9.3; 46.3 µs	Ti^{4+} F-centers	[21]
Topaz $Al_2SiO_4(F,OH)_2$	460 (1.23)	320	"10^{-3}" ms	Heavy elements	[41]
Wadeite $K_2ZrSi_3O_9$	480 (1.13)	235 (1.14), 300 (0.69)		TiO_6	[29,42]

Table 4. Summary of luminescent properties of synthetic compounds containing extrinsic Ti for which BSL has been documented. For all, the perceived luminescence color is blue.

Crystals	Emission nm (FWHM eV)	Excitation nm (FWHM eV)	Lifetime Decay	Proposed Cause	Reference
$Ti:BaZrO_3$	408 (0.76)	274 (0.83)		Ti^{4+}	[43,44]
$Ti:BaSnSi_3O_9$	425 (0.70)	245 (1.04)		Ti^{4+}	[45]
$Ti:CaZrO_3$	427	260		Ti	[35]
$Ti:Ca_3Al_4ZnO_{10}$	370 (1.45)	265 (0.89)		Ti^{4+}	[36]
$Ti:La_2Sn_2O_7$	434	265		Ti	[35]
$Ti:Lu_2O_3$	380 (0.43)	240 (1.65)		Ti^{4+}	[46]
$Ti:Mg_5SnB_2O_{10}$	430 (1.42)	260 (1.12)		TiO_6	[47,48]
$Ti:MgSnO_4$	473	265	2.5 µs	Ti	[35]
ZrO_2 monoclinic	470	235, 290, and 375	2.46 µs	Ti^{4+}	[18,49]

Note that in Table 3 (natural minerals) titanium is the most often cited cause for BSL. Titanate groups are mentioned as the probable cause, but also Ti^{3+}, or the association of Ti with intrinsic defects. Finally, totally different causes are proposed for some, such as "heavy elements", with specific reference to Tl and Pb.

As expected, titanium is considered the activator in all the Ti-doped synthetic materials exhibiting BSL as it is the intended dopant used to trigger luminescence. There again, we note this element appearing as TiO_6 and Ti^{4+}.

Note that there are many more cases of blue luminescence attributed to titanium or titanate groups in the literature consulted (more than 25), but excitation data and/or proof of the presence of traces of Ti is missing in many compounds to establish that it is indeed BSL (see Table S1 in Supplementary Materials).

In addition to experimental work, recent advances in the theory of luminescence modelling in solids have proved useful to tackle the origin of luminescence when it is controversial. In particular, some attention has been paid to the role of titanium in oxides [18,44]. These studies confirmed without ambiguity that the blue luminescence in monoclinic ZrO_2 and $BaTiZrO_3$ is due to Ti^{4+} instead of F-centers.

On this bibliographic basis, it is apparent that titanate groups are likely to be the cause of BSL. We therefore proceeded with checking that the luminescence characteristics of BSL materials not described so far match those of the references materials selected or described in the literature. We actually started by checking that these compounds do contain traces of titanium.

2. Materials and Methods

We surveyed hundreds of gems and mineral specimens for BSL. Out of those, ten have been selected to represent either an adequate reference (benitoite, synthetic corundum, synthetic spinel) or as representative of BSL. They are listed in Table 5. They all fluoresce blue under shortwave UV (SWUV–254 nm, Figure 1). All samples were tested using Raman and other spectroscopic techniques to verify their nature. As hydrozincite is the only polycrystalline material, we assessed its detailed nature by X-ray diffraction. It contains a small amount of cerussite, which is not blue luminescing. All samples are part of the gemology teaching collection at Nantes University.

We include three references to be able to compare data obtained with the same instrument in the same conditions. This avoids variability factors (different lamps, correction or software) that may exist from one luminescence spectrometer to the other.

Table 5. Ten minerals or gem samples exhibiting BSL, selected for this study.

Minerals	Chemical Formula	Nature	Absorption Color	Sample
Benitoite	$BaTiSi_3O_9$	Natural	Blue	3684
Corundum: synthetic sapphire	Al_2O_3	Synthetic	Colorless	3697
Dumortierite	$Al_7BO_3Si_3O_{18}$	Natural	Blue	3977
Silicate glass: "Fondu du Jura"	65.7%w SiO_2, 9.1%w Na_2O, 25.2%w Al_2O_3	Synthetic	Blue	1846
Hydrozincite	$Zn_5(CO_3)_2(OH)_6$	Natural	White	4101
Beryl Morganite	$Be_3Al_2(SiO_3)_6$	Natural	Colorless	1865
Synthetic Opal	SiO_2, nH_2O	Synthetic	White with play of color	2029
Pezzotaite	$CsAl_2Si_6O_{18}$	Natural	Purple-Pink	966
Flame fusion synthetic Spinel	$MgAl_2O_4$	Synthetic	Colorless	1386
Tourmaline: Elbaite	$Na(Li,Al)_{1.5}Al_6(Si_6O_{18})(BO_3)_3(OH)_3$	Natural	Colorless	3056

The luminescence data were acquired with a Horiba JobinYvon Fluorolog-3 fluorimeter. The light source is a 450 W xenon lamp, the detector a Hamamatsu R13456 photo-multiplier. The emission spectra were carried out with an excitation at 254 nm, covering the range 300 to 900 nm with an excitation spectral bandwidth ("slit") of 4 nm and sampling every nanometer with an integration time of 1 s per point. The excitation spectra were acquired covering 240 to 400 nm with an emission spectral bandwidth ("slit") of 4 nm and sampling every nanometer with an integration time of 1 s per point. The acquisition was done via the Fluor-Essence software. The low-temperature luminescence experiments were performed in situ in a vacuum system with a conduction-cooling device at the temperature of liquid nitrogen (77 K) and with the same spectral parameters.

Fluorescence decay time is an important parameter to characterize the cause of luminescence. A fluorophore which is excited by a photon will drop to the ground state with a certain probability based on the decay rates through a number of different (radiative and/or nonradiative) decay pathways. To observe fluorescence, one of these pathways must be by spontaneous emission of a photon, for ex for BSL. BSL lifetime decay curves are fitted with Origin software with the following single exponential formula:

$$y = y0 + A1 * \exp\left(\frac{-(x - x0)}{t1}\right) \tag{1}$$

where y0 is the y offset, A1 is the amplitude, x the time in seconds, x0 the x offset and t1 is the decay time measured.

Trace element data were obtained by laser ablation inductively coupled plasma mass spectrometry (LA-ICP-MS) using a G2 Excimer Laser Ablation System (at 50% of its maximum energy) coupled to a Varian quadrupole 820-MS system. Analyses consist for each sample of 5 lines of 60 spots. Each spot was obtained with an energy density of 4.54 J/cm^2 with a repetition rate of 10 Hz, spot size of 110 µm and a speed of 10 µm/s. The analysis consists of 30 s background acquisitions, 30 s data acquisition and 60 s time after each line (cell wash-out, gas stabilization, computer processing and move to next sample). We used two standard references: NIST 610 and NIST 612 measured at the beginning of the sequence and between each sample or two samples in a row [50]. The concentration of elements was calculated using the GLITTER software and calibrated using ^{27}Al [51].

3. Results

3.1. Inductively-Coupled Plasma Mass Spectrometry by Laser Ablation (LA-ICP-MS)

We investigated with LA-ICP-MS the elements that had been proposed as activators for BSL in the literature: chiefly Ti but also S, Tl and Pb. Results are reported in Table 6.

Table 6. Impurities concentrations (ppmw—parts per million weight) of the possible activators of BSL in the ten samples selected obtained by LA-ICP-MS techniques. avg (average), SD (standard deviation), nd (below limit of detection).

	Benitoite	Synthetic Sapphire	Dumortierite	Glass	Hydrozincite
	avg-SD	avg-SD	avg-SD	avg-SD	avg-SD
^{33}S	201.072–36.852	nd	1064.6–341.6	2136.8–624.5	10.1–1.0
^{47}Ti	Constituant	23.5–1.5	3658.2–127.2	1274.2–46.5	31.2–3.3
^{205}Tl	nd	nd	0.074–0.014	0.08–0.01	nd
^{208}Pb	0.354–0.021	1.3–0.1	11.6–0.9	2895.0–236.4	438.2–115.2
	Beryl (Morganite) avg-SD	Synthetic opal avg-SD	Pezzotaite avg-SD	synthetic spinel avg-SD	Tourmaline (Elbaite) avg-SD
^{33}S	703.216–225.848	254.6–36.5	396.3–131.5	nd	630.5–195.8
^{47}Ti	23.374–1.104	28.5–1.1	508.4–23.8	23.5–1.5	38.9–1.5
^{205}Tl	7.100–0.576	nd	2.59–0.14	nd	0.074–0.009
^{208}Pb	5.230–0.361	1.27–0.06	8.4–0.5	1.27–0.09	108.2–5.9

In all samples most of the activators proposed for blue luminescence are detected, but in contrasting amounts. Tl is detected in the ppm range or is below detection limits, making it a less likely candidate for BSL activation. Also, sulfur is not detected in one of the documented examples of BSL, synthetic sapphire. Pb is present in some samples but often at ppm level. On the other hand, titanium is always detected above the 20 ppm level in LA-ICP-MS on average, including in the reference materials. Nevertheless, extreme care must be taken when interpreting luminescence results based on a single criterion, in particular chemistry. Indeed, many emissions are caused by activators present only at the ppm level such Cr^{3+} in Ga_2O_3 [52] or the uranyl molecular ion in opal [53].

3.2. Photoluminescence Properties

3.2.1. Emission Spectra

Beyond the presence of the supposed activator, we must ascertain that all BSL described present similar spectroscopic characteristics for their emission. Figure 2 illustrates the emission spectra of our ten samples.

BSL is caused by an emission band with an apparent maximum between 414 and 463 nm (2.68 to 3.00 eV), as one would expect for a blue color perception. Also, the emission is always broad, ranging from about 75 to 100 nm (about 0.6 eV) FWHM. The emission is not always a single symmetrical band, so other contributions by other elements or maybe other electronic transitions could be part of this emission feature. In addition, the results are presented in nm for ease of comparison with other publications, but this induces a slight asymmetry of the spectral features. Nevertheless, overall, the emission of the new BSL materials have an emission profile which is close to that of the reference materials.

3.2.2. Excitation Spectra

Excitation spectra are critical to assign the cause of luminescence. They describe the absorptions responsible for the emission (Figure 3).

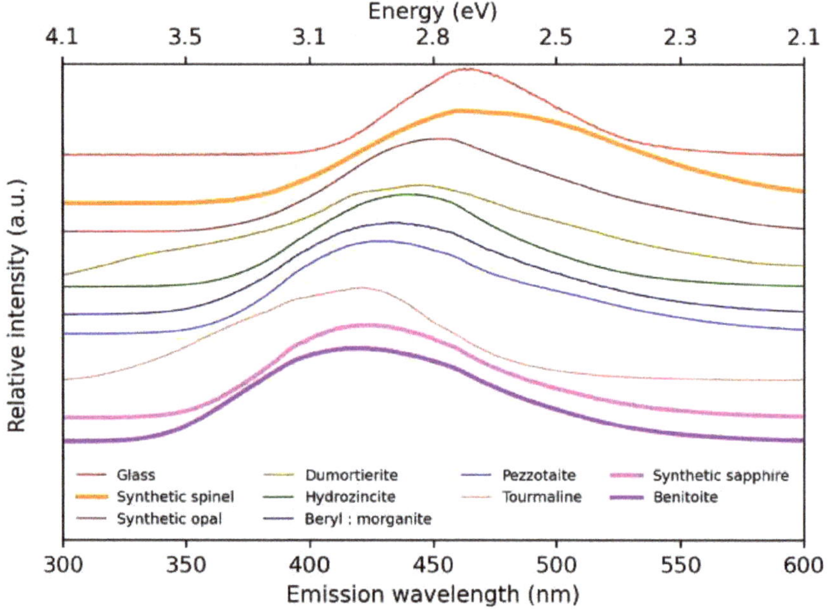

Figure 2. Emission spectra of our ten samples under λ_{ex} = 254 nm at room temperature. Apparent maxima of the broad emission band are located in the 414 to 463 nm range. Thicker lines mark reference materials. Spectra are shifted vertically for clarity.

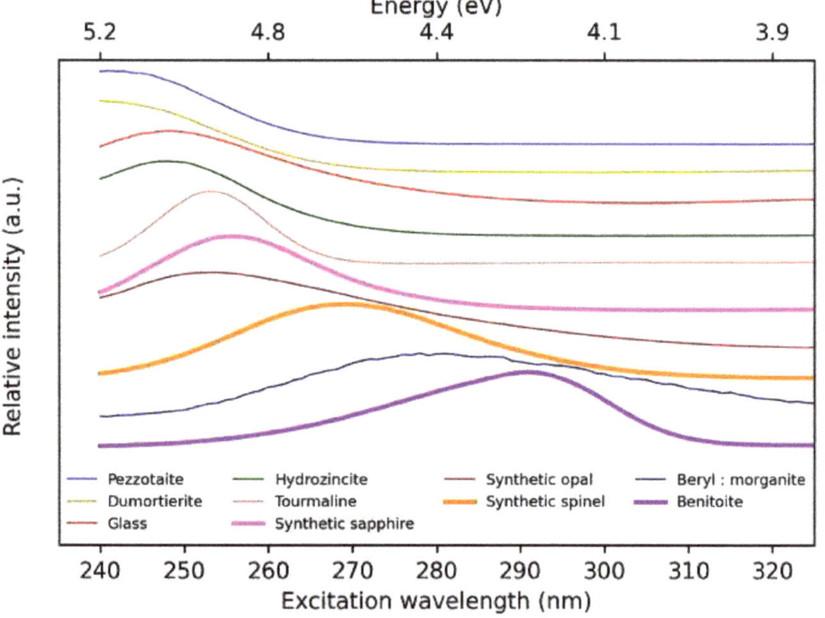

Figure 3. Excitation spectra of our ten samples for a luminescence excited by shortwave ultraviolet. Apparent maxima of the broad excitation band are located in the 240 to 290 nm range, as expected. Thicker lines mark reference materials. Spectra are shifted vertically for clarity.

Here also, the excitation spectral feature is not always a single symmetrical band. Two components seem present, one in the 240–250 nm (4.96–5.16 eV) range and the other in the 270–290 nm (4.27–4.59 eV) range. This is consistent with Gaft observations in several titanates that there are sometimes two excitation bands, one nearer 240 nm, and the other towards 290 nm [54]. These authors attribute the 240 nm excitation band to the allowed $^1A_{1g}$-$^1T_{2u}$ transition, and the weaker band at 290 nm to the formally forbidden $^1A_{1g}$-$^1T_{1u}$ transition. Nevertheless, overall, the excitation spectra of the reference and new BSL minerals are clearly comparable and consistent, if they are not identical.

3.2.3. Lifetime Decay of Luminescence

The third parameter to be taken into account when studying luminescence is the lifetime decay. Rarely studied in mineralogy, it is nevertheless essential to calculate it in order to determine which defect is at the origin of the luminescence. Similar activators should have similar lifetimes. The lifetime decay results of the ten BSL samples are presented in Figure 4.

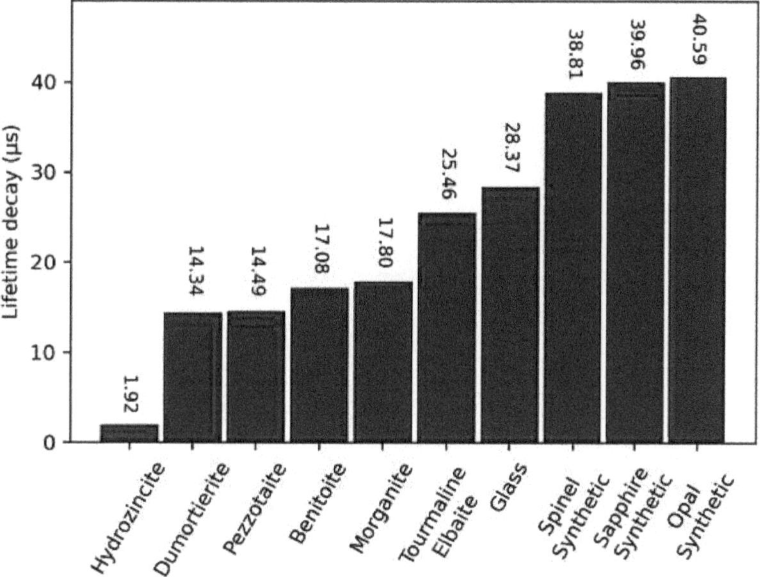

Figure 4. Lifetime decay of the ten samples showing BSL.

The values range from about 2 to approximately 40 μs on our instrument. The data for the reference materials are intercalated within the other new BSL, such as dumortierite, morganite, tourmaline and synthetic opal. Despite the apparent spread, this remains consistent for the same phenomenon across a wide variety of host atomic structures.

4. Discussion

All of the BSL minerals studied here show a broad emission band of about 100 nm FWHM wide (0.31 eV) between 414 and 463 nm (2.99 to 2.68 eV respectively) and a broad excitation band between 240 and 290 nm (4.27–5.16 eV) about 50 nm wide (0.89 eV), together with a decay lifetime of the order of 10–40 microseconds (Table 7). This data is consistent with titanate activator. We must first eliminate the other possible BSL activators invoked in the literature.

Table 7. Emission/excitation maxima of the ten samples showing BSL and associated lifetime decay.

Mineral	Emission Maxima (nm–eV)	Excitation Maxima (nm–eV)	Lifetime Decay (µs)
Benitoite	414–2.99	291–4.26	17.08
Corundum: Synthetic Sapphire	425–2.92	256–4.84	39.96
Dumortierite	446–2.78	Near 240–Near 5.16	14.34
Pb silicate glass	463–2.68	248–5.00	28.37
Hydrozincite	439–2.82	248–5.00	1.92
Beryl Morganite	433–2.86	278–4.46	17.8
Synthetic Opal	452–2.74	253–4.90	40.59
Pezzotaite	426–2.91	Near 240–Near 5.16	14.49
Flame fusion Synthetic Spinel	461–2.69	269–4.61	38.81
Tourmaline: Elbaite	420–2.95	253–4.90	25.46

The bibliographical search identifies four potential activators for the origin of BSL: titanium, lead, thallium or sulphur.

The luminescence of thallium in halogen compounds KCl, KBr, KI is close to that observed for BSL in terms of emission and excitation spectra [55]. However, Tl^+ lifetime is about 280 ns in KH_2PO_4, which is very different from our BSL values above 2 µs [56]. In addition, Tl is often below the limit of detection in BSL materials.

$(S_2)^-$ is very rarely mentioned as a source of blue luminescence in minerals [14], although it is rather known to be a common activator for emission in the orange-red spectral range [45,57,58]. There is little support for $(S_2)^-$ as activator as decay times are not provided and literature on this subject is scarce. Moreover, there is no relation between the S content that we measured and blue luminescence intensity.

Pb luminescence is mostly associated with emission in relatively longwave UV (mostly in 320–380 nm range) [54,59–63]. Also, Pb emission lifetimes appear to be much shorter than those observed in BSL, such as 190 to 300 ns at room temperature [55,64]. It also seems to be very dependent on Pb environment ranging from a few ns to microseconds in KCl, KBr [65]. It should be noted that some luminescence in the yellow range is attributed to Pb as in cerussite ($PbCO_3$) or $SrTiO_3$ [66]. Thus, lead does not appear to be a source of the microsecond lifetime blue luminescence. In the specific case of our hydrozincite, it contains titanium, which is not the case of hydrozincite studied by [12] for which emission was attributed to Pb^{2+}. Furthermore, our sample contains a small amount of cerussite, which explains the high Pb concentration in ICPMS measurements, actually a contamination by a non-BSL phase.

Titanate groups are the most likely activators associated with BSL; it has the same position maxima and width in emission and excitation spectra and same order of magnitude of decay times (Table 7), as confirmed by running references in the same manner as the BSL "unknowns" have.

We demonstrated that titanate is at the origin of BSL in dumortierite, whose origin has always been supposed to be related to titanium, but no detailed literature is available on the subject [29]. Note that many pink or blue quartz, colored by varieties of dumortierite, show this luminescence. We were also able to see the BSL in a synthetic opal. This is more surprising as blue luminescence in natural opals and amorphous silica is known to be associated with the presence of non-bonding oxygen or oxygen deficiency in silica (ODC) tetrahedra [54,67,68]. However, the common blue luminescence of opals is mostly triggered in longwave (365 nm) in contrast to that caused by the presence of titanate which is SW-excited only. It is logical that synthetic opal offers different luminescence behavior having a slightly different structure [69]. It is possible that TiO_2 substitutes partially for ZrO_2 the material between SiO_2 sphere– in this synthetic opal [70].

We describe for the first time blue luminescence in beryl, in this case morganite. The similarity of the structure of beryl and pezzotaite makes it likely that BSL in the gems would have similar causes. It is what we find and then, we ascribed its blue luminescence to TiO_6 center presence in both of them. Titanium probably substitutes for Al in its octahedral site in either mineral. This is a rare luminescence in beryl/pezzotaite as it necessitates the presence of traces of Ti and the virtual absence of iron. BSL is limited in beryl to near colorless to pink varieties (goshenite, morganite) and it is not found in more common, iron-containing beryl varieties.

It is not surprising that titanate is found in a wide variety of minerals. Titanium is the ninth most abundant constituent of the earth's crust and mantle [71]. Naturally, it can be found as a component of numerous minerals such as rutile/anatase/brookite (TiO_2), ilmenite ($FeTiO_3$), sphene ($CaTi(SiO_4)O$) and many others. However, titanium is present in the majority of soils, rocks and sediments in very small quantities (a few percent to ppm). In many natural crystalline compounds, Ti has two main valences, Ti^{3+} and Ti^{4+} and is mainly found in octahedral coordination [72]. It is rarely found as TiO_4/TiO_5 [73–75]. As with many transition elements, the mere presence of a few ppm is a potential source of unique physical, in this case optical, properties. For decades, titanium has been known to be a cause of luminescence under cathode rays, ultraviolet light, flame excitation or heating in many compounds, so it is a well-established and likely luminophore or activator [76].

Based on the work of Sidike et al. 2010 and Satoh et al. 2017, correlations have been observed between the Ti-O distance and the emission wavelength (Figure 5) [16,77]. However, very few studies determine directly the real Ti-O distance in Ti-doped compounds. The Ti-O distance mentioned in many publications is the metal-oxygen distance in the titanium-free compound. When present as an impurity, the determination of the exact (average) Ti-O distance would require using synchrotron techniques, for example. Nevertheless, we plotted the position of the emission versus estimated Ti-O bond length (Figure 5) using measured or estimated Ti-O bond lengths available in the literature.

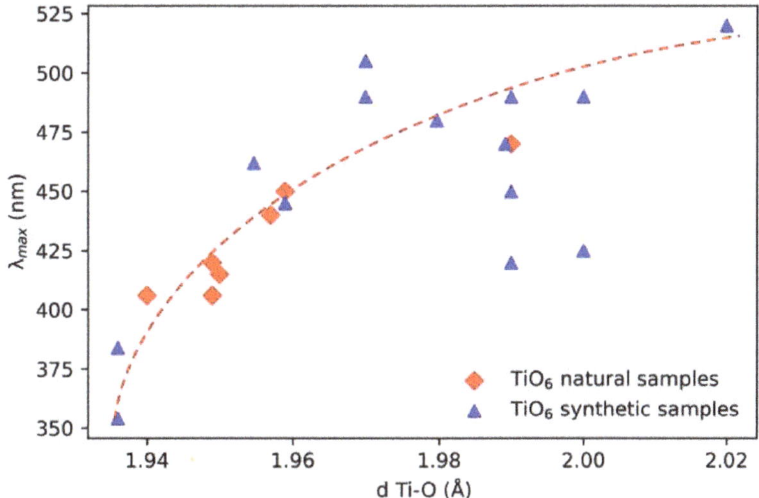

Figure 5. Position of the emission versus estimated Ti-O bond length in Ti-bearing compounds. Most of the Ti-O distances are estimated by structural analysis (DRX) or synchrotron. A visual guide, not a fit, in interrupted red indicates a trend. Data provided in SI.

A visual guide, not a fit, indicates a correlation trend between the length of the Ti-O distance and the luminescence emission maximum. Counter-examples are curiously located only in the 1.98 to 2.00 Å range. This means that emission is shifted from blue to green as Ti-O lengthens leading the blue-green, green, yellow luminescence due to titanate groups at longer Ti-O, superior to 2 Å although alternative possibilities have been proposed (see below).

Our assignment of BSL to the titanate group is based on four measurements. The coincidence for seven minerals (plus three references) is striking; yet it is always possible in luminescence studies that there is an alternate cause, due to a coincidence of spectral properties between two activators. Despite the difference in lifetime measurements, the resemblance between the blue emission of Pb^{2+} in hydrozincite (without titanium) and hydrozincite with titanium merits further investigation. To complete the present work, it would be interesting to probe further with time resolved luminescence, and also perform temperature studies.

The lack of information on the luminescence of minerals prevents us from mentioning further possible candidates for BSL. There are a number of articles regarding the luminescence of titanium-containing materials which indicate potential BSL, but often either the effective color or the excitation wavelengths are missing. Thus, the list of BSL materials might be much longer, but often only the emission data is provided, which does not stop many authors from interpreting the source of a simple blue luminescence as being due to titanates without chemical analyses proving the presence of Ti as a trace. Many blue-white luminescences are ascribed to TiO_6 in fluorescence database (such as Fluomin), for example murmanite $Na_4Ti_4(Si_2O_7)_2O_4\,4H_2O$, a potential candidate if more information is provided. The visual description is clearly insufficient to identify BSL in hibonite $Ca_2(Al,Ti)_{24}O_{38}$ [78]. A non-exhaustive list of many luminescent compounds satisfying at least in part BSL criteria is provided in SI.

Titanate related luminescence is not restricted to BSL. Some minerals emit a green to a yellow luminescence with the same excitation spectrum and similar lifetimes to BSL. This is recorded for natural minerals such as andalusite [54], topaz [41], cassiterite [79] lorenzite, talc, uvarovite [29], baghdadite [16], chrysoberyl: alexandrite [54] and enstatites [80]. This is also true of several chemical compounds such as $MgSn_2O_4$, zirconium titanate doped with lanthanum or $BaTi(PO_4)_2$ [81–83] In some of these compounds, titanium is present in TiO_5 groups, which may explain the shift of color [84–86]. It has also been proposed that there is a link between the shift of the emission wavelength and the amount of titanium in the solid, (in stannates [34]). Finally, some authors mention the presence of oxygen vacancies in TiO_5 polyhedra as an explanation for emissions in the green to yellow range [35,87,88]. These activators have also been proposed for BSL, so yellow and green emissions may be induced by titanate groups with different environments from those leading to BSL.

5. Conclusions

We have demonstrated for the first time that BSL is due to octahedral titanate groups in seven natural minerals and synthetic materials: dumortierite, hydrozincite, beryl, pezzotaite, some elbaites, some lead aluminosilicate glasses and some synthetic opals. This was established by comparing the luminescence properties with reference materials for which the origin of BSL has been firmly established as being due to TiO_6. BSL minerals are characterized by a broad emission band in the spectral range perceived as blue, between 400 and 500 nm (3.1 to 2.48 eV), respectively. This band is relatively broad (about 100 nm–near 0.6 eV). It is excited between 250 and 300 nm, with an excitation band about 50 nm wide (0.6 eV). Finally, the decay lifetime is of the order of ten microseconds, spreading from 2 to 40 microseconds.

It is apparent from a literature search that many more minerals or materials might possess a titanate-induced BSL. Yet, because of the lack of details regarding luminescence parameters (excitation spectrum and decay time, for example) one cannot prove the titanate origin of these emissions (see Tables S1 and S2 in Supplementary Materials). In addition, recent progress in luminescence modelling has started to confirm titanate origin where the origin of BSL was ambiguous. There is reason to believe that, in the future, the titanate activation of BSL will be demonstrated for an increasing number of minerals (natural or synthetic).

Supplementary Materials: The following supporting information can be downloaded at: https://www.mdpi.com/article/10.3390/min13010104/s1, Table S1: Mineral and synthetic compounds on which Ti-related blue luminescence (possibly BSL) has been described or inferred [1,27,29,39,79,83,88–101]. Table S2: Data associated with Figure 5, mineral & formula, Ti-O distance, emission maxima (in nm and eV), bibliographic sources for luminescence and Ti-O distance [1,9,16,29,40,77,89–92,102–119].

Author Contributions: Conceptualization, M.V.; Data curation, M.V.; Formal analysis, M.V.; Investigation, M.V. and C.L.; Methodology, M.V., E.F. and C.L.; Resources, E.F.; Supervision, E.F. and S.J.; Validation, E.F. and S.J.; Visualization, M.V., E.F., T.C., C.L. and S.J.; Writing–original draft, M.V., E.F. and C.L.; Writing–review and editing, M.V., E.F. All authors have read and agreed to the published version of the manuscript.

Funding: This research received no external funding.

Data Availability Statement: Not applicable.

Acknowledgments: The Authors thank Carole La for her help with LA-ICP-MS analyses. Cassandre Moinard performed many luminescence measurements. Glenn Waychunas provided initial information and samples on this topic.

Conflicts of Interest: The authors declare no conflict of interest.

References

1. Tang, H.; Berger, H.; Schmid, P.E.; Lévy, F.; Burri, G. Photoluminescence in TiO_2 anatase single crystals. *Solid State Commun.* **1993**, *87*, 847–850. [CrossRef]
2. Marshall, D.J. *Cathodoluminescence of Geological Materials*; Unwin-Hyman: Boston, MA, USA, 1988; 146p.
3. Garcia-Guinea, J.; Garrido, F.; Lopez-Arce, P.; Correcher, V.; de la Figuera, J. Spectral Green cathodoluminescence emission from surfaces of insulators with metal-hydroxyl bonds. *J. Lumin.* **2017**, *190*, 128–135. [CrossRef]
4. Takahashi, N.; Tsujimori, K.; Kayam, M.; Nishido, H. Cathodoluminescence petrography of P-type jadeitites from the New Idria serpentinite body, California. *J. Mineral. Petrol. Sci.* **2017**, *112*, 291–299. [CrossRef]
5. Blasse, G. The Luminescence of Closed-Shell Transition-Metal Complexes. New Developments. In *Luminescence and Energy Transfer*; Springer: Berlin/Heidelberg, Germany, 1980; pp. 1–41.
6. Hemphill, W.R.; Tyson, R.M.; Theisen, A.F. Spectral luminescence properties of natural specimens in the scheelite-powellite series, and an assessment of their detectivity with an airborne Fraunhofer line discriminator. *Econ. Geol.* **1988**, *83*, 637–646. [CrossRef]
7. Tyson, R.M.; Hemphill, W.R.; Theisen, A.F. Effect of the W:Mo ratio on the shift of excitation and emission spectra in the scheelite-powellite series. *Am. Mineral.* **1988**, *73*, 1145–1154.
8. Bode, J.H.G.; Van Oosterhout, A.B. Defect luminescence of ordered perovskites A_2BWO_6. *J. Lumin.* **1975**, *10*, 237–242. [CrossRef]
9. Gaft, M.; Nagli, l.; Waychunas, G.; Weiss, D. The nature of blue luminescence from natural benitoite $BaTiSi_3O_9$. *Phys. Chem. Miner.* **2004**, *31*, 365–373. [CrossRef]
10. Waychunas, G.A. Apatite luminescence. *Rev. Mineral. Geochem.* **2002**, *48*, 701–742. [CrossRef]
11. Nikbakht, T.; Kakuee, O.; Lamehi-rachti, M. An efficient ionoluminescence analysis of turquoise gemstone as a weakly luminescent mineral. *Spectrochim. Acta Part A Mol. Biomol. Spectrosc.* **2017**, *179*, 171–177. [CrossRef]
12. Gaft, M.; Seigel, H.; Panczer, G.; Reisfeld, R. Laser-induced time-resolved luminescence spectroscopy of Pb^{2+} in minerals. *Eur. J. Mineral.* **2002**, *14*, 1041–1048. [CrossRef]
13. Panczer, G.; Gaft, M.; De Ligny, D.; Boudeulle, M.; Champagnon, B. Luminescent centres in pezzottaite, $CsBe_2LiAl_2Si_6O_{18}$. *Eur. J. Mineral.* **2010**, *22*, 605–612. [CrossRef]
14. Chang, I.F.; Onton, A. Optical properties of photochromatic sulfur-doped chlorosodalite. *J. Electron. Mater.* **1973**, *2*, 17–46. [CrossRef]
15. Carvalho, J.M.; Rodrigues, L.C.; Hölsä, J.; Lastusaari, M.; Nunes, L.A.; Felinto, M.C.; Brito, H.F. Influence of titanium and lutetium on the persistent luminescence of ZrO_2. *Opt. Mater. Express* **2012**, *2*, 331–340. [CrossRef]
16. Sidike, A.; Kobayashi, S.; Zhu, H.-J.; Kusachi, I.; Yamashita, N. Photoluminescence of baratovite and katayamalite. *Phys. Chem. Miner.* **2010**, *37*, 705–710. [CrossRef]
17. Lupei, A.; Lupei, V.; Ionescu, C.; Tang, H.G.; Chen, M.L. Spectroscopy of Ti^{3+}: α-Al_2O_3. *Opt. Commun.* **1986**, *59*, 36–38. [CrossRef]
18. Lafargue-dit-Hauret, W.; Schira, R.; Latouche, C.; Jobic, S. Theoretical calculations meet experiment to explain the luminescence properties and the presence of defects in m-ZrO_2. *Chem. Mater.* **2021**, *33*, 2984–2992. [CrossRef]
19. Takahashi, Y.; Kitamura, K.; Iyi, N.; Inoue, S.; Fujiwara, T. Blue photoluminescence of germania-stabilized benitoite. *J. Ceram. Soc. Jpn.* **2008**, *116*, 1143–1146. [CrossRef]
20. Pathak, N.; Ghosh, P.S.; Gupta, S.K.; Mukherjee, S.; Kadam, R.; Arya, A. An insight into the various defects-induced emission in $MgAl_2O_4$ and their tunability with phase behavior: Combined experimental and theoretical approach. *J. Phys. Chem. C* **2016**, *120*, 4016–4031. [CrossRef]

21. Sawai, S.; Uchino, T. Visible photoluminescence from $MgAl_2O_4$ spinel with cation disorder and oxygen vacancy. *J. Appl. Phys.* **2012**, *112*, 103523. [CrossRef]
22. Raj, S.S.; Gupta, S.K.; Grover, V.; Muthe, K.P.; Natarajan, V.; Tyagi, A.K. $MgAl_2O_4$ spinel: Synthesis, carbon incorporation and defect-induced luminescence. *J. Mol. Struct.* **2015**, *1089*, 81–85. [CrossRef]
23. Mikhailik, V.B.; Kraus, H.; Balcerzyk, M.; Czarnacki, W.; Moszynski, M.; Mykhaylyk, M.S.; Wahl, D. Low-temperature spectroscopic and scintillation characterisation of Ti-doped Al_2O_3. *Nucl. Instrum. Methods Phys. Res. Sect. A Accel. Spectrometers Detect. Assoc. Equip.* **2005**, *546*, 523–534. [CrossRef]
24. Chen, W.; Tang, H.; Shi, C.; Deng, J.; Shi, J.; Zhou, Y.; Yin, S. Investigation on the origin of the blue emission in titanium doped sapphire: Is F^+ color center the blue emission center? *Appl. Phys. Lett.* **1995**, *67*, 317–319. [CrossRef]
25. Mikhailik, V.B.; Di Stefano, P.C.F.; Henry, S.; Kraus, H.; Lynch, A.; Tsybulskyi, V.; Verdier, M.A. Studies of concentration dependences in the luminescence of Ti-doped Al_2O_3. *J. Appl. Phys.* **2011**, *109*, 053116. [CrossRef]
26. Pallotti, D.K.; Passoni, L.; Maddalena, P.; Di Fonzo, F.; Lettieri, S. Photoluminescence mechanisms in anatase and rutile TiO_2. *J. Phys. Chem. C* **2017**, *121*, 9011–9021. [CrossRef]
27. Norrbo, I.; Gluchowski, P.; Hyppanen, I.; Laihinen, T.; Laukkanen, P.; Makela, J.; Lastusaari, M. Mechanisms of tenebrescence and persistent luminescence in synthetic hackmanite $Na_8Al_6Si_6O_{24}(Cl,S)_2$. *ACS Appl. Mater. Interfaces* **2016**, *8*, 11592–11602. [CrossRef]
28. Andrade LH, C.; Lima, S.M.; Novatski, A.; Neto, A.M.; Bento, A.C.; Baesso, M.L.; Boulon, G. Spectroscopic assignments of Ti^{3+} and Ti^{4+} in titanium-doped OH− free low-silica calcium aluminosilicate glass and role of structural defects on the observed long lifetime and high fluorescence of Ti^{3+} ions. *Phys. Rev. B* **2008**, *78*, 224202. [CrossRef]
29. Gorobets, B.S.; Rogojine, A.A. *Luminescent Spectra of Minerals: Reference-Book*; All-Russia Institute of Mineral Resources (VIMS): Moscow, Russia, 2002; 302p.
30. Taher, M.A.; Asadollahzadeh, H.; Fazelirad, H. Determination of trace amounts of iron by a simple fluorescence quenching method. *Anal. Methods* **2015**, *7*, 6726–6731. [CrossRef]
31. Al-kady, A.S.; Gaber, M.; Hussein, M.M.; Ebeid, E.Z.M. Structural and fluorescence quenching characterization of hematite nanoparticles. *Spectrochim. Acta Part A Mol. Biomol. Spectrosc.* **2011**, *83*, 398–405. [CrossRef]
32. Amthauer, G.; Rossman, G.R. Mixed valence of iron in minerals with cation clusters. *Phys. Chem. Miner.* **1984**, *11*, 37–51. [CrossRef]
33. Fritsch, E.; Waychunas, G.A. *Gemstones. M. Robbins, Fluorescence*; Geoscience Press: Phoenix, AZ, USA, 1994; pp. 163–165.
34. Macke, A.J.H. Luminescence and energy transfer in the ordered perovskite system $La_2MgSn_{1-x}Ti_xO_6$. *Phys. Status Solidi* **1977**, *39*, 117–123. [CrossRef]
35. Macke, A.J.H. Investigations on the luminescence of titanium-activated stannates and zirconates. *J. Solid State Chem.* **1976**, *18*, 337–346. [CrossRef]
36. Wu, Q.; Hu, Z. Na_2TiGeO_5—A self-light-emitting phosphor with the stable structure and tunable emission resulted from Cr^{3+}-doped for FEDs. *J. Am. Ceram. Soc.* **2019**, *102*, 2727–2736.
37. Takahashi, Y.; Masai, h.; Fujiwara, T.; Kitamura, K.; Inoue, S. Afterglow in synthetic bazirite, $BaZrSi_3O_9$. *J. Ceram. Soc. Jpn.* **2008**, *116*, 357–360. [CrossRef]
38. Smith, A.L. Some new complex silicate phosphors containing calcium, magnesium, and beryllium. *J. Electrochem. Soc.* **1949**, *96*, 287. [CrossRef]
39. Evans, B.D.; Pogatshnik, G.J.; Chen, Y. Optical properties of lattice defects in α-Al_2O_3. *Nucl. Instrum. Methods Phys. Res. Sect. B: Beam Interact. Mater. At.* **1994**, *91*, 258–262. [CrossRef]
40. Bausa, L.E.; Vergara, I.; Jaque, F.; Sole, J.G. Ultraviolet laser excited luminescence of Ti-sapphire. *J. Phys. Condens. Matter* **1990**, *2*, 9919. [CrossRef]
41. Marques, C.; Santos, L.; Falcao, A.N.; Silva, R.C.; Alves, E. Luminescence studies in colour centres produced in natural topaz. *J. Lumin.* **2000**, *87*, 583–585. [CrossRef]
42. Iida, Y.; Sawamura, K.; Iwasaki, K.; Nakanishi, T.; Iwakura, F.; Nakajima, Y.; Yasumori, A. Persistent Luminescence Properties of Ti^{4+}-doped $K_2ZrSi_3O_9$ Wadeite. *Sens. Mater.* **2020**, *32*, 1427–1433. [CrossRef]
43. Moon, C.; Nishi, M.; Miura, K.; Hirao, K. Blue long-lasting phosphorescence of Ti-doped $BaZrO_3$ perovskites. *J. Lumin.* **2009**, *129*, 817–819. [CrossRef]
44. Cavignac, T.; Jobic, S.; Latouche, C. Modeling Luminescence Spectrum of $BaZrO_3$: Ti Including Vibronic Coupling from First Principles Calculations. *J. Chem. Theory Comput.* **2022**, *18*, 7714–7721. [CrossRef]
45. Konijnendijk, W.L. Luminescence of $BaSnSi_3O_9$: Ti^{4+} compared to $BaZrSi_3O_9$: Ti^{4+}. *Inorg. Nucl. Chem. Lett.* **1981**, *17*, 129–132. [CrossRef]
46. Zeng, G.; Dong, Q.; Bao, W. Preparation and Photoluminescence Properties of Ti-Doped Lu_2O_3 Powder. *J. Appl. Spectrosc.* **2016**, *83*, 460–465. [CrossRef]
47. Kawano, T.; Yamane, H. Synthesis, crystal structure analysis, and photoluminescence of Ti^{4+}-doped $Mg_5SnB_2O_{10}$. *Chem. Mater.* **2010**, *22*, 5937–5944. [CrossRef]
48. Konijnendijk, W.L.; Blasse, G. On the luminescence of Ti^{4+} in $Mg_5SnB_2O_{10}$ and $Mg_3ZrB_2O_8$. *Mater. Chem. Phys.* **1985**, *12*, 591–599. [CrossRef]

49. Cong, Y.; Li, B.; Yue, S.; Fan, D.; Wang, X.J. Effect of oxygen vacancy on phase transition and photoluminescence properties of nanocrystalline zirconia synthesized by the one-pot reaction. *J. Phys. Chem. C* **2009**, *113*, 13974–13978. [CrossRef]

50. Jochum, K.P.; Weis, U.; Stoll, B.; Kuzmin, D.; Yang, Q.; Raczek, I.; Enzweiler, J. Determination of reference values for NIST SRM 610–617 glasses following ISO guidelines. *Geostand. Geoanalytical Res.* **2011**, *35*, 397–429. [CrossRef]

51. Van Achterbergh, E.; Ryan, C.G.; Griffin, W.L. Glitter: On-line interactive data reduction for the laser ablation inductively coupled plasma mass spectrometry microprobe. In Proceedings of the Ninth Annual VM Goldschmidt Conference, Cambridge, MA, USA, 22–27 August 1999; p. 7215.

52. Sun, R.; Ooi, Y.K.; Dickens, P.T.; Lynn, K.G.; Scarpulla, M.A. On the origin of red luminescence from iron-doped β-Ga$_2$O$_3$ bulk crystals. *Appl. Phys. Lett.* **2020**, *117*, 052101. [CrossRef]

53. Gaillou, E.; Fritsch, E.; Massuyeau, F. Luminescence of gem opals: A review of intrinsic an extrinsic emission. *Aust. Gemmol.* **2012**, *24*, 365–373.

54. Gaft, M.; Reisfeld, R.; Panczer, G. *Modern Luminescence Spectroscopy of Minerals and Materials*; Springer: Berlin/Heidelberg, Germany, 2015; 606p.

55. Curtice, R.E.; Scott, A.B. The luminescence of thallium (I) halo complexes. *Inorg. Chem.* **1964**, *3*, 1383–1387. [CrossRef]

56. Ogorodnikov, I.N.; Pustovarov, V.A.; Puzikov, V.M.; Salo, V.I.; Voronov, A.P. A luminescence and absorption spectroscopy study of KH$_2$PO$_4$ crystals doped with Tl$^+$ ions. *Opt. Mater.* **2012**, *34*, 1522–1528. [CrossRef]

57. Kirk, R.D. Role of sulfur in the luminescence and coloration of some aluminosilicates. *J. Electrochem. Soc.* **1954**, *101*, 461. [CrossRef]

58. Stoliaroff, A.; Schira, R.; Blumentritt, F.; Fritsch, E.; Jobic, S.; Latouche, C. Point Defects Modeling Explains Multiple Sulfur Species in Sulfur-Doped Na$_4$(Al$_3$Si$_3$O$_{12}$)Cl Sodalite. *J. Phys. Chem. C* **2021**, *125*, 16674–16680. [CrossRef]

59. Folkerts, H.F.; Blasse, G. Two types of luminescence from Pb^{2+} in alkaline-earth carbonates with the aragonite structure. *J. Phys. Chem. Solids* **1996**, *57*, 303–306. [CrossRef]

60. Lin, J.; Su, Q. Luminescence of Pb^{2+} and energy transfer from Pb^{2+} to rare earth ions in silicate oxyapatites. *Phys. Status Solidi* **1996**, *196*, 261–267. [CrossRef]

61. Folkerts, H.F.; Blasse, G. Luminescence of Pb^{2+} in several calcium borates. *J. Mater. Chem.* **1995**, *5*, 273–276. [CrossRef]

62. Folkerts, H.F.; Hamstra, M.A.; Blasse, G. The luminescence of Pb^{2+} in alkaline earth sulfates. *Chem. Phys. Lett.* **1995**, *246*, 135–138. [CrossRef]

63. Asano, S.; Yamashita, N. Effet du champ magnétique sur la luminescence de l'ion Pb^{2+} dans les luminophores CaO, CaS, CaSe et MgS. *Phys. Status Solidi* **1981**, *108*, 549–558. [CrossRef]

64. Jary, V.; Nikl, M.; Mihokova, E.; Bohacek, P.; Trunda, B.; Polak, K.; Studnicka, V.; Mucka, V. Photoluminescence of Pb^{2+}-doped SrHfO$_3$. *Radiat. Meas.* **2010**, *45*, 406–408. [CrossRef]

65. Tsuboi, T. Temporal Evolution of Luminescence by Pb^{2+}-Doped KBr and KCl Phosphors. *Electrochem. Solid-State Lett.* **2000**, *3*, 200. [CrossRef]

66. Folkerts, H.F.; Blasse, G. Luminescence of Pb^{2+} in SrTiO$_3$. *Chem. Mater.* **1994**, *6*, 969–972. [CrossRef]

67. Skuja, L. Optically active oxygen-deficiency-related centers in amorphous silicon dioxide. *J. NON-Cryst. Solids* **1998**, *239*, 16–48. [CrossRef]

68. Yu, D.P.; Hang, Q.L.; Ding, Y.; Zhang, Z.G.; Bai, Z.G.; Wang, J.J.; Zou, Y.H.; Qian, Q.; Xiong, G.C.; Feng, S.Q. Amorphous silica nanowires: Intensive blue light emitters. *Appl. Phys. Lett.* **1998**, *73*, 3076–3078. [CrossRef]

69. Nassau, K. *Gems Made by Man*, 1st ed.; Chilton Book Co.: Radnor, PA, USA, 1980; p. 364.

70. Gauthier, J.P. Observation directe par microscopie électronique à transmission de diverses variétés d'opale. II. Opale synthétique. *J. De Microsc. Et De Spectrosc. Electron.* **1986**, *11*, 37–52.

71. Mason, B. *Principle of Geochemistry*, 3rd ed.; John Wiley and Sons: New York, NY, USA, 1952; p. 276.

72. Waychunas, G.A. Synchrotron radiation XANES spectroscopy of Ti in minerals; effects of Ti bonding distances, Ti valence, and site geometry on absorption edge structure. *Am. Mineral.* **1987**, *72*, 89–101.

73. Morrison, G.; Christian, M.S.; Besmann, T.M.; Zur Loye, H.C. Flux Growth of Uranyl Titanates: Rare Examples of TiO$_4$ tetrahedra and TiO$_5$ square bipyramids. *J. Phys. Chem. A* **2020**, *124*, 9487–9495. [CrossRef]

74. Jiang, N.; Su, D.; Spence, J.C.H. Determination of Ti coordination from pre-edge peaks in Ti K-edge XANES. *Phys. Rev. B* **2007**, *76*, 214117. [CrossRef]

75. Ichihashi, Y.; Yamashita, H.; Anpo, M.; Souma, Y.; Matsumura, Y. Photoluminescence properties of tetrahedral titanium oxide species in zeolitic materials. *Catal. Lett.* **1998**, *53*, 107–109. [CrossRef]

76. Nichols, E.L. The luminescence of titanium oxide. *Phys. Rev.* **1923**, *22*, 420. [CrossRef]

77. Satoh, Y.; Takemoto, M. Charge transfer-type fluorescence of Ti-doped Ca$_{14}$Al$_{10}$Zn$_6$O$_{35}$. *J. Lumin.* **2017**, *185*, 141–144. [CrossRef]

78. Keil, K.; Fuchs, L.H. Hibonite [Ca$_2$(Al,Ti)$_{24}$O$_{38}$] from the Leoville and Allende chondritic meteorites. *Earth Planet. Sci. Lett.* **1971**, *12*, 184–190. [CrossRef]

79. Hall, M.R.; Ribbe, P.H. An electron microprobe study of luminescence centers in cassiterite. *Am. Mineral. J. Earth Planet. Mater.* **1971**, *56*, 31–45.

80. Bloise, A.; Pingitore, V.; Miriello, D.; Apollaro, C.; Armentano, D.; Barrese, E.; Oliva, A. Flux growth and characterization of Ti-and Ni-doped enstatite single crystals. *J. Cryst. Growth* **2011**, *329*, 86–91. [CrossRef]

81. Zhang, S.; Zhang, P.; Liu, X.; Yang, Z.; Huang, Y.; Seo, H.J. Synthesis, structure, and luminescence of Eu^{3+}-activated La$_4$Ti$_3$O$_{12}$ nanoparticles with layered perovskite structure. *J. Am. Ceram. Soc.* **2019**, *102*, 1784–1793. [CrossRef]

82. Freitas, G.F.G.; Nasar, R.S.; Cerqueira, M.; Melo, M.; Longo, E.; Varela, J.A. Luminescence in semi-crystalline zirconium titanate doped with lanthanum. *Mater. Sci. Eng. A* **2006**, *434*, 19–22. [CrossRef]
83. Blasse, G.; Dirksen, G.J. The luminescence of barium titanium phosphate, $BaTi(PO_4)_2$. *Chem. Phys. Lett.* **1979**, *62*, 19–20. [CrossRef]
84. Li, Y.; Zhou, Y.; Xiao, J.; Yang, D.; Dai, L.; Yang, Y.; Zhao, L. A rare-earth-free self-activated phosphor: Li_2TiSiO_5 with TiO_5 square pyramids. *N. J. Chem.* **2020**, *44*, 5828–5833. [CrossRef]
85. Ding, J.; Li, Y.; Wu, Q.; Long, Q.; Wang, Y.; Wang, Y. A novel self-activated white-light-emitting phosphor of Na_2TiSiO_5 with two Ti sites of TiO_5 and TiO_6. *RSC Adv.* **2016**, *6*, 8605–8611. [CrossRef]
86. Takahashi, Y.; Kitamura, K.; Iyi, N.; Inoue, S. Visible orange photoluminescence in a barium titanosilicate $BaTiSi_2O_7$. *Appl. Phys. Lett.* **2006**, *88*, 151903. [CrossRef]
87. De Haart, L.G.J.; De Vries, A.J.; Blasse, G. On the photoluminescence of semiconducting titanates applied in photoelectrochemical cells. *J. Solid State Chem.* **1985**, *59*, 291–300. [CrossRef]
88. Huang, Y.; Tsuboi, T.; Seo, H.J. A high efficient $Ba_2TiP_2O_9$ phosphor for X-ray and UV excitation. *Ceram. Int.* **2013**, *39*, 861–864. [CrossRef]
89. Bouma, B.; Blasse, G. Dependence of luminescence of titanates on their crystal structure. *J. Phys. Chem. Solids* **1995**, *56*, 261–265. [CrossRef]
90. De Figueiredo, A.T.; Longo, V.M.; de Lazaro, S.; Mastelaro, V.R.; De Vicente, F.S.; Hernandes, A.C.; Longo, E. Blue-green and red photoluminescence in $CaTiO_3$: Sm. *J. Lumin.* **2007**, *126*, 403–407. [CrossRef]
91. Ma, Q.; Zhang, A.; Lue, M.; Zhou, Y.; Qiu, Z.; Zhou, G. Novel class of aeschynite structure $LaNbTiO_6$-based orange-red phosphors via a modified combustion approach. *J. Phys. Chem. B* **2007**, *111*, 12693–12699. [CrossRef]
92. Lin, Y.; Nan, C.W.; Wang, J.; He, H.; Zhai, J.; Jiang, L. Photoluminescence of nanosized $Na_{0.5}Bi_{0.5}TiO_3$ synthesized by a sol–gel process. *Mater. Lett.* **2004**, *58*, 829–832. [CrossRef]
93. Parsons, I.; Steele, D.A.; Lee, M.R.; Magee, C.W. Titanium as a cathodoluminescence activator in alkali feldspars. *Am. Mineral.* **2008**, *93*, 875–879. [CrossRef]
94. Cathelineau, T. Catapleiite from Mont Saint-Hilaire, Québec, Canada. *J. Gemmol.* **2020**, *37*, 237–239. [CrossRef]
95. Silva Santos, H.; Norrbo, I.; Laihinen, T.; Sinkkonen, J.; Mäkilä, E.; MCarvalho, J.; Lastusaari, M. Synthesis and Features of Luminescent Bromo-and Iodohectorite Nanoclay Materials. *Appl. Sci.* **2017**, *7*, 1243. [CrossRef]
96. Yamashita, T.; Ueda, K. Blue photoluminescence in Ti-doped alkaline-earth stannates. *J. Solid State Chem.* **2007**, *180*, 1410–1413. [CrossRef]
97. Xia, M.; Zhang, Y.; Li, M.; Zhong, Y.; Gu, S.; Zhou, N.; Zhou, Z. High thermal stability and blue-violet emitting phosphor $CaYAlO_4$: Ti^{4+} with enhanced emission by Ca^{2+} vacancies. *J. Rare Earths* **2020**, *38*, 227–233. [CrossRef]
98. Wong, W.C.; Danger, T.; Huber, G.; Petermann, K. Spectroscopy and excited-state absorption of Ti^{4+}: $Li_4Ge_5O_{12}$ and Ti^{4+}: Y_2SiO_5. *J. Lumin.* **1997**, *72*, 208–210. [CrossRef]
99. Lin, C.M.; Lai, Y.-S.; Chen, J.S. Photoluminescence behavior of Ti-doped Zn_2SiO_4 thin film phosphors. *ECS Trans.* **2006**, *1*, 1. [CrossRef]
100. Haldar, P.K.; Dey, S.; Mukhopadhyay, S.; Mukhopahyay, S.; Parya, T.K. Structural and optical properties of Ti^{4+} doped sintered $ZnAl_2O_4$ ceramics. *Interceram-Int. Ceram. Rev.* **2014**, *63*, 382–385. [CrossRef]
101. Alajerami, Y.S.M.; Hashim, S.; Hassan, W.M.S.W.; Ramli, A. The effect of titanium oxide on the optical properties of lithium potassium borate glass. *J. Mol. Struct.* **2012**, *1026*, 159–167. [CrossRef]
102. Fischer, K. Verfeinerung der Kristallstruktur von Benitoit BaTi [Si309]. *Z. Für Krist.-Cryst. Mater.* **1969**, *129*, 222–243. [CrossRef]
103. Sandomirskii, P.A.; Simonov, M.A.; Belov, N.V. Crystal structure of baratovite, $KLi_3Ca_7Ti_2(Si_6O_{18})_2F_2$. In *Doklady Akademii Nauk*; Russian Academy of Sciences: Moscow, Russia, 1976; pp. 615–618.
104. Kato, T.; Murakami, N. The crystal structure of katayamalite. *Mineral. J.* **1985**, *12*, 206–217. [CrossRef]
105. Sundberg, M.R.; Lehtinen, M.; Kivekas, R. Refinement of the crystal structure of ramsayite (lorenzenite). *Am. Mineral.* **1987**, *72*, 173–177.
106. Speer, J.A.; Gibbs, G.V. The crystal structure of synthetic titanite, $CaTiOSiO_4$, and the domain textures of natural titanites. *Am. Mineral.* **1976**, *61*, 238–247.
107. Merlino, S.; Pasero, M.; Artioli, G.; Khomyakov, A.P. Penkvilksite, a new kind of silicate structure: OD character, X-ray single-crystal (1 M), and powder Rietveld (2O) refinements of two MDO polytypes. *Am. Mineral.* **1994**, *79*, 1185–1193.
108. Horn, M.S.C.F.; Schwebdtfeger, C.F.; Meagher, E.P. Refinement of the structure of anatase at several temperatures. *Z. Krist.-Cryst. Mater.* **1972**, *136*, 273–281.
109. Monarumit, N.; Wongkokua, W.; Satitkune, S. Oxidation state of Ti atoms and Ti-O bond length on natural sapphire gem-materials probed by X-ray absorption spectroscopy. In *Key Engineering Materials*; Trans Tech Publications Ltd.: Wollerau, Switzerland, 2017; pp. 585–589.
110. Sasaki, S.; Prewitt, C.T.; Bass, J.D.; Schulze, W.A. Orthorhombic perovskite $CaTiO_3$ and $CdTiO_3$: Structure and space group. *Acta Crystallogr. Sect. C: Cryst. Struct. Commun.* **1987**, *43*, 1668–1674. [CrossRef]
111. Yakubovich, O.V.; Kireev, V.V. Refinement of the crystal structure of $Na_2Ti_3O_7$. *Crystallogr. Rep.* **2003**, *48*, 24–28. [CrossRef]
112. Ziadi, A.; Thiele, G.; Elouadi, B. The crystal structure of Li_2TiSiO_5. *J. Solid State Chem.* **1994**, *109*, 112–115. [CrossRef]
113. Nyman, H.; O'Keeffe, M.; Bovin, J.O. Sodium titanium silicate, Na_2TiSiO_5. *Acta Crystallogr. Sect. B: Struct. Crystallogr. Cryst. Chem.* **1978**, *34*, 905–906. [CrossRef]

114. Longo, J.M.; Kierkegaard, P. The crystal structure of $NbOPO_4$. *Acta Chem. Scand.* **1966**, *20*, 72–78. [CrossRef]

115. Robertson, A.; Fletcher, J.G.; Skakle, J.M.S.; West, A.R. Synthesis of $LiTiPO_5$ and $LiTiAsO_5$ with the α-Fe_2PO_5 structure. *J. Solid State Chem.* **1994**, *109*, 53–59. [CrossRef]

116. Culbertson, C.M.; Flak, A.T.; Yatskin, M.; Cheong, P.H.Y.; Can, D.P.; Dolgos, M.R. Neutron total scattering studies of group ii titanates ($ATiO_3$, A^{2+}= Mg, Ca, Sr, Ba). *Sci. Rep.* **2020**, *10*, 1–10. [CrossRef]

117. Tordjman, P.I.; Masse, E.; Guitel, J.C. Structure cristalline du monophosphate $KTiPO_5$. *Z. Krist.* **1974**, *139*, 103–115. [CrossRef]

118. Golobič, A.; Škapin, S.D.; Suvorov, D.; Meden, A. Solving structural problems of ceramic materials. *Croat. Chem. Acta* **2004**, *77*, 435–446.

119. Yoneda, Y.; Noguchi, Y. Nanoscale structural analysis of $Bi_{0.5}Na_{0.5}TiO_3$. *Jpn. J. Appl. Phys.* **2020**, *59*, SPPA01. [CrossRef]

Article

Study of 405 nm Laser-Induced Time-Resolved Photoluminescence Spectroscopy on Spinel and Alexandrite

Wenxing Xu [1,*], Tsung-Han Tsai [2,*] and Aaron Palke [2]

[1] Independent Researcher, Cheerstrasse 13, 6014 Lucerne, Switzerland
[2] Gemological Institute of America, 50 W 47th Street, New York, NY 10036, USA
* Correspondence: wenxingx@gmail.com (W.X.); stsai@gia.edu (T.-H.T.)

Abstract: Research on photoluminescence spectroscopy on Cr-doped gem materials has demonstrated great success regarding the identification of gemstones in terms of building rapid test systems. In this study, 405 nm photoluminescence spectroscopy was used to measure the luminescence decay profiles of dozens of natural and lab-grown spinel (including heated spinel) and alexandrite. Spinel and alexandrite are both capable of producing photoluminescence with a long lifetime: spinel between 9 and 23 microseconds and alexandrite from 25 to 53 microseconds. The photoluminescence lifetime and exponential parameters of the half-life demonstrated notable differences in the ranges of decay times between natural, heated, and lab-grown versions of these materials.

Keywords: time-resolved spectroscopy; spinel; alexandrite; photoluminescence lifetime; fluorescence decay; gemstone testing

Citation: Xu, W.; Tsai, T.-H.; Palke, A. Study of 405 nm Laser-Induced Time-Resolved Photoluminescence Spectroscopy on Spinel and Alexandrite. *Minerals* **2023**, *13*, 419. https://doi.org/10.3390/min13030419

Academic Editor: Jordi Ibanez-Insa

Received: 10 February 2023
Revised: 8 March 2023
Accepted: 14 March 2023
Published: 16 March 2023

1. Introduction

Studies of photoluminescence (PL) spectroscopy on colored gemstones have been successful in identifying different gemstone species such as corundum, spinel, emerald, and alexandrite, which can be used in rapid testing systems [1]. Time-resolved PL spectroscopy can detect events within the environment of a fluorophore. The events that can be measured are the types of decay, indicated by a decrease in PL after excitation [2]. This technique has been widely used in the analysis of biomolecular structures and the nitrogen-vacancy defects of diamond materials (e.g., [3,4]).

Spinel crystallizes in a cubic crystal system, belonging to the mineral class of "oxides and hydroxides", and possesses the idealized chemical composition of $MgAl_2O_4$, wherein Mg is a divalent cation at the tetrahedrally coordinated site and Al is trivalent and situated at the octahedrally coordinated site. Red spinel is a variant colored by Cr. The PL spectra of spinel are widely used to distinguish natural from flux lab-grown spinel or heated spinel (e.g., [5,6]). However, flux lab-grown and heated spinel still show overlapping Cr PL spectra and require additional chemical testing or gemological observation to separate them.

Chrysoberyl, $BeAl_2O_4$, is an orthorhombic mineral crystallizing in the space group Pbmn [7]. Chrysoberyl and spinel both possess close-packed oxygen structures, wherein chrysoberyl's is hexagonally close-packed (hcp) and spinel's is cubic close-packed (ccp). Chrysoberyl has a lower structural symmetry than spinel because Be^{2+} is a significantly smaller cation than Al^{3+} [8]. Beryllium occupies tetrahedra and Al occupies two types of slightly distorted interstitial octahedral sites: B1 octahedra with Ci symmetry and B2 octahedra with Cs symmetry [9]. These octahedra have different volumes; additionally, B2 has an Al–O distance of 1.936 Å, which is larger than B1, with an Al–O distance of 1.890 Å [10]. Alexandrite is a type of chromium-doped bearing chrysoberyl in which Cr^{3+} is substituted in the Al^{3+} sites.

This study attempted to measure the PL lifetime of spinel and alexandrite. Furthermore, decay profiles were used to characterize their different variants: natural, flux lab-grown and heated spinel, and natural and lab-grown alexandrite. Time-resolved PL spectroscopy has the potential to be applied as an additional luminescence-testing technique for gemstone identification.

Spinel has the most characteristic photoluminescence spectra ranging from 650 to 750 nm. Gaft et al. assigned luminescence bands at 677, 685, 697, 710, and 718 nm to Cr^{3+} [11]. The most intensive luminescence duplet of the studied sample, which was located at 685 and 687 nm, belonged to the R1 and R2 lines. The R lines were purely electronic and were collectively assigned to the most intense centers. The R lines arose from the state of the Cr^{3+} ion, which corresponded to an ideal short-range order due to the spin-forbidden transition [12,13]. The most intensive peak of natural spinel was located at 685 nm, while those of the flux lab-grown and heated natural spinel were located at 687 nm, showing broader peak widths (Figure 1).

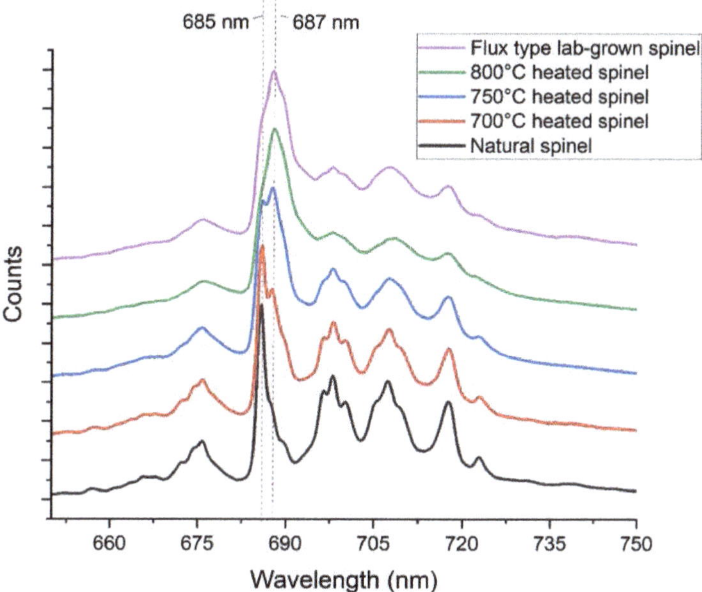

Figure 1. Photoluminescence spectra of the natural, heated, and flux lab-grown spinel.

Several powder- or single-crystal diffraction studies on the $MgAl_2O_4$ spinel after quenching from high temperature determined that high temperatures cause the spinel to exhibit order–disorder behavior (e.g., [14–21]). A Mg cation exchanges its site with an Al cation; this formula can be described as $(Mg_{1-x}Al_x)M(Al_{2-x}Mg_x)O_4$, wherein x is between 0 and 1 [15]. Flux lab-grown spinel is crystallized from low-temperature melting at around 1000 °C; therefore, it presents the same degree of disorder in its crystal structure as the heated natural spinel. Since the luminescence properties, radiative transition characteristics, and emissions at selective site excitation depend on the symmetry of the luminescence center environment [11], the Cr^{3+} photoluminescence spectra of the natural heated and lab-grown spinel showed corresponding modifications, which were caused by their respective heating temperatures [12,13,22–28].

There are two types of Cr^{3+} luminescence centers present in alexandrite: R-lines at approximately 679 with 677.3 nm and at 694.4 with 691.7 nm, which are accompanied by N-lines of Cr–Cr pairs located at 644, 650, 653, 667, 669, 678, 680, 690, 694, 702, 707, and 716 nm [11,29,30] (Figure 2). The luminescence band located at 702 nm may be attributed to V^{2+} [11]. The R-lines in chrysoberyl were divided into the Rm (mirror, B2 octahedron) and Ri (inversion, B1 octahedron) lines, which were located at 678 nm (Rm-line) and

690 nm (Ri-line). Numerous experiments have indicated the uneven distribution of cations between the two non-equivalent octahedral sites in chrysoberyl's structure [6,31,32]. The most intensive luminescence duplet at 679 nm was caused by the high Cr^{3+} content at the B2 site with Cs symmetry, where about 70% of the chromium is present [33]. Ferric iron prefers the more symmetrical B1 site, whereas the trivalent chromium cations responsible for the color of alexandrite occupy the B2 site [34].

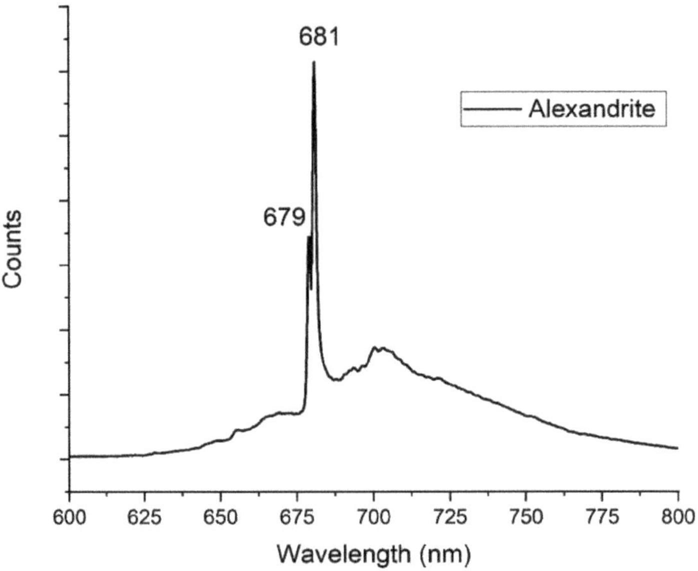

Figure 2. Photoluminescence spectra of alexandrite.

The photoluminescence lifetime is the time measured for the excited energy to decay exponentially to N/e (36.8%) of the original state via the loss of energy through fluorescence or non-radiative processes. The fluorescence lifetime is affected by external factors such as temperature, polarity, and the presence of fluorescence quenchers, and also depends on the internal structure of the fluorophore complexity [35]. In this study, all the PL decay was measured of the gem-quality mineral material at room temperature.

2. Materials and Methods

In this study, eight gem-quality, faceted, and well-formed octahedral, rough, natural, red-to-pink spinel were selected that originated from Burma and Sri Lanka. To avoid the use of chemically homogenous materials, we deliberately selected spinel from different origins. Additionally, four flux-type lab-grown spinel were selected. Both natural and flux-type lab-grown spinel had the stoichiometry ratio of $MgO:Al_2O_3$ 1:1.

To create heated natural spinel, three rough spinel samples were cut in half and heated in air. The goal of the heat experiments was to generate fluorescence with different ratios between the 685 and 687 nm peaks. Changes in the photoluminescence spectra after heat treatment have previously been observed and attributed to the structural rearrangement during tempering [5,6,22,36]: the peaks at 685 nm and 687 nm changed the relative intensities in the range between 700 and 800 °C; above 800 °C, the R-line peak completely shifted to 687 nm, and there were no further changes in the photoluminescence spectra. Subsequently, the heating temperatures were set at three levels: 700 °C, 750 °C, and 800 °C for 24 h.

Moreover, six natural alexandrite and five lab-grown alexandrite samples with mixed origins were studied. The alexandrite samples were all of a faceted gem quality. Further information on the samples is listed in Table 1.

Table 1. List of the selected spinel and alexandrite samples.

Sample	Type	Shape	Weight (Carats)	Color
NatSp1	Natural spinel	Rough	0.89	pink
NatSp2	Natural spinel	Rough	0.75	pink
NatSp3	Natural spinel	Rough	0.92	pink
NatSp4	Natural spinel	Faceted	3.56	Purple red
NatSp5	Natural spinel	Faceted	5.12	Red
NatSp6	Natural spinel	Rough	1.18	Pink
NatSp7	Natural spinel	Rough	1.32	Pink
NatSp8	Natural spinel	Rough	1.15	Pink
FLGSp1	Lab-grown spinel	Faceted	2.14	Red
FLGSp2	Lab-grown spinel	Polished Slab	0.53	Red
FLGSp3	Lab-grown spinel	Polished Slab	0.56	Red
FLGSp4	Lab-grown spinel	Polished Slab	0.76	Red
NatAx1	Natural alexandrite	Faceted	1.23	Dark green to purple
NatAx2	Natural alexandrite	Faceted	1.43	Dark green to purple
NatAx3	Natural alexandrite	Faceted	1.35	Dark green to purple
NatAx4	Natural alexandrite	Faceted	1.56	Dark green to purple
NatAx5	Natural alexandrite	Faceted	1.11	Dark green to purple
NatAx6	Natural alexandrite	Faceted	1.89	Dark green to purple
LGAx1	Lab-grown alexandrite	Faceted	2.01	Green to red purple
LGAx2	Lab-grown alexandrite	Faceted	1.98	Green to red
LGAx3	Lab-grown alexandrite	Faceted	2.56	Green to purple
LGAx4	Lab-grown alexandrite	Faceted	2.98	Green to red
LGAx5	Lab-grown alexandrite	Faceted	1.34	Green to purple

In this study, we selected a 405 nm laser (FC-D-405-50 mW, CNI) as the excitation source. The custom made 405 nm spectroscopy probe (SPC-R405, Spectra Solution) guided the laser and utilized optics with a 9 mm working distance to create an approximately 100 um focal spot to excite and collect the PL signal from the sample. Three optical filters were selected to isolate the excitation from the emission signal including a laser clean-up filter (LL01-405, Semrock) that precisely defined the excitation wavelength, a dichroic beam splitter (FF409-Di03, Semrock) to reflect the laser to the sample while passing the PL signal, and a long pass filter (LP02-407RU-25) to provide additional cut-off to the excitation. Finally, a compact miniature spectrometer (Avaspec-mini, Avantes) with 1.2 nm spectral resolution and 400 to 900 nm sensing range was used to collect the PL signal. Two 100 um optical fibers were used to guide the laser to the spectroscopy probe and to relay the spectra to the spectrometer. The spectroscopy probe was pre-aligned to the center of a black color anodized metal aperture sample stage with a 0.7 mm diameter.

In order to analyze the PL decay, a series of PL spectra using a millisecond integration time were continuously collected. The laser was shut down shortly after the beginning of the spectra collection to initiate the PL decay. The collected spectra were stored to the random-access memory (RAM) of the spectrometer to accurately label the time stamp on each spectrum with a minimized sampling interval. The sampling interval between each spectrum was observed between 0.6 to 1.6 ms, mainly caused by the electronic spectrometer. The spectra collection and the time stamp labeling was automatically performed by the Avasoft 8.8 software provided by the manufacturer of the spectrometer. This PL lifetime spectroscopy can serve as a low-cost (<$10,000) alternative setup compared to other sophisticated time-resolved techniques such as time-correlated single photon counting (TCSPC) and time-gated spectroscopy.

3. Results

3.1. Photoluminescence Decay Spectra and Duration of Decay

3.1.1. Orientational Homogeneity

First, a series of orientational homogeneity tests were conducted. Notably it is important to determine whether the results of the duration of decay and decay profile were repeatable for each sample regardless of their orientation as well as show that the 405 PL device produced sufficiently repeatable results. The natural spinel sample NatSp3 is an octahedral-form rough crystal. With the laser vertically oriented with respect to each crystal facet, the decay spectra were collected six times.

The R-line band at 685 nm was the most persistent and was the last peak to disappear. The intensities of the entire spectral band series decreased at the same speed, and the intensity ratios between each band remained unchanged until they declined under the background (Figure 3). The decay curve of the 685 nm band intensity was used to represent each decay event. All of the curves were normalized to 100. The six decay curves of each measurement illustrated in Figure 4 were almost identical. The repeat tests show that the type of the photoluminescence decay was independent of orientation, and the decay spectra, the duration of decay, and the decay curve's profile were all reproducible by the 405 PL device. The duration of decay values of each measurement with respect to sample NatSp3 are depicted in Figure 4; they resulted in an average PL lifetime of 21 ms with a standard error of 0.8 ms.

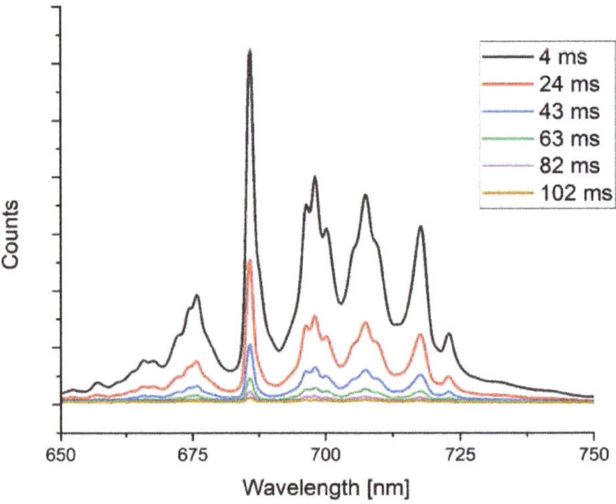

Figure 3. Time-resolved fluorescence decay spectra of natural spinel.

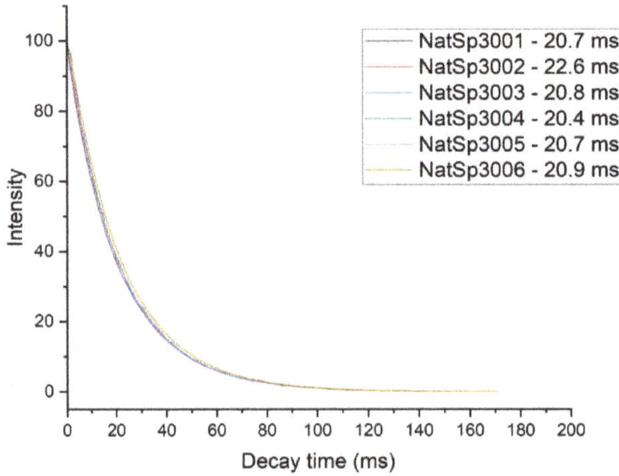

Figure 4. Six photoluminescence decay curves at 685 nm of the rough spinel NatSp3.

Although the 405 PL device was not specifically built for time-resolved testing, in our opinion, it is sufficient for this preliminary study because the luminescence of spinel lasts long enough. Accordingly, a 3%–4% standard error is acceptable.

3.1.2. Comparison Natural Spinel, Heated Spinel, and Flux Lab-Grown Spinel

Heated natural spinel and flux lab-grown spinel are different from unheated natural spinel and are characterized by a dominant R-line band at 687 nm, with a broader peak width, which revealed an overlapping of both R-line bands at 685 and 687 nm (Figure 1). The same interpretation was made by [6], which was quantified via Gaussian deconvolution. Interestingly, the luminescence decay spectra proved the existence of double peaks. The 687 nm band showed a more rapidly decreasing speed than the 685 nm band. Accordingly, the decay spectra of the flux lab-grown and heated natural spinel began with dominant bands at 687 nm, but as the luminescence decayed, the dominant peak position shifted to around 685 nm with a simultaneous decrease in the peak width. In the middle of the decay process, double peaks could also be observed (Figure 5).

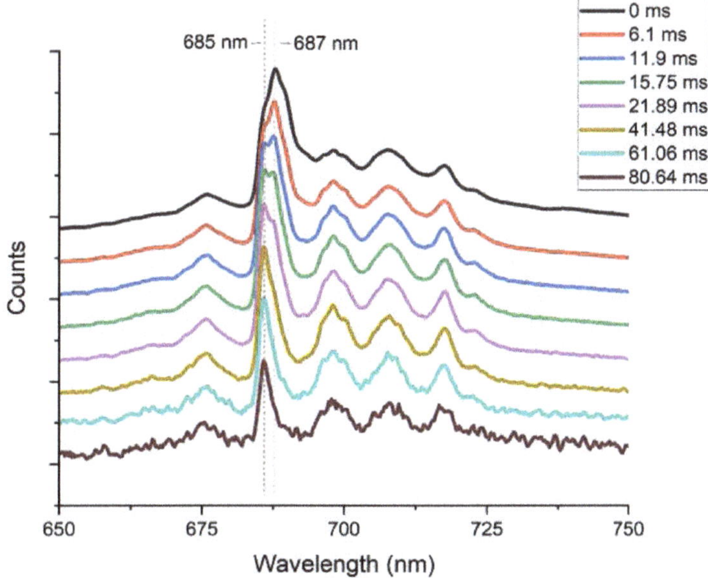

Figure 5. Time-resolved fluorescence decay spectra of the flux lab grown and heated spinel.

Regarding the PL lifetime, there was separation in the range of decay times between the natural spinel, heated spinel, and flux lab-grown spinel (see Table 2). The eight natural, unheated spinel samples had a PL lifetime between 12.8 ms and 23.4 ms, with an average of 18.7 ms and a standard error of 3.1 ms. The three heated natural spinel samples all showed decreased durations of decay compared to their unheated counterparts, with an average duration of decay of 14 ms; nevertheless, these values were still within the range of natural spinel's duration of decay. This also revealed that the crystal-structure-disordering process caused a reduction in the fluorescence-related duration of decay under the same chemical environment. The four flux lab-grown spinel samples had an average duration of decay of 8.87 ± 2.41 ms. Aizawa et al. showed very a similar lifetime of lab-grown spinel of 10 ms [37]. In comparison, the PL lifetime values of the natural unheated and heated spinel were generally longer than the flux lab-grown spinel. Due to the overlap between the heated natural spinel and flux lab-grown spinel using only single fluorescence spectra, this result is meaningful (Figure 6).

Table 2. Summary of the decay duration and exponential parameter half-life of the spinel and alexandrite samples. Half-life A1 ($t_{1/2}$) and half-life A2 ($t_{1/2}$) corresponded to the one-phase and two-phase exponential function fitting parameters, respectively. A1 and A2 are the coefficients to the exponential function.

Type	Sample	Lifetime (ms)	Exponential Fitting Parameters					
			Half-Life A1 ($t_{1/2}$) (ms)	Standard Error	Half-life A2 ($t_{1/2}$) (ms)	Standard Error	A1	A2
Natural spinel	NatSp1	17.5	11.8	0.11				
	NatSp2	18.7	13.2	0.07				
	NatSp3	20.5	14.4	0.05				
	NatSp4	23.4	15.2	0.13				
	NatSp5	12.8	8.7	0.14				
	NatSp6	19.6	13.5	0.06				
	NatSp7	20.3	14.3	0.05				
	NatSp8	16.9	11.8	0.07				
	Average	18.7	12.9					
	Standard error	3.1	2.1					
Natural heated spinel	NatSp6 800	11.0	14.8	1.87	5.57	0.33	0.28	0.73
	NatSp7 750	15.5	15.2	1.17	6.4	0.65	0.54	0.47
	NatSp8 700	15.4	12.8	0.08	4.34	0.13	0.78	0.21
	Average	14.0	14.3					
	Standard error	2.6	1.2					
Flux lab-grown spinel	FLGSp1	7.2	8.4	0.12	2.22	0.06	0.52	0.47
	FLGSp2	12.3	14.5	1.61	5.89	0.41	0.36	0.65
	FLGSp3	8.9	10.1	0.19	3.1	0.08	0.52	0.48
	FLGSp4	7.1	42.1	160.9	4.35	0.34	0.05	1.02
	Average	8.9	18.8					
	Standard error	2.4	17.4					
Natural alexandrite	NatAx1	30.5	0.5	0	19.73	0.5		
	NatAx2	25.7	0.7	0.03	17.56	1.08		
	NatAx3	28.7	0.2	0.01	20.75	0.89		
	NatAx4	43.9	0.5	0.02	24.37	0.22		
	NatAx5	27.5	0.4	0.04	17.41	0.38		
	NatAx6	26.3	2.1	1.63	14.03	0.72		
	Average	30.4	0.7		18.97			
	Standard error	6.8	0.7		3.51			
Lab-grown alexandrite	LGAx1	49.3	0.9	0.01	34.67	1.85		
	LGAx2	31.8	0.8	0.03	16.79	0.41		
	LGAx3	32.1	4.2	8.05	20.05	0.83		
	LGAx4	44.3	2.7	2.8	31.03	0.53		
	LGAx5	52.8	0.5	0.03	31.14	0.49		
	Average	42.1	1.8		26.74			
	Standard error	9.7	1.6		7.82			

3.1.3. Alexandrite's Decay Spectra and Duration of Decay

The 405 nm laser-excited decay spectra enabled us to detect two different Cr^{3+} centers, which presented as two distinguishable decay processes. The eleven alexandrite samples showed consistent properties. The first decay step occurred at the dominant duplet R-lines at 681 and 679 in the excitation state (Figure 2), which were caused by about 70% of the Cr^{3+} content at the B2 site [33]. These bands diminished into the background rapidly within the first 5 ms. After the first decay step, a series of Cr^{3+} peaks in the B1 site bands appeared, with major bands at 690 and 696 nm (Figure 7). The decay process of the B1 site bands lasted much longer, contributing to the majority of the duration of decay of the whole process. The 696 nm band disappeared, thus marking it as the last emission band. Walling et al. also determined a two-step photoluminescence decay process of alexandrite crystal, a 2.3 ms lifetime of the mirror site emission, and the inversion-site emission had a longer lifetime of 48 ms.

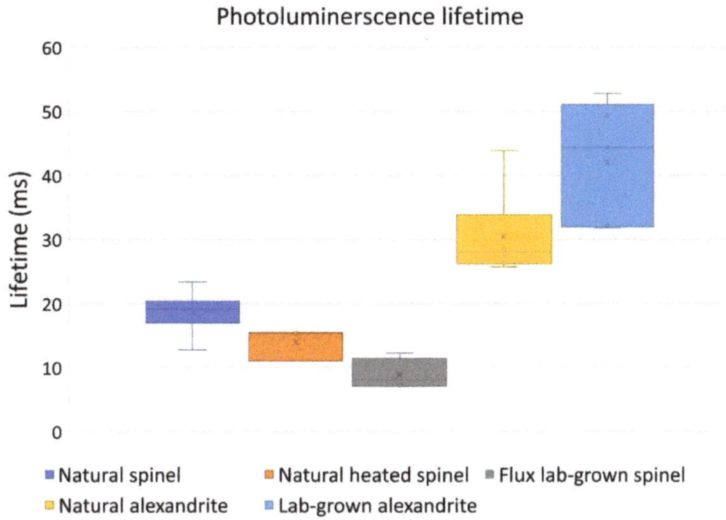

Figure 6. PL lifetime of the analyzed spinel and alexandrite samples.

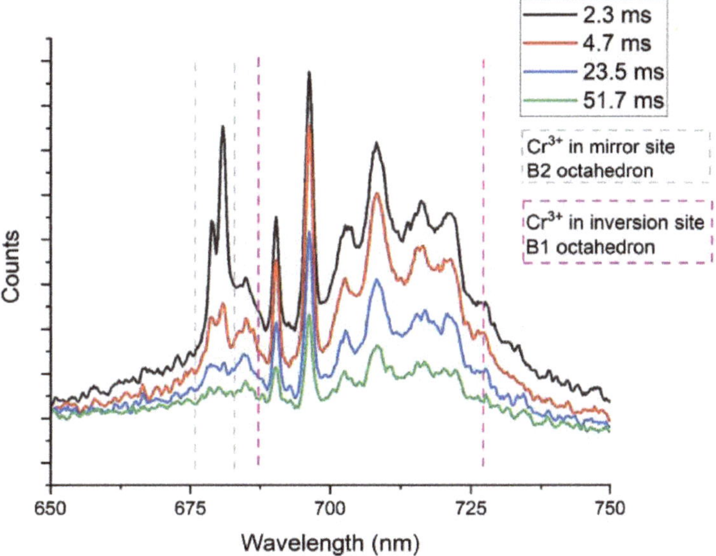

Figure 7. Time-resolved fluorescence decay spectra of alexandrite.

Due to the excessively short decay time of the B2 site, there were no more than five measurements that could be collected on the first stage of decay. The PL lifetime is represent by the B1 site decay. The lifetime of the six natural alexandrite samples was between 25.7 and 43.9 ms with an average of 30.43 ms. In comparison, the lifetimes of the lab-grown alexandrite samples were higher with an overlap with the latter values between 31.8 and 52.8 ms. The PL lifetime of the natural and lab-grown alexandrite showed a significant separation between the two groups with a limited overlap zone (Figure 6).

3.2. Decay Curve Fitting with Exponential Function

To quantify the curve profile, an exponential decay function was applied to fit the curves. An exponential decay curve fits the following equation:

$$y = e^{-t/\tau} \tag{1}$$

where t is the time and τ is the decay constant. The half-life of the decay is related to the decay constants in the following way:

$$t_{1/2} = In(2)\tau \tag{2}$$

where $t_{1/2}$ is the half-life. The half-life of a decay curve quantifies the decay speed of each measurement.

Two-phase exponential decay functions are applied when a one-phase exponential decay function cannot fit the curve's shape.

The formula for a one-phase exponential decay function is as follows:

$$y = A1 \times e^{-\frac{t}{\tau}} + y0 \tag{3}$$

The formula for a two-phase exponential decay function is as follows:

$$y = A1 \times e^{-\frac{t}{\tau 1}} + A2 \times e^{-\frac{t}{\tau 2}} + y0 \tag{4}$$

3.2.1. Spinel Decay Curve Fitting

The decay curves of all the tested spinel samples are presented in Figure 8 and Table S1. Not only was the PL lifetime, but the decay curve shapes were also different between the natural spinel and flux lab-grown spinel, and the curves of the natural spinel displayed a more obvious gentle slope than lab-grown spinel. Heated spinel samples NatSp6, 7, and 8 illustrated a visible increase in the slope in comparison with their unheated part, located between the natural and flux lab-grown spinel curves.

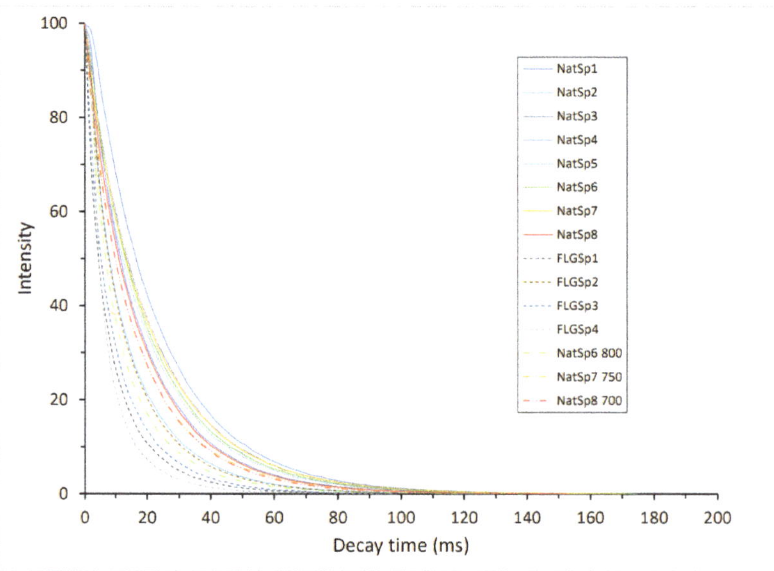

Figure 8. The photoluminescence decay curves at 685 nm of spinel: natural spinel versus the heated and flux lab-grown spinel samples.

The decay half-lives of the fitting parameters ($t_{1/2}$) together with their standard errors are listed in Table 2. The fit of the natural spinel's decay curves with respect to the one-phase exponential function was suitable for all samples. However, the flux lab-grown and heated spinel all needed to be fit using the two-phase exponential function and are listed under half-life $t_{1/2}$ A1 and half-life $t_{1/2}$ A2 in Table 2. This result coincides with the observation that the dominant fluorescence band at 687 nm band showed a higher decay speed in the two spinel groups.

The fitting results revealed that only one decay path was present in natural spinel and followed the exponential decay process, while in the lab-grown and heated spinel crystal structures, two decay paths existed, and the two decay paths were differentiated by their decay speeds. The site exchange between a Mg cation and an Al cation created the second decay path. Mg^{2+} at the octahedrally coordinated and Al^{3+} at the tetrahedrally coordinated sites will create two inequivalent lattice cells, whose presence is more conducive to photon transmission. These paths can be termed as "normal" and "inverse" paths, respectively. Switching between the Mg^{2+} cation and Al^{3+} cation increases the proportion of the "inverse" decay path.

The coefficients A1 and A2 in the two-phase exponential fitting function refer to the proportions of "normal" decay and "inverse" decay, respectively. A1/A2 decreased correspondingly: 78/21, 54/47, and 28/73 via the three spinel samples heated at 700 °C, 750 °C, and 800 °C. This corresponds to spinel's order–disorder behavior theory, in which, with respect to $(Mg_{1-x}Al_x)M(Al_{2-x}Mg_x)O_4$, x increases through heating [15]. However, this does not mean that the A1/A2 ratio can be directly transferred to x, but proves that the decay curve fitting results are valid and reflect the order–disorder characteristics of the spinel crystals.

Through exponential fitting, it was concluded that the natural, unheated spinel had an average half-life ($t_{1/2}$) of 13.6 ± 1.9 ms. The two-phase exponential fitting of the heated spinel data split two decay processes: a slow decay with a half-life from 12.8 to 15.1 ms and a quick decay with a half-life from 4.3 to 6.4 ms. The slow decay's parameters were very close to the unheated samples, and it corresponded to a "normal" decay path. The quicker decay corresponded to "inverse" decay.

The exponential fitting of the two components of the flux lab-grown samples also resulted in one "slow" and one "quick" decay. However, their half-life values showed a different combination compared to the heated natural sample.

In summary, natural spinel is characterized by one-phase exponential decay, whereas heated and flux lab-grown spinel follow two-phase exponential decay.

3.2.2. Alexandrite Decay Curve Fitting

The 696 nm band decay curves of the alexandrite samples are displayed in Figure 9 and Table S2. As a result of the two-component decay process with greatly different decay speeds between the B1 and B2 sites, both decay curves were entirely bent, presenting an obvious angular appearance. The decay curves of the natural and lab-grown alexandrite can be divided by their visibility at the angular area (see in Figure 9), while natural alexandrite started their second step by 1%–4% intensity, the lab-grown alexandrite curves displayed a higher bending point between 5 and 10% intensity. This indicates that in the lab-grown material, there were proportionally significantly more Cr sites at the B1 site than in the natural material structure.

The whole decay curve can be visually split up into a two-component exponential decay process. This began with short lifetime B2 site decay followed by long lifetime B1 site decay. The B2 site decay fitting parameters are listed under half-life A1, which resulted in a large standard error. We note that due to the minimal sampling interval of our system, the photoluminescence lifetimes shorter than 5 ms may not have been confidently measured. Based on our measurement, the B2 site decay time was approximately 2 ms.

Alternatively, the characterization of the natural and lab-grown alexandrite decay curves focused on fitting the B1 site's decay process. The use of a one-phase exponential decay function was suitable for determining the B1 site's decay profile. The fitting parameters' decay half-life ($t_{1/2}$), together with their standard errors, are listed under half-life A2 in Table 2.

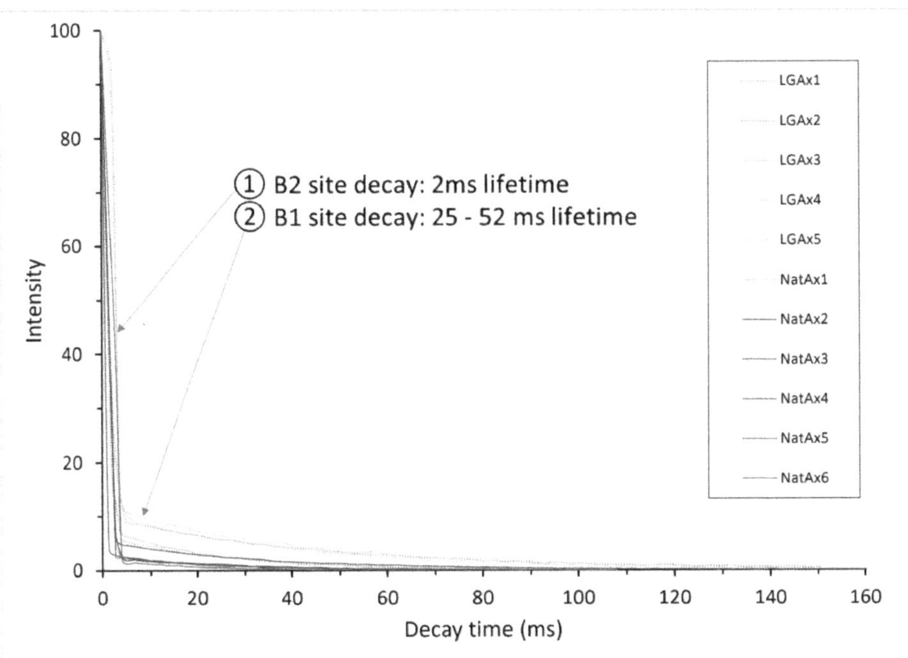

Figure 9. Photoluminescence decay curves at 696 nm of the natural and lab-grown alexandrite samples. Indication of the two decay processes that are related to Cr^{3+} in the mirror site (B2) and Cr^{3+} in the inverse site (B1). B1 site decays begin with a different intensity of individual samples.

The evaluation of the half-life ($t_{1/2}$) showed the same resolution between the natural and lab-grown alexandrite compared to the PL lifetime. The decay half-life values of the natural alexandrite's B1 site were between 14 and 24 ms, with an average of 18.9 ms, whereas those of the lab-grown alexandrite samples were between 16.8 and 34.7 ms, with an average 26.7 ms. On average, the lab-grown alexandrite's decay speed was lower than the natural one. Among the five lab-grown samples, three resulted in half-lives over 30 ms.

In contrast to spinel, the natural and lab-grown alexandrite did not show any differences in any emission, Raman, or fluorescence spectra. Ollier et al. (2015) [38] pointed out that the Fe^{3+} ions were efficiently substituted in the mirror site, and had a strong impact on the Cr^{3+}-related lifetime of the mirror site. This explains the differences in the half-life parameter between the natural and lab-grown alexandrite observed herein. Natural alexandrite from different origins contains between 600 and 10,000 ppm of iron (e.g., [39–42]). Generally, lab-grown gem material contains less iron in comparison to its natural varieties. In some lab-grown alexandrite, impurities of iron can be under the detection limit of ICPMS or EDXRF. In our case, we were able to use the laser-induced time-resolved photoluminescence decay profile to distinguish the low-iron-content lab-grown alexandrite from the high-iron content natural alexandrite. Furthermore, the PL decay curve indicated significant higher Cr^{3+} contents in the mirror site of the lab-grown alexandrite compared to the natural alexandrite, which can be used as a second characteristic criterion. Plotting the PL lifetime versus the intensity value at a 7.05 ms decay time where the junction point between the B1 and B2 site decay processes in Figure 10 significantly enhanced the separation between the two groups.

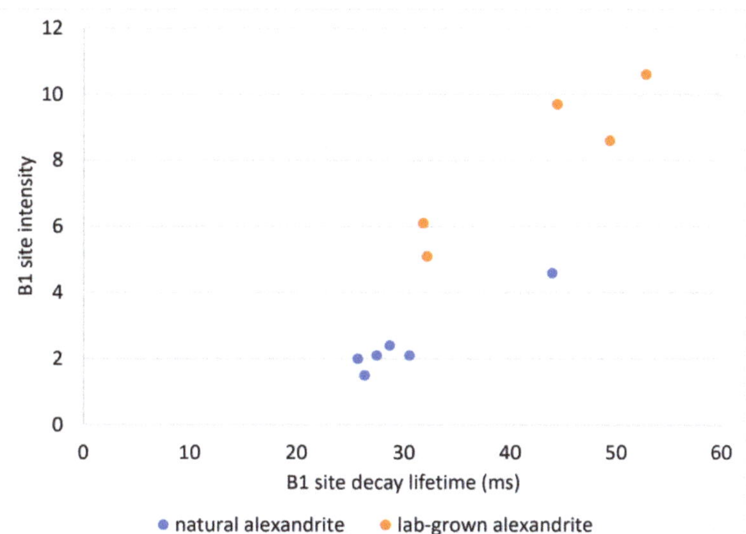

Figure 10. Plotting the B1 site decay lifetime with the B1 site starting intensity. This resulted in a better separation between the natural and lab-grown alexandrite.

4. Discussion

Spinel and alexandrite both exhibited a long photoluminescence lifetime up to 23.4 ms and 52.8 ms. Another Cr-doped gemstone species showing a long fluorescence lifetime is ruby. Similar fluorescence decay experiments demonstrated ruby's fluorescence lifetime to be about 3.5 ms [43], which is 1/5 to 1/10 that of spinel and alexandrite. There are crystallographic commonalities among ruby, spinel, and chrysoberyl: (1) they all possess a close-packed oxygen structure; and (2) Cr^{3+} is substituted for Al^{3+} in the octahedral sites. The major difference between their structural complexities is that two thirds of ruby's octahedral sites are occupied while an eighth of the tetrahedral holes and half of the octahedral holes are occupied in spinel and alexandrite crystals [7]. Structural complexity governs the number of decay paths available to electrons and the duration for which electrons remain in an excited state. The case of heated spinel demonstrated that the modification of the disorder of the lattice on different levels dramatically changes the decay profiles. This property provides a potential application for investigations into gemstone treatments.

Another aspect influencing a gemstone's fluorescence lifetime is chemistry, specifically, the aluminum substitution elements of chromium versus the iron content, for which the duration of decay of alexandrite is partially related to its iron concentration, as has been demonstrated in another study [22]. Iron concentration is a very important element in the comparison of natural and lab-grown gemstone variants, but also gem-materials from different origins. A typical example is the use of iron to distinguish marble-type rubies from metamorphic original rubies (e.g., [44–47]). The correlation between chemical patterns and luminescence lifetime profiles requires further investigation with more statistical data.

This experimental, primary study based on the 405 nm PL device is highly instructive. Through its completion, we have obtained a better understanding of the capabilities and applications of laser-introduced time-resolved photoluminescence in terms of gemstone identification, treatment determination, and, potentially, origin determination. Further investigations (e.g., using nanosecond measurements and more sophisticated instrumentation) will increase the availability of methods for distinguishing different groups of gemstones. Unlike traditional gemstone-testing methods, the quantitative time-resolved luminescence method is capable of assessing large quantities of data. Ideally, together with the automation of data acquisition and analysis software, this method can fulfill the requirements of gemstone-testing analyses.

5. Conclusions

Using 405 nm laser-induced, time-resolved photoluminescence decay spectroscopy to study dozens of reference samples of natural and lab-grown spinel and alexandrite as well as heated spinel, we determined that both spinel and alexandrite possess long photoluminescence lifetime properties of up to 23.4 ms and 52.8 ms. Detailed investigation into the samples' decay spectra profiles clearly separated the natural, heated natural, and flux lab-grown spinel by their duration of decay and exponentially fitted half-lives ($t_{1/2}$). Natural and lab-grown alexandrite also possess different half-life ($t_{1/2}$) patterns due to their diverse impurity chemical components.

Supplementary Materials: The following supporting information can be downloaded at: https://www.mdpi.com/article/10.3390/min13030419/s1, Table S1: Decay curves of the analyzed spinel samples; Table S2: Decay curves of the analyzed alexandrite samples.

Author Contributions: Conceptualization, W.X.; Methodology, W.X. and T.-H.T.; Software, W.X. and T.-H.T.; Formal analysis, W.X.; Investigation, W.X.; Resources, W.X., T.-H.T. and A.P.; Data curation, W.X. and T.-H.T.; Writing—original draft preparation, W.X.; Writing—review and editing, W.X. and T.-H.T.; Project administration, W.X. and A.P. All authors have read and agreed to the published version of the manuscript.

Funding: This research received no external funding.

Data Availability Statement: All data generated or used during the study appear in the submitted article.

Conflicts of Interest: The authors declare no conflict of interest.

References

1. Tsai, T.H.; D'Haenens-Johansson, U.F.S. Rapid gemstone screening and identification using fluorescence spectroscopy. *Appl. Opt.* **2021**, *60*, 3412–3421. [CrossRef]
2. Nevin, A.; Cesaratto, A.; Bellei, S.; D'Andrea, C.; Toniolo, L.; Valentini, G.; Comelli, D. Time-resolved photoluminescence spectroscopy and imaging: New approaches to the analysis of cultural heritage and its degradation. *Sensors* **2014**, *14*, 6338–6355. [CrossRef]
3. Gupta, A.; Hacquebard, L.; Childress, L. Efficient signal processing for time-resolved fluorescence detection of nitrogen-vacancy spins in diamond. *J. Opt. Soc. Am.* **2016**, *33*, B28–B34. [CrossRef]
4. Jones, D.C.; Alexandrov, Y.; Curry, N.; Kumar, S.; Lanigan, P.M.P.; McGuiness, C.D.; Dale, M.W.; Twitchen, D.J.; Fisher, D.; Neil, M.A.A. Multidimensional spectroscopy and imaging of defects in synthetic diamond: Excitation-emission-lifetime luminescence measurements with multiexponential fitting and phasor analysis. *J. Phys. D Appl. Phys.* **2021**, *54*, 045303. [CrossRef]
5. Saeseaw, S.; Wang, W.; Scarratt, K.; Emmett, J.L.; Douthit, T.R. Distinguishing Heated Spinels from Unheated Natural Spinels and from Synthetic Spinels: A Short Review of Ongoing Research. 2009. Available online: http://www.giathai.net/pdf/Heated_spinel_Identification_at_May_25_2009.pdf (accessed on 7 January 2014).
6. Widmer, R.; Malsy, A.K.; Armbruster, T. Effects of heat treatment on red gemstone spinel: Single-crystal X-ray, Raman, and photoluminescence study. *Phys. Chem. Miner.* **2015**, *42*, 251–260. [CrossRef]
7. Okrusch, M.; Matthes, S. Oxide und Hydroxide. In *Mineralogie*; Springer-Lehrbuch; Springer: Berlin/Heidelberg, Germany, 2014. [CrossRef]
8. Shannon, R.D. Revised Effective Ionic Radii and Systematic Studies of Interatomic Distances in Halides and Chalcogenides. *Acta Crystallogr. A* **1976**, *32*, 751–767. [CrossRef]
9. Hawthorne, F.C.; Huminicki, D.M.C. The Crystal Chemistry of Beryllium. *Rev. Mineral. Geochem.* **2002**, *50*, 333–403. [CrossRef]
10. Weber, S.U.; Grodzicki, M.; Lottermoser, W.; Redhammer, G.J.; Topa, D.; Tippelt, G.; Amthauer, G. 57Fe Mössbauer Spectroscopy, X–Ray Single Crystal Diffractometry and Electronic Structure Calculations on Natural Alexandrite. *Phys. Chem. Miner.* **2007**, *34*, 507–515. [CrossRef]
11. Gaft, M.; Reisfeld, R.; Panczer, G. *Modern Luminescence Spectroscopy of Minerals and Materials*; Springer: Berlin/Heidelberg, Germany, 2015; 606p.
12. Mikenda, W.; Preisinger, A. N–Lines in the luminescence spectra of Cr^{3+}–doped spinels (i) identification of n–lines. *J. Lumin.* **1981**, *26*, 53–66. [CrossRef]
13. Mikenda, W.; Preisinger, A. N–Lines in the luminescence spectra of Cr^{3+}–doped spinels (ii) origins of n–lines. *J. Lumin.* **1981**, *26*, 67–68. [CrossRef]

14. Yamanaka, T.; Takéuchi, Y. Order-disorder transition in $MgAl_2O_4$ spinel at high temperatures up to 1700 °C. *Z Kristallogr.* **1983**, *165*, 65–78. [CrossRef]

15. Peterson, R.C.; Lager, G.A.; Hitterman, R.L. A time-of-flight powder diffraction study of $MgAl_2O_4$ at temperatures up to 1273 k. *Am. Miner.* **1991**, *76*, 1455–1458.

16. Redfern, S.A.T.; Harrison, R.J.; O'Neill, H.; Wood, D.R.R. Thermodynamics and kinetics of cation ordering in $MgAl_2O_4$ spinel up to 1600 °C from in situ neutron diffraction. *Am. Miner.* **1999**, *84*, 299–310. [CrossRef]

17. Andreozzi, G.B.; Princivalle, F.; Skogby, H.; Della Giusta, A. Cation ordering and structural variations with temperature in $MgAl_2O_4$ spinel: An X-ray single crystal study. *Am. Miner.* **2000**, *85*, 1164–1171. [CrossRef]

18. Andreozzi, G.B.; Princivalle, F. Kinetics of cation ordering in synthetic $MgAl_2O_4$ spinel. *Am. Miner.* **2002**, *87*, 838–844. [CrossRef]

19. Méducin, F.; Redfern, S.A.T. Study of cation order-disorder in spinel by in situ neutron diffraction up to 1600 K and 3.2 GPa. *Am. Miner.* **2004**, *89*, 981–986. [CrossRef]

20. Princivalle, F.; Martignago, F.; Dal Negro, A. Kinetics of cation ordering in natural $Mg(Al, Cr^{3+})_2O_4$ spinels. *Am. Miner.* **2006**, *91*, 313–318. [CrossRef]

21. Barpanda, P.; Behera, S.K.; Gupta, P.K.; Pratihar, S.K.; Bhattacharya, S. Chemically induced disorder order transition in magnesium aluminium spinel. *J. Eur. Ceram. Soc.* **2006**, *26*, 2603–2609. [CrossRef]

22. Cynn, H.; Sharma, S.K.; Cooney, T.F.; Nicol, M. High-temperature Raman investigation of order-disorder behavior in the $MgAl_2O_4$ spinel. *Phys. Rev.* **1992**, *B45*, 500–502. [CrossRef]

23. Slotznick, S.P.; Shim, S.H. In situ Raman spectroscopy measurements of $MgAl_2O_4$ spinel up to 1400 °C. *Am. Miner.* **2008**, *93*, 470–476. [CrossRef]

24. Malsy, A.K.; Karampelas, S.; Schwarz, D.; Klemm, L.; Armbruster, T.; Tuan, D.A. Orange–red to orange–pink gem spinels from a new deposit at Lang Chap (Tan Huong-Truc Lau). Vietnam. *J. Gemmol.* **2012**, *33*, 19–27. [CrossRef]

25. Wood, D.L.; Imbusch, G.F.; Macfarlane, R.M.; Kisliuk, P.; Larkin, D.M. Optical spectrum of Cr^{3+} ions in spinels. *J. Chem. Phys.* **1968**, *48*, 5255–5263. [CrossRef]

26. Derkosch, J.; Mikenda, W. N-lines in the luminescence spectra of Cr^{3+}-doped spinels: (IV) excitation spectra. *J. Lumines* **1983**, *28*, 431–441. [CrossRef]

27. Strek, W.; Derén, P.; Jezowska-Trzebiatowska, B. Optical properties of Cr^{3+} in $MgAl_2O_4$ spinel. *Physica* **1988**, *B152*, 379–384. [CrossRef]

28. Phan, T.L.; Yu, S.C.; Phan, M.H.; Han, T.P.J. Photoluminescence properties Cr^{3+}-doped $MgAl_2O_4$ natural spinel. *J. Korean Phys. Soc.* **2004**, *45*, 63–66.

29. Powell, R.C.; Xi, L.; Gang, X.; Quarles, G.J.; Walling, J.C. Spectroscopic properties of alexandrite crystals. *Phys. Rev. B* **1985**, *32*, 2788. [CrossRef]

30. Malickova, I.; Bacik, P.; Fridrichová, J.; Hanus, R.; Štubňa, J.; Milovská, S.; Škoda, R. Detailed luminescence spectra interpretation of selected oxides: Spinel from Myanmar and chrysoberyl—Var. alexandrite from Tanzania. *Acta Geol. Slovaca* **2020**, *12*, 69–74.

31. Yeom, T.H.; Choh, S.H.; Honhg, K.S.; Yeom, H.Y.; Park, Y.H.; Yu, Y.M. Nuclear Quadrupole Interactions of 27Al in Alexandrite Single Crystal. *Z. Nat.* **1998**, *53*, 568–572. [CrossRef]

32. Yeom, T.H.; Choh, S.H. Cr^{3+} (I) and Cr^{3+} (II) Centers in Alexandrite Single Crystal. *J. Appl. Phys.* **2001**, *90*, 5946–5950. [CrossRef]

33. Bordage, A.; Rossano, S.; Horn, A.H.; Fuchs, Y. Site partitioning of Cr^{3+} in the trichroic alexandrite $BeAl_2O_4:Cr^{3+}$ crystal: Contribution from X-ray absorption spectroscopy. *J. Phys.* **2012**, *24*, 225401–225409. [CrossRef]

34. Urusov, V.S.; Gromalova, N.A.; Vyatkin, S.V.; Rusakovn, V.S.; Maltsev, V.V.; Eremin, N.N. Study of structural and valence state of Cr and Fe in chrysoberyl and alexandrite with EPR and Mössbauer spectroscopy. *Mosc. Univ. Geol. Bull.* **2011**, *66*, 102–110. [CrossRef]

35. Berezin, M.Y.; Achilefu, S. Fluorescence Lifetime Measurements and Biological Imaging. *Chem. Rev.* **2010**, *110*, 2641–2684. [CrossRef] [PubMed]

36. Minh, N.V.; Yang, I.-S. A Raman study of cation-disorder transition temperature of natural MgAl2O4 spinel. *Vibr. Spec.* **2004**, *35*, 93–96. [CrossRef]

37. Aizawa, H.; Ohishi, N.; Ogawa, S.; Watanabe, E.; Katsumata, T.; Komuro, S.; Morikawa, T.; Toba, E. Characteristics of chromium doped spinel crystals for a fiber-optic thermometer application. *Rev. Sci. Instrum.* **2002**, *73*, 3089. [CrossRef]

38. Ollier, N.; Fuchs, Y.; Cavani, O.; Horn, A.H.; Rossano, S. Influence of impurities on Cr^{3+} luminescence properties in Brazilian emerald and alexandrite. *Eur. J. Mineral.* **2015**, *27*, 783–792. [CrossRef]

39. Schmetzer, K. *Russian Alexandrites*; Schweizerbart Science: Stuttgart, Germany, 2010.

40. Schmetzer, K.; Malsy, A.-K. Natural Alexandrite with an Irregular Growth Pattern: A Case Study. *J. Gemmol. Assoc. Hong Kong* **2011**, *32*, 68–75. Available online: http://www.gahk.org/journal/Alexandrite_for%20web_v1.pdf (accessed on 7 January 2014).

41. Schmetzer, K.; Malsy, A.-K. Alexandrite and colour-change chrysoberyl from the Lake Manyara alexandrite-emerald deposit in northern Tanzania. *J. Gemmol.* **2011**, *32*, 179–209. [CrossRef]

42. Sun, Z.; Palke, A.C.; Muyal, J.; DeGhionno, D.; McClure, S.F. Geographic Origin Determination of Alexandrite. *Gems Gemol.* **2019**, *55*, 4. [CrossRef]

43. Chandler, D.E.; Majumdar, Z.K.; Heiss, G.J.; Clegg, R.M. Ruby Crystal for Demonstrating Time- and Frequency-Domain Methods of Fluorescence Lifetime Measurements. *J. Fluoresc.* **2006**, *16*, 793–807. [CrossRef]

44. Schwarz, D.; Schmetzer, K. Rubies from the Vatomandry area, eastern Madagascar. *J. Gemmol.* **2001**, *27*, 409–416. [CrossRef]

45. Sorokina, E.S.; Litvinenko, A.K.; Hofmeister, W.; Häger, T.; Jacob, D.E.; Nasriddinov, Z.Z. Rubies and sapphires from Snezhnoe, Tajikistan. *G&G* **2015**, *51*, 160–175. [CrossRef]
46. Stone-Sundberg, J.; Thomas, T.; Sun, Z.; Guan, Y.; Cole, Z.; Equall, R.; Emmett, J.L. Accurate reporting of key trace elements in ruby and sapphire using matrix-matched standards. *G&G* **2017**, *53*, 438–451. [CrossRef]
47. Palke, A.C.; Saeseaw, S.; Renfro, N.D.; Sun, Z.; McClure, S.F. Geographic Origin Determination of Ruby. *Gems Gemol.* **2019**, *55*, 4. [CrossRef]

Article

Application of High-Temperature Copper Diffusion in Surface Recoloring of Faceted Labradorites

Qingchao Zhou [1,2,3,*], Chengsi Wang [1,3] and Andy-Hsitien Shen [1,3,*]

[1] Gemmological Institute, China University of Geosciences, Wuhan 430074, China; wangcs@cug.edu.cn
[2] School of Jewelry, West Yunnan University of Applied Sciences, Baoshan 679100, China
[3] Hubei Province Gem & Jewelry Engineering Technology Research Center, Wuhan 430074, China
* Correspondence: qczhou@wyuas.edu.cn (Q.Z.); shenxt@cug.edu.cn (A.-H.S.)

Abstract: Owing to the high market values of natural sunstones in Oregon, a kind of artificially diffused red feldspar exhibited at the Tucson Exhibition at the beginning of this century, whose color origin is the same as that of natural sunstone (copper nanoparticles). However, the details of the artificial diffusion process are less disclosed, there is no systematic method to obtain such gemstones. In this paper, we developed the high-temperature copper diffusion process for the surface recoloring of faceted labradorites, which are partly buried in the diffusant. By optimizing the experimental parameters of high-temperature copper diffusion, we successfully recolored the faceted labradorites to red and light red. The gemological and spectroscopic characteristics of the recolored faceted labradorite were further characterized. The red and light-red faceted labradorites exhibited the unique surface plasmon resonance absorption peaks of copper nanoparticles near 580 nm, which is the origin of red color. The typical inclusions formed in the faceted labradorite is in the shape of "fire cloud". The interface of red and light-red faceted labradorite that is in contact with the diffusant is less contaminated, we believe that the contamination could be further reduced or eliminated by optimizing the high-temperature copper diffusion process. The way that the sample is in contact with the diffusant partly is versatile and promising in the surface treatment of materials that have already been processed.

Keywords: labradorite; faceted; copper diffusion; recoloring

Citation: Zhou, Q.; Wang, C.; Shen, A.-H. Application of High-Temperature Copper Diffusion in Surface Recoloring of Faceted Labradorites. *Minerals* **2022**, *12*, 920. https://doi.org/10.3390/min12080920

Academic Editors: Stefanos Karampelas and Emmanuel Fritsch

Received: 15 June 2022
Accepted: 19 July 2022
Published: 22 July 2022

Publisher's Note: MDPI stays neutral with regard to jurisdictional claims in published maps and institutional affiliations.

1. Introduction

Feldspar is one of the most important rock-forming minerals in the earth's crust, and those with superior quality can be used as gemstones. The feldspar gemstones on the market mainly include moonstone, sunstone, amazonite, iridescence labradorite, and other species with special optical effects. Among them, the sunstone shows strong golden and red metallic luster under the light owning to its sheet metal inclusions. Furthermore, there is a special kind of sunstone that not only has the common daylight effect, but also can show the red and green body colors like ruby and emerald at the same time. Because of its special optical characteristics, this type of gems is favored by many gem carving artists and collectors. As the state stone of Oregon, the rarity and beauty make natural sunstones a high market price. In terms of this kind of natural sunstone, the origins that have been confirmed by gemmologists are Oregon in the United States and Afar region in Ethiopia [1–3].

The research on the coloring mechanism of natural sunstone has never stopped since the discovery of this kind of gemstone in 1908. However, due to the technical limitations, the idea of the existence of copper nanoparticles in natural sunstone has not been directly verified for more than 30 years since then [4–6]. Until 2019, Wang et al. from our research group directly observed the microscopic morphology of copper nanoparticles in natural sunstone by means of FIB-TEM [7]. This problem that has puzzled people for more than 100 years was finally solved.

It is also because of the high price and huge market potential of natural sunstones that an artificially diffused red andesine feldspar was unveiled at the Tucson Exhibition in the United States at the beginning of this century [8–10]. The gem sellers claimed that the red andesine feldspar is natural from a new origin, which aroused the attention of many gem research institutes all over the world. Since the diffused red andesine feldspar has the same coloring mechanism (copper nanoparticles) with natural sunstone, it took nearly ten years of exploration to confirm that this kind of "new origin red andesine feldspar" is a scam [11–16].

Presently, except for the beryllium diffusion of sapphire, most of the diffusion treatment for recoloring is limited to the shallow surface of the gemstone. Nevertheless, in the process of high-temperature diffusion treatment, the diffusant will inevitably cause contamination or damage to the gemstone surface, which need to be removed by a secondary polishing process [17]. The secondary polishing will also remove the colored areas on the surface that are introduced by diffusion. In our previous study, we have proposed a "partly burying" method for the high-temperature copper diffusion of labradorite and andesine feldspars, and systematically investigated the Cu^+-Na^+ ions exchange and in situ formation process of copper nanoparticles in labradorite [18]. By using the "partly burying" method, we can bury only a part of the labradorite in the diffusant, and the surface of labradorite that is not in contact with the diffusant can avoid the contaminations.

In this paper, we further explored the application of high-temperature copper diffusion in surface recoloring of faceted labradorites, the inconspicuous pavilion of the faceted labradorite was preferably adopted as the partly burying areas in diffusant. The red and light red faceted labradorites were prepared by adjusting the parameters of high-temperature copper diffusion. In addition, we characterized the absorption spectra of the red and light red faceted labradorites to evaluate the color changes, all the samples exhibit unique surface plasmon resonance (SPR) absorption peaks of copper nanoparticles near 580 nm. The "fire cloud" like inclusions and minor surface contaminations of the faceted labradorite after copper diffusion were systematically characterized by microscopic examinations. The "partly burying" method proposed for the surface recoloring of faceted labradorites in this paper may also have great application prospects in other faceted gemstones.

2. Materials and Methods

2.1. Materials

The labradorite crystals were purchased from Oregon mine, which is available on their official website. All reagents were used as received without further purification: ZrO_2 (zirconium (IV) oxide 99%, Aladdin), CuO (copper (II) oxide 99%, Aladdin), activated carbon (C 98%, Aladdin).

2.2. Methods

2.2.1. High-Temperature Copper Diffusion of Faceted Labradorites

Six labradorite rough stones from Oregon were selected for cutting and polishing. The information of the six faceted labradorite gems is listed in Table 1.

Table 1. Information of faceted labradorites before and after copper diffusion.

Sample Number and Description	Images before Copper Diffusion	Images after Copper Diffusion
CP-1 oval cut, 2.08 ct 10.5 × 7.2 × 4.3 mm	 pale yellow	 gray black (partly red)

Table 1. *Cont.*

Sample Number and Description	Images before Copper Diffusion	Images after Copper Diffusion
CP-2 oval cut, 2.09 ct 10.4 × 7.6 × 4.1 mm	pale yellow	light red
CP-3 emerald cut, 1.94 ct 11.7 × 5.5 × 3.9 mm	pale yellow	light red
CP-4 pear cut, 2.55 ct 11.3 × 8.2 × 5.2 mm	pale yellow	light red
CP-5 pear cut, 2.77 ct 12.6 × 7.4 × 5.7 mm	pale yellow	red
CP-6 cushion cut, 1.93 ct 8.9 × 7.0 × 4.0 mm	pale yellow	light red

CP-1: Weigh 0.05 g of CuO powder and 5 g of zirconia powder as diffusant, which is ground with an agate mortar to mix uniformly. The faceted labradorite CP-1 is planted in the diffusant. The heating program of the tube furnace for copper diffusion was set at 1050 °C for 6 h and 1150 °C for 3 h, the heating and cooling rate of the tube furnace is 6 °C/min.

CP-2 and CP-3: Weigh 0.05 g of CuO powder and 5 g of zirconia powder as diffusant, which is ground with an agate mortar to mix uniformly. The faceted labradorites CP-2 and CP-3 are planted in the diffusant. An additional 5 g of activated carbon was added to the tube furnace. The heating program of the tube furnace for copper diffusion was set at 1050 °C for 6 h and 1150 °C for 3 h, the heating and cooling rate of the tube furnace is 6 °C/min.

CP-4: Weigh 0.025 g of CuO powder and 5 g of zirconia powder as diffusant, other conditions are the same as CP-2.

CP-5: Weigh 0.1 g of CuO powder and 5 g of zirconia powder as diffusant, the heating program of the tube furnace for copper diffusion was set at 1050 °C for 6 h and 1150 °C for 6 h, other conditions are the same as CP-2.

CP-6: Weigh 0.01 g of CuO powder and 5 g of zirconia powder as diffusant, other conditions are the same as CP-2.

2.2.2. Characterizations and Measurements

Images of all the samples were captured in a light box (D55 light source) under identical conditions to compare color changes. All the samples were tested for fluorescence reactions under long-wave (365 nm) ultraviolet (LW-UV). Microscopic observations were performed with a Leica M205A. Ultraviolet-visible (UV-Vis) absorption spectra were characterized by using a Lambda 650S spectrometer. The wavelength covers the ultraviolet to near-infrared band from 300 to 900 nm, the data interval is 1 nm, and the slit width is 5 nm. Fluorescence spectra were characterized by using a Jasco FP8500 fluorescence spectrometer. The photomultiplier tube (PMT) voltage was fixed at 600 V for all samples to compare the fluorescence intensity. The emission spectra (340–750 nm) was measured with the excitation wavelength of 320 nm at a response time of 0.5 s and a scan speed of 1000 nm/min. For all samples, the parameter settings of the instrument remain unchanged. Raman spectra were characterized by using a micro confocal Raman spectrometer (HORIBA LabRAM HR Evolution) in air using a 532 nm laser, 6.25 mW of laser power, and 9–15 cm^{-1} resolution. The light was focused on an approximately 2 μm spot. Spectra were collected in the wavenumber range of 45–4450 cm^{-1}. Major and trace element analyses were conducted by laser ablation inductively coupled plasma mass spectrometry (LA-ICP-MS) at the State Key Laboratory of Geological Processes and Mineral Resources, China University of Geosciences, Wuhan. Detailed operating conditions for the laser ablation system and the ICP-MS instrument and data reduction are the same as described by Liu et al. [19]. All data were acquired on sample in single spot ablation mode at a spot size of 44 μm in this paper. Each analysis incorporated a background acquisition of approximately 20–30 s (gas blank) followed by 50 s of data acquisition from the sample. The Agilent Chemstation was utilized for the acquisition of each individual analysis. Element contents were calibrated against multiple-reference materials (BCR-2G, BIR-1G and BHVO-2G) without applying internal standardization.

3. Results and Discussion

3.1. High-Temperature Copper Diffusion of Faceted Labradorites

Figure 1 illustrates the copper diffusion of faceted labradorite, which is partly buried in the diffusant. The labradorite was first cut and polished to nearly colorless and transparent faceted gemstones. Then, the diffusant was prepared by uniformly mixing the CuO and ZrO$_2$ powder and then transferred to an alumina crucible. ZrO$_2$ is selected as the dispersant of CuO because of its high melting point of ~2700 °C, and it hardly reacts with CuO and labradorite crystals during the high-temperature diffusion process. Finally, the pavilion of faceted labradorite was planted in the diffusant, the alumina crucible planted with the faceted labradorite was placed in a tube furnace for high-temperature copper diffusion experiments.

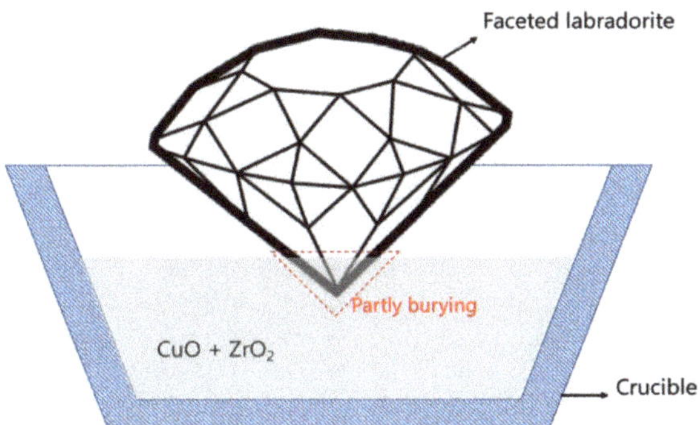

Figure 1. Schematic illustration of the copper diffusion of faceted labradorites.

3.2. Faceted Labradorites before Copper Diffusion

3.2.1. Gemological Characteristics

Table 1 lists the pictures of the six faceted labradorites that have been cut and polished. Compared with the original rough stones, the color effect of the faceted labradorites has been greatly improved. The rough labradorites used for cutting are not large, so the size of the six faceted labradorite is also relatively small (1.93–2.77 ct). In order to maximize the weight of faceted labradorites, it can be seen from the pictures that many obvious defects (inclusions and cracks) were not avoided during the cutting process. As shown in Figure 2, some typical inclusions can be observed in the six faceted labradorites, such as the "fried egg"-like inclusions, yellow dotted inclusions, white flocs, and healing fissures. It should be noted that these obvious inclusions and healing fissures should be avoided in the cutting process if it is to be processed into commercial grade faceted labradorite.

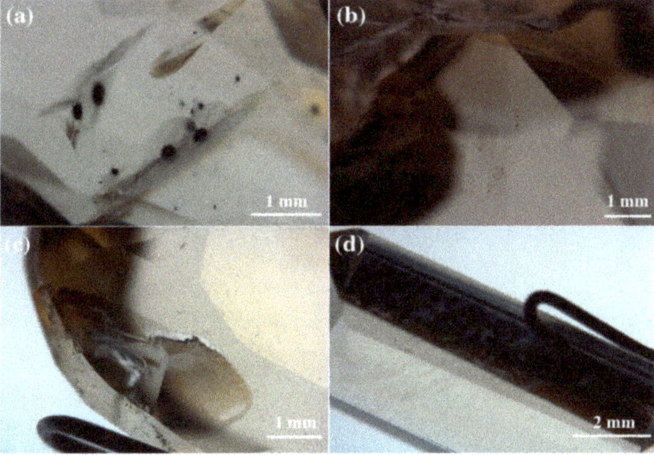

Figure 2. Typical inclusions in the faceted labradorites before high-temperature copper diffusion, (**a**) "fried egg"-like inclusions, (**b**) yellow dotted inclusions, (**c**) healing fissures, and (**d**) white flocs.

3.2.2. Spectroscopic Characteristics

Figure 3a shows the absorption spectra of the above six faceted labradorites before copper diffusion. Due to the different thickness of the faceted labradorites, the horizontal baseline of the absorption spectra is different, but all the faceted labradorites exhibit three absorption peaks at 381 nm, 421 nm, and 447 nm. These three absorption peaks are attributed to the inherent Fe^{3+} ions in natural labradorites [5]. According to the LA-ICP-MS data, the average concentration of element Fe in labradorite is 0.44 wt% (calculated as FeO).

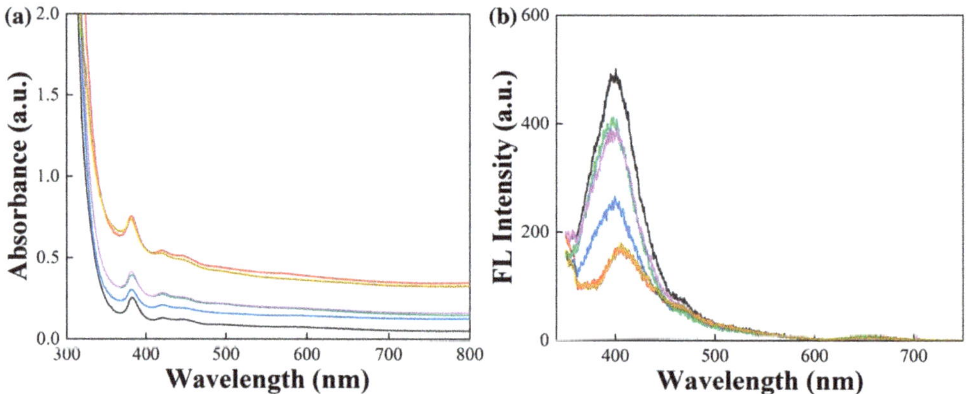

Figure 3. UV-Vis absorption spectra (**a**) and fluorescence emission spectra (**b**) of the faceted labradorites.

Figure 3b shows the fluorescence emission spectra of the above six faceted labradorites before copper diffusion. It is obvious that all the six faceted labradorites exhibit a weak emission peak (100–500 cps) around 400 nm under the excitation of 320 nm. We are currently unsure of the attribution of this fluorescence emission peak but tend to believe that it is caused by Cu^+, because the labradorite with more Cu^+ after high-temperature copper diffusion exhibits a strong fluorescence emission peak near this wavelength. The concentration of copper in the untreated labradorite is 1–10 ppmw determined by LA-ICP-MS.

3.3. Recolored Faceted Labradorites by High-Temperature Copper Diffusion

3.3.1. The Parameters of High-Temperature Copper Diffusion

The surface of the faceted labradorite was recolored by high-temperature copper diffusion. In order to have a more direct comparison with the sample before copper diffusion, the images of the faceted labradorite after high-temperature copper diffusion are summarized in Table 1. Among them, the sample CP-1 was treated in the diffusant with CuO mass concentration of 1% for 9 h (1050 °C for 6 h + 1150 °C for 3 h). It can be seen from the picture that the sample CP-1 is only turned red partly after the copper diffusion, most areas of the sample turned gray black. In our previous research, the surface of the entire rough labradorite can be treated to red by using this copper diffusion scheme [18]. It can be concluded that the surface of the faceted labradorite is different from the natural rough labradorite, and the surface of the natural rough labradorite is obviously more favorable for the formation of copper nanoparticles during high-temperature copper diffusion. The possible reason is that the surface of the natural rough labradorite corresponds to the relatively fragile surface of the crystal. During the entire process of the rough labradorite from the formation to the mining, it will be subjected to a variety of external forces. Under the action of external forces, the relatively fragile parts of the natural labradorite will break easily. From the perspective of crystal structure, there are more defects on these relatively fragile surfaces. On the surface with more defects, the diffusion rate of Cu^+ ions will be greatly accelerated during the high-temperature copper diffusion, which explains the reason why copper nanoparticles are more likely to be formed on the natural surface of labradorite instead of the artificial facets.

In order to recolor the surface of the faceted labradorites, we further added a certain quality of activated carbon to the tube furnace during the high-temperature copper diffusion, which will accelerate the formation of copper nanoparticles. During the high-temperature copper diffusion, the addition of activated carbon can consume the oxygen in the tube furnace, thereby promoting the transformation of CuO to Cu_2O at high temperatures [20]. This transition increases the concentration of Cu^+ ions in the diffusant, which facilitates the subsequent process of Cu^+ diffusion and copper nanoparticle formation.

As shown in the pictures in Table 1, the surfaces of sample CP-2 and CP-3 treated by high-temperature copper diffusion turned to light red after adding the activated carbon to the tube furnace. The purpose of adding two samples is to verify the reproducibility of the copper diffusion, and the influence of small differences among different samples can be also excluded. Moreover, the contamination and damage of faceted labradorites caused by diffusant can be greatly reduced after adding activated carbon to the tube furnace.

In addition, we further recolor the faceted labradorite into different shades of red by adjusting the CuO concentration in the diffusant during the high-temperature copper diffusion. As shown in the pictures in Table 1, when the mass concentration of CuO in the diffusant is reduced (0.2–0.5%), the color of the faceted labradorites CP-4 and CP-6 treated by copper diffusion becomes lighter, when the mass concentration of CuO in the diffusant is increased (2%), the color of the faceted labradorite CP-5 treated by copper diffusion becomes darker. The copper concentration of the faceted labradorites treated by high-temperature copper diffusion was tested by LA-ICP-MS. The copper concentration on the shallow surface of the CP-5 sample with the darkest color was 2780 ppm. The copper concentration on the shallow surface of samples CP-4, CP-6, CP-2, and CP-3 were 451, 329, 893, and 878 ppm, respectively. The above results show that more copper can be incorporated to the shallow surface of the labradorite by increasing the concentration of CuO in the diffusant.

3.3.2. Gemological Characteristics of the Recolored Faceted Labradorites

The recolored faceted labradorites were first studied by using classical gemological equipment and methods, namely, the refractometer, longwave ultraviolet light (LW-UV) and microscope. All samples showed more or less equal values for refractive indices, birefringence, and very weak pleochroism (Table 2). Under LW-UV, the pale red faceted labradorites showed weak orange fluorescence, and the red stone (CP-5) showed relatively stronger orange fluorescence. The gemological properties of some labradorites and andesines reported in other references are listed in Table 2 as a comparison.

Table 2. Comparison of gemological properties of labradorite and andesine reported in other references [8,9,13].

	Treated Labradorite from Oregon	Sunstone from Oregon	Andesine from Congo	Labradorite from Congo	Andesine from Tibet
	this paper	Krzemnicki, 2004 [9]	Fritsch, 2002 [8]	Krzemnicki, 2004 [9]	Abduriyim, 2009 [13]
RI	RIα 1.561–1.563 RIβ 1.570–1.572	RIα 1.560–1.565 RIβ 1.570–1.572	RIα 1.551 RIβ 1.560	RIα 1.553–1.555 RIβ 1.562–1.563	RIα 1.550–1.552 RIβ 1.555–1.557
DR	0.009–0.010	0.007–0.010	0.009	0.007–0.011	0.009–0.010
color	pale red and red	pale red and pale green	red	red and green	reddish orange, orange-red, and deep red
pleochroism	very weak	very weak (red sample) to distinct (green sample)	very weak	very weak (red sample) to distinct (green sample)	weak
transparency	transparent	transparent	transparent	transparent	transparent to translucent
LW-UV	weak to orange	none	none	weak to distinct orange	weak chalky orange
observations	"fire cloud" like inclusions, tiny particles with metallic luster	slight "schiller" effect	"schiller" effect	milky turbidity green labradorite: red under incandescent light	color zoning, twin lamellae, turbid milky clouds and particles, lath-like hollow channels, pipe-like growth tubes, irregular dislocations, fractures, uncommon tiny native copper grains or platelets

3.3.3. Optical Microscopic Characteristics of the Recolored Faceted Labradorites

The microscopic characteristics of the faceted labradorite after high-temperature copper diffusion were analyzed by optical microscopy, including the contaminations in the contact interface between faceted labradorite and the diffusant, the inclusions distributed in the shallow surface of faceted labradorite. The microscopic characteristics and relevant pictures are summarized in Table 3.

Table 3. Microscopic images of recolored faceted labradorites by high-temperature copper diffusion.

Sample Number	The Area in Contact with the Diffusant	Inclusions on the Shallow Surface
CP-1	Many black spots, the contamination is serious	Dense micron-scale black inclusions
CP-2	White attachments, the contamination is minor	"Fire cloud" like inclusions
CP-3	White attachments, the contamination is minor	"Fire cloud" like inclusions, tiny particles with metallic luster

Table 3. *Cont.*

Sample Number	The Area in Contact with the Diffusant	Inclusions on the Shallow Surface
CP-4	White attachments, the contamination is minor	"Fire cloud" like inclusions
CP-5	Few white attachments, and the contamination is very slight	"Fire cloud" like inclusions
CP-6	White attachments, the contamination is minor	"Fire cloud" like inclusions, tiny particles with metallic luster

It can be seen from the picture in Table 3 that serious contamination (many black spots) was present in the surface area of the recolored labradorite where the CP-1 sample was in contact with the diffusant. Emmett analyzed that this phenomenon is due to the reaction of CuO in the diffusant with the oxide components (CaO, SiO$_2$, etc.) on the surface of the labradorite to form a solid solution [17]. In the absence of activated carbon in the tube furnace, there will be a large amount of CuO in the diffusant at high temperature, which is the main reason for the contamination of labradorite during the copper diffusion. Moreover, there are many black dots distributed in the shallow surface area that are not in contact with the diffusant. By measuring under an optical microscope, the size of these tiny black particles is 1–6 μm, most of them are tiny rods, the short-axis direction is 1–3 μm, and the long-axis direction is 3–6 μm. The gray black color of the faceted labradorite after copper diffusion is attributed to these micron-scale inclusions, which then characterized them by micro-Raman spectroscopy. As shown in Figure 4, the Raman spectrum of these black micron-scale inclusions matches that of CuO, appearing Raman peaks at 291, 338, 623, and 1100 cm^{-1}. We inferred that these micron-scale CuO particles are aggregated by the Cu^{2+} diffused to the shallow surface of the labradorite.

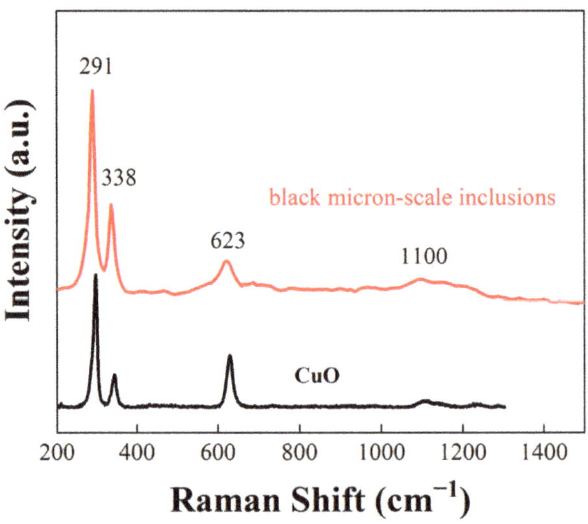

Figure 4. Raman spectra of black micron-scale inclusions in recolored faceted labradorite CP-1 and standard CuO (532 nm laser excitation).

To avoid the formation of black micron-scale CuO inclusions in the labradorite, it is necessary to find a way to reduce the concentration of CuO in the diffusant. In our previous study, we have analyzed the thermodynamic reasons for the stable existence of CuO in the diffusant in air atmosphere. In order to improve the conversion efficiency of CuO to Cu$_2$O at high temperature over 1050 °C, the most effective way is to reduce the oxygen fugacity in the reaction conditions, that is, adding substances that can consume oxygen in the tube furnace. In this way, the transformation rate of CuO→Cu$_2$O is increased as soon as the temperature reaches 1050 °C. On the one hand, the transformation of CuO to Cu$_2$O can reduce the reaction of CuO in the diffusant with the labradorite so as to reduce the contamination on the surface. On the other hand, the transformation of CuO to Cu$_2$O can increase the concentration of Cu$^+$ in the diffusant so as to promote the subsequent formation of copper nanoparticles on the shallow surface of labradorite. In addition, we maintained the temperature at 1050 °C for 6 h during the copper diffusion experiment, at which the reaction rate of CuO with labradorite is low but the transformation of CuO→Cu$_2$O had already begun.

From the pictures in Table 3, it can be seen that black micron-scale inclusions are no longer observed on the shallow surface of samples CP-2 and CP-3, which indicates that the addition of activated carbon accelerates the transformation of CuO→Cu$_2$O in the diffusant. The copper in the diffusant mostly exists in the form of Cu$_2$O. During the subsequent high-temperature copper diffusion, the Cu$^+$ diffused into the labradorite is further reduced to Cu^0 and aggregated to form copper nanoparticles. The red inclusions formed on the shallow surface of labradorite is in the shape of "fire cloud". The red inclusions of sample CP-3 are further enlarged to show tiny particles with metallic luster, which are supposed to be the copper particles. In addition, the interfaces of the two labradorites that are in contact with the diffusant no longer have black spots contaminations, but some white attachments.

This kind of surface contamination should be further eliminated by optimization of the copper diffusion process.

By decreasing the concentration of CuO in the diffusant, we prepared the light red samples CP-4 and CP-6 by copper diffusion. As can be seen from the pictures listed in the Table 3, the surface contamination of these two samples is comparable to that of CP-2 and CP-3. The light red in color corresponds to the less "fire cloud" like red inclusions. Similarly, by increasing the concentration of CuO in the diffusant, we prepared the dark red sample CP-5 by copper diffusion. The surface contamination of CP-5 was the least among all samples. Except for a small area, the whole surface of the faceted labradorite is covered with "fire cloud" like red inclusions. The copper diffusion scheme revealed that properly increasing the concentration of CuO in the diffusant can not only make the color of the labradorite darker, but also reduce the surface contamination to a certain extent. In our opinion, the increased concentration of CuO in the diffusant increases the amount of Cu^+ at the interface between the diffusant and labradorite, thereby reducing the contact between ZrO_2 and labradorite.

3.3.4. Spectroscopic Characteristics of the Recolored Faceted Labradorite

Figure 5 shows the UV-Vis absorption spectrum and fluorescence emission spectrum of the recolored faceted labradorite CP-1 by copper diffusion. It can be seen from the figure that the typical SPR absorption peak of copper nanoparticles near 580 nm is very weak, which indicates that few copper nanoparticles were formed on the shallow surface of the faceted labradorite. The transmittance of the recolored faceted labradorite CP-1 decreases due to the contaminations on the surface and the micron-scale CuO formed on the shallow surface. The fluorescence emission peak around 395 nm of the faceted labradorite is also very weak (1400 cps), which indicates that the amount of Cu^+ ions diffused into the labradorite is small [21,22]. More copper tends to diffuse into the faceted labradorite in the form of Cu^{2+} instead of Cu^+, so the subsequent formation of copper nanoparticles cannot be satisfied.

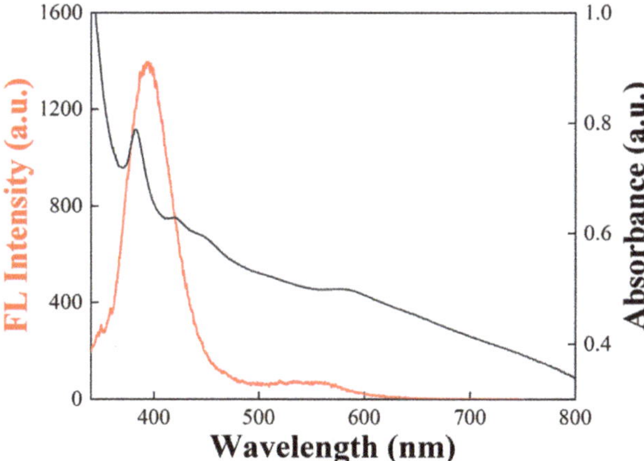

Figure 5. UV-Vis absorption (black line) and fluorescence emission spectra (red line) of recolored faceted labradorite CP-1 after high-temperature copper diffusion.

As shown in Figure 6a, the SPR absorption peaks of recolored faceted labradorites are stronger than that of sample CP-1 owing to the addition of activated carbon in tube furnace during high-temperature copper diffusion. The SPR absorption peak of sample CP-5 is the strongest, which corresponds to the most intense red color. From the fluorescence spectra in Figure 6b, it can be seen that the recolored faceted labradorites with the addition of activated carbon exhibit strong fluorescence emission peaks near 395 nm (2700–3500 cps).

The measurement conditions of the fluorescence spectra for all samples in this paper are maintained the same. Presently, it seems that the typical SPR absorption peak of copper nanoparticles and the fluorescence emission peaks of Cu^+ appear simultaneously in the recolored faceted labradorite by high-temperature copper diffusion. Moreover, the recolored faceted labradorites by high-temperature copper diffusion are easily distinguishable from the natural Oregon sunstones and will not confuse the consumers. On the one hand, the color of the recolored faceted labradorites is only distributed on the surface; on the other hand, the recolored faceted labradorites exhibit strong fluorescence emission under the excitation of 320 nm ultraviolet light.

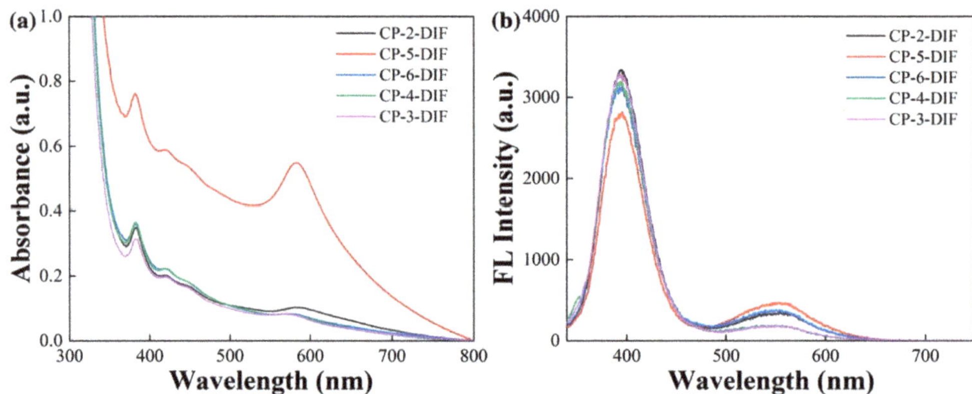

Figure 6. UV-Vis absorption spectra (**a**) and fluorescence emission spectra (**b**) of recolored faceted labradorites CP-2, CP-3, CP-4, CP-5, CP-6 by high-temperature copper diffusion.

4. Conclusions

Based on the way that the sample is in contact with the diffusant partly, we here successfully recolored six faceted labradorites by using high-temperature copper diffusion. Among them, a faceted labradorite is gray black with a little red, a faceted labradorite is red, and the remaining four faceted labradorites are light red. Spectroscopic data illustrate that the SPR absorption peak of copper nanoparticles near 580 nm can be observed in the UV-Vis absorption spectra of all six faceted labradorites, the gray black one is the weakest and the red one is the strongest. Fluorescence spectral data show that the red and light red faceted labradorites exhibit stronger fluorescence emission of Cu^+ near 395 nm than the gray black one. Micro-Raman spectrum shows that the gray black color of the faceted labradorite is attributed to the distribution of dense gray black CuO inclusions (1–3 μm) on the shallow surface. Differently, the typical "fire cloud"-like inclusions are observed on the shallow surface of both red and light red faceted labradorites. The surface contamination of gray black faceted labradorites encountered the diffusant is more serious because of the chemical reaction between CuO in the diffusant and the metal oxide composition in labradorite. By introducing the activated carbon to the tube furnace, the surface contamination of the red and light red faceted labradorites is greatly reduced because the CuO in the diffusant was converted to Cu_2O. We believe that the surface contamination could be further reduced or eliminated by optimizing the high-temperature copper diffusion process. More importantly, the way that the sample is in contact with the diffusant partly is versatile to the surface recoloring of other faceted gemstones.

Author Contributions: Conceptualization, Q.Z.; methodology, Q.Z.; validation, Q.Z. and C.W.; writing—original draft preparation, Q.Z.; writing—review and editing, A.-H.S. All authors have read and agreed to the published version of the manuscript.

Funding: This paper is CIGT contribution CIGTWZ-2022021. The authors acknowledge the financial support of a Grant (CIGTXM-04-S202146) from Center for Innovative Gem Testing Technology, China University of Geosciences (Wuhan). This research was funded by the Special Basic Cooperative Research Programs of Yunnan Provincial Undergraduate Universities' Association (grant NO. 202101BA070001-028).

Data Availability Statement: Data is available from the corresponding authors.

Acknowledgments: The authors thank Jia Liu for her help in measurements.

Conflicts of Interest: The authors declare no conflict of interest.

References

1. Johnston, C.L.; Gunter, M.E.; Knowles, C.R. Sunstone labradorite from the Ponderosa mine, Oregon. *Gems Gemol.* **1991**, *27*, 220–233. [CrossRef]
2. Kiefert, L.; Wang, C.; Sintayehu, T.; Link, K. Sunstone labradorite-bytownite from Ethiopia. *J. Gemmol.* **2019**, *36*, 694–696. [CrossRef]
3. Sun, Z.; Renfro, N.D.; Palke, A.C.; Breitzmann, H.; Muyal, J.; Hand, D.; Rossman, G.R. Sunstone plagioclase feldspar from Ethiopia. *Gems Gemol.* **2020**, *56*, 184–188.
4. Farfan, G.; Xu, H. Pleochroism in Calcific Labradorite from Oregon: Effects from Size and Orientation of Nano-and Micro-Precipitates of Copper and Pyroxene. *Chem. Geol.* **2008**, *161*, 59–88.
5. Hofmeister, A.M.; Rossman, G.R. Exsolution of Metallic Copper from Lake County Labradorite. *Geology* **1985**, *13*, 644–647. [CrossRef]
6. Nishida, N.; Kimata, M. Identification of microinclusions in the nanometer dimension by electron microprobe analysis. The chromonophores of colored sunstones. *Jpn. Mag. Mineral. Petrol. Sci.* **2002**, *31*, 268–274. (In Japanese)
7. Wang, C.S.; Shen, A.H.; Palke, A.C. Color origin of the Oregon sunstone. In Proceedings of the 36th International Gemmological Conference IGC, Nantes, France, 27–31 August 2019; pp. 71–74.
8. Fritsch, E. Gem News International: Red andesine feldspar from Congo. *Gems Gemol.* **2002**, *38*, 94–95.
9. Krzemnicki, M.S. Red and green labradorite feldspar from Congo. *J. Gemmol.* **2004**, *29*, 15–23. [CrossRef]
10. Laurs, B.L. Gem News International: Gem plagioclase reportedly from Tibet. *Gems Gemol.* **2005**, *41*, 356–357.
11. Lan, Y.; Chen, C.; Lu, T.J.; Wang, W.W. Characteristics of Surrounding Rocks and Surface Residues of "Red Feldspar from Tibet". *J. Gems Gemmol.* **2011**, *13*, 1–5. (In Chinese)
12. Wang, W.W.; Lan, Y.; Lu, T.J.; Jiang, W.; Chen, C.; Li, Q.; Chen, Z.; Xie, J. Documental Report of Geological Field Investigation on "Red Feldspar" in Tibet, China. *J. Gems Gemmol.* **2011**, *13*, 1–5. (In Chinese)
13. Abduriyim, A. The characteristics of red andesine from the Himalaya Highland, Tibet. *J. Gemmol.* **2009**, *31*, 283–298. [CrossRef]
14. Fontaine, G.H.; Hametner, K.; Peretti, A.; Günther, D. Authenticity and provenance studies of copper-bearing andesines using Cu isotope ratios and element analysis by fs-LA-MC-ICPMS and ns-LA-ICPMS. *Anal. Bioanal. Chem.* **2010**, *398*, 2915–2928. [CrossRef] [PubMed]
15. Rossman, G.R. The Chinese Red Feldspar Controversy: Chronology of Research Through July 2009. *Gems Gemol.* **2011**, *47*, 16–30. [CrossRef]
16. Zhou, Q.C.; Wang, C.S.; Shen, A.H. The Fluorescence Characteristics and Identification of Copper Diffusion-Treated Red Feldspar. In Proceedings of the 38th International Gemmological Conference IGC, Online, 20–21 November 2021; pp. 32–33.
17. Emmett, J.; Douthit, T. Copper Diffusion of Plagioclase: GIA, News from Research. Available online: www.gia.edu/research-resources/news-from-research (accessed on 10 September 2009).
18. Zhou, Q.C.; Wang, C.S.; Shen, A.H. Copper Nanoparticles Embedded in Natural Plagioclase Mineral Crystals: In Situ Formation and Third-Order Nonlinearity. *J. Phys. Chem. C* **2022**, *126*, 387–395. [CrossRef]
19. Liu, Y.; Hu, Z.; Gao, S.; Günther, D.; Xu, J.; Gao, C.; Chen, H. In situ analysis of major and trace elements of anhydrous minerals by LA-ICP-MS without applying an internal standard. *Chem. Geol.* **2008**, *257*, 34–43. [CrossRef]
20. Rakhshani, A.E. Preparation, Characteristics and Photovoltaic Properties of Cuprous Oxide—A Review. *Solid-State Electron.* **1986**, *29*, 7–17. [CrossRef]
21. Boutinaud, P.; Parent, C.; Le Flem, G.; Pedrini, C.; Moine, B. Spectroscopic Investigation of the Copper (I)-Rich Phosphate $CuZr_2(PO_4)_3$. *J. Phys. Condens. Matter* **1992**, *4*, 3031. [CrossRef]
22. Boutinaud, P.; Garcia, D.; Parent, C.; Faucher, M.; Le Flem, G. Energy Levels of Cu^+ in the Oxide Insulators $CuLaO_2$ and $CuZr_2(PO_4)_3$. *J. Phys. Chem. Solids* **1995**, *56*, 1147–1154. [CrossRef]

Article

The Heat Treatment of Pink Zoisite

Clemens Schwarzinger

Institute for Chemical Technology or Organic Materials, Johannes Kepler University Linz, 4040 Linz, Austria; clemens.schwarzinger@jku.at

Abstract: Natural pink zoisites owe their color to a high concentration of manganese paired with low concentrations of other coloring elements such as vanadium or titanium. Upon conventional heating, such stones typically suffer from the reduction of Mn^{3+} to the colorless Mn^{2+} species alongside the destruction of the brownish yellow color that is related to titanium. We have processed manganese containing zoisites under the high pressure of pure oxygen which allowed the manganese to remain oxidized, while the brownish yellow color component was still successfully removed. Depending on the vanadium level, the treated gems show a pink to purplish pink color. Detection of this treatment is not easy as the temperature is too low to result in a change in internal features, but a combination of UV-Vis-NIR spectroscopy and trace element chemistry provided by LA-ICP-MS give evidence of such treatment.

Keywords: zoisite; tanzanite; heat treatment; mineral chemistry; LA-ICP-MS; UV-Vis spectroscopy

Citation: Schwarzinger, C. The Heat Treatment of Pink Zoisite. *Minerals* **2022**, *12*, 1472. https://doi.org/10.3390/min12111472

Academic Editors: Stefanos Karampelas and Emmanuel Fritsch

Received: 21 October 2022
Accepted: 18 November 2022
Published: 21 November 2022

Publisher's Note: MDPI stays neutral with regard to jurisdictional claims in published maps and institutional affiliations.

1. Introduction

The mineral zoisite, $Ca_2Al_3[O\,|\,OH\,|\,SiO_4\,|\,Si_2O_7]$ was first found in the 18th century in a location in Austria called Saualpe (pig alp) and thus named Saualpite [1]. In honor of Karl Sigmund Zois, Freiherr von Edelstein, a rich patron of natural sciences who has funded the research and travels that led to the finding of this new mineral, it was renamed zoisite in 1805 [2]. While zoisite is known as an opaque mineral in several colors, such as the pinkish thulite [3] or the green Anyolithe, which is best known as a host mineral of large rubies from Tanzania [4], the gem and jewelry industry only became aware of this material in the late 1960s when gem-quality blue to violet crystals were found close to Merelani, in the Laletema hills of Tanzania [5–7]. As the name zoisite was considered to be an obstacle in marketing, the name tanzanite was coined for this blue variety by Henry B. Platt of Tiffany & Company [8,9]. The deposit, its geology, and mineralogy have been reviewed extensively by many authors and shall not be discussed here in detail [8,10–12] as this work only focuses on the pinkish variety of zoisite. Besides this famous location, gem-quality zoisite has so far only been found in Alchuri, Pakistan, mostly of brownish green color [13] and a small find of crystals with a light purple hue is reported from Afghanistan [14].

Most natural zoisite that comes out of the single most important locality in Tanzania is colored brownish (>70%), with a small fraction of about 5% being natural blue and about 1% having a natural pink color (Figure 1). Those numbers are according to a major gem rough dealer who has dealt with tanzanites since 1970 and has seen and handled hundreds of thousands of them [15]. Smith reports a general opinion of only 1% being naturally blue but based on his own experiences in 2009, he estimates that more than 10% are natural blues [16]. The overall color of a zoisite is very hard to describe due its strong trichroism [17]. Each axis can have a very distinct color and when a gem is cut from such a piece of rough, the orientation plays a major role in assigning the final face-up color. The cause of color has been the subject of many studies and is still discussed controversially. It is known that there are two different sites for Al^{3+} in zoisite which are distorted octahedrons [18]. Transition elements such as V, Cr, Fe, or Mn can replace aluminum and induce color. Vanadium has been found to be the main reason for blue coloration in tanzanite [18], while chromium results in green coloration [19]. Langer et al. have reported the UV-Vis spectra of pink zoisite and stated that the reason for the pink color is Mn(III) ions replacing Al in the crystal lattice. They also provide theoretical calculations of the band positions and a comparison with synthetic materials. Their interpretation is in accordance with many other examples where Mn(III) causes pink coloration [20] as well as the data provided on the minerals.gps.caltech.edu webpage by George Rossman [21]. Hainschwang and Notari also describe the manganese content to be distinctively higher in pink zoisites than in those of other colors, but point out that the color mechanism is not quite understood yet [22]. A good online article can be found at the Spectroscopy for Gemologists website, describing a 0.97 ct pink zoisite, presenting UV-Vis, FTIR, Photoluminescence spectra [23]. However, Wikipedia lists Mn(II), which is basically colorless [24], as a substitute for calcium [25].

Figure 1. Natural pink zoisite crystals with orange, purple, or no secondary color. Those samples from Merelani, Tanzania, were not part of the study and measure 25 mm × 25 mm × 9 mm, 18 mm × 11 mm × 7 mm, and 36 mm × 10 mm × 7 mm.

In this study, the term pink zoisite is used when at least one axis displays an intense pink coloration. Unfortunately, the pink axis is in many cases visible along the thinner side of a crystal (e.g., Figure 1 center), which is another reason why pink gemstones are rare and typically rather small. Cut stones are typically below 1 ct and anything above has to be considered rare; the largest gem the author has been offered so far is 8.22 ct.

While the heat treatment of brownish zoisite to turn it into blue tanzanite is well known and a common practice in the trade [26,27], the cause of color modification is still not clear and discussed controversial. Chemical analyses clearly suggest that vanadium plays an essential role in the blue coloration of zoisite but its oxidation state and changes during heat treatment are not fully understood.

So far, no method has been described that enables the treatment of zoisites with a certain chemical composition to turn them into pink or pinkish colors. It is the aim of this paper to describe the necessary prerequisites in terms of untreated material, how it is treated, and finally, the influence of the treatment on the spectroscopic properties.

2. Materials and Methods

These results were obtained as part of a larger ongoing project on the coloration of zoisites. In total, more than 1000 samples have been investigated; the number of samples directly used in this study was around 20. All samples are from Merelani, Tanzania and were obtained from three independent dealers, two of them located in the USA and one in Germany. The samples were either used without any pretreatment such as polishing windows or cut into gemstones by the author. A description of all samples is given in Table 1 and a photo provided in Figure 2.

Densities were determined with a Kern EBM 200-3V balance equipped with a Kern YDB-01 set for density determination. Refractive indices were measured on a RHG-181 gem refractometer using diiodomethane as contact liquid.

Heat treatment was performed either in a Simon–Mueller oven with the samples embedded in activated charcoal, or in a steel autoclave that was pressurized to 30 bar with pure oxygen. Heating rates were 2 °C min^{-1} up to the final temperature, holding for 30 min and then cooling with again 2 °C min^{-1}.

Elemental concentrations were measured by Laser ablation—inductively coupled plasma mass spectrometry (LA-ICP-MS) using a Cetac LSX-213 G2+ laser ablation system (Teledyne Cetac Technologies, Omaha, NE, USA) connected to a Thermo Fisher Scientific X-Series II ICP MS system (Thermo Fisher Scientific, Waltham, MA, USA). Tuning and calibration were carried out using NIST 610, 612, and 614 glass standards to maximize signal intensity while minimizing oxide ratios; reference values were taken from Jochum et al. [28]. ^{29}Si was used as an internal standard and all samples corrected to a value of 18.5%. The laser ablation was operated with a 100 μm spot size, a laser fluence of 9.2 J cm^{-2}, and a repetition rate of 10 Hz. 230 shots were made on each spot and the ablated products transferred to the ICP-MS with flow rates of two times 600 mL.min^{-1} He gas. Each sample was measured at three different spots. All concentrations are given in ppmw.

UV-Vis-NIR spectra were collected on a Shimadzu UV-3600i plus spectrometer (Shimadzu, Tokyo, Japan) equipped with a multi-purpose large sample compartment MPC-603A containing an integrating sphere, polarizing filter, and an accessory to measure small samples (~2 mm beam diameter). Spectra were measured from 240 to 1600 nm with a slit width of 20 nm, data interval of 2 nm, and the scan rate set to high (70 s per spectrum). Rough samples were oriented along crystallographic faces if present, otherwise a polarized computer screen was used to get the best possible separation of color axes. Optical pathlengths could not be determined as most samples had a rough surface that prohibited exact measurement. Cut gems were measured from the culet in the direction of the table to minimize reflections due to cutting and also oriented with a PC screen to separate the color axes as good as possible. Quantitative color description has been done using the L*a*b color space tool of the Shimadzu LabSolutions UV-Vis software (Vers. 1.11).

Table 1. Description of samples used for chemical analysis.

Sample	Color	Weight/ct	Dimensions/mm
1	pinkish purple	0.81	5.5 × 4.4 × 4.4
2	orange	0.67 (1.95) *	4.4 × 4.4 × 3.7
3	orangey pink	0.4 (1.34)	4.3 × 4.3 × 2.4
4	pink	1.56	7 × 5.6 × 4.4
5	purplish pink	0.83	8.6 × 4.8 × 2.7
6	purplish pink	2.05	8.6 × 6.8 × 5.1
7	bluish purple	0.92	7.1 × 4.9 × 4.1
8	orange	3.56	11 × 6.5 × 5.6
9	orange	1.05	7.5 × 4.4 × 4.5
10	orange	3.63	12.6 × 5.9 × 5.8
11	pink	3.44	10.4 × 7.9 × 3.6
12	red	6.71	14.4 × 10.1 × 7
13	pink	8.44	17.9 × 7.9 × 6.5
14	pink	1.79 (3.2)	10.1 × 6.2 × 3.9
15	pink	6.55	13 × 11.6 × 5.1
16	orangey pink	0.59 (1.97)	5 (diameter) × 3.4

* Values in brackets are before cutting.

Figure 2. Photo of the samples described in Table 1, samples 1 to 8 top row from left to right, samples 9 to 16 bottom row left to right; samples 2 and 16 have been heat treated under pressure with oxygen at 460 °C, sample 11 at 500 °C; samples 1 and 7 have been heat treated in charcoal and ambient atmosphere at 550 °C.

3. Results and Discussion

3.1. Chemical Composition of Pink Zoisites

Standard gemological properties have been determined for one natural pink zoisite, one unheated piece of rough (polished on three sides), two oxygen-treated pink, and one conventionally-heated blue gemstone. Refractive indices were measured on the table by rotating the samples 5–6 times; the values presented in Table 2 are the ranges observed for the lower reading (1) and the higher reading (2). Recording of RIs along the three crystallographic axes was not possible as the gems have not been oriented along those axes to yield the best possible optical performance. All stones are doubly refractive and the refractive index values are in good agreement with literature data, although a bit on the lower end [6,29]. Densities with about 3.18 g cm^{-3} are lower than the 3.35 g cm^{-3} typically reported for tanzanite [6,30] but in good agreement with the range of 3.15–3.35 cm^{-3} listed for zoisite [31] or the theoretical value of 3.09 g cm^{-3} [32]. All values are listed in Table 2.

In the first step of this study, the chemistry of pink, pinkish, and orangey zoisite crystals or cutting rough was determined by LA-ICP-MS. The data presented in Table 3 are selected from a larger dataset and include 1 pinkish red sample, 7 pink, 2 orangey pinks, 4 orange, and 2 purple ones. The manganese concentration of the measured samples lies between 60 and 1000 ppm with the majority ranging between 150 and 300 ppm. Whereas as a rule of thumb it can be said that the concentration of manganese correlates with the saturation of the pink color component, no general deduction of the color is possible from this value alone. As can be seen from the two purplish samples 6 and 7, which both have very high Mn levels, a high level of vanadium always results in a blue color component. In general, the vanadium values for the tested stones are between 50 and 1400 ppm, with the purest pinks having concentrations below 150 ppm and purplish and orange stones having the highest concentrations. Magnesium does not have any known influence on color, its concentrations are widespread between below 100 and 3000 ppm. Iron does also not seem to play an important role in the coloring mechanism; its overall concentrations range from 40 up to 2400 ppm with some samples of similar pink color having levels of 40–850 ppm. Chromium concentrations are generally very low, from below 1 ppm up to 100 ppm; the highest values are found in the samples that are orange. Finally, titanium, which in preliminary studies as well as literature [12] has been found to correlate with a UV-Vis band that results in a yellow to brown color, has been found in levels from 30 to 360 ppm.

Table 2. Standard gemological properties.

Color	Treatment	Weight	RI Range 1	RI Range 2	Double Refractive	Density/g cm^{-3}
pink	none	1.79 ct	1.680–1.700	1.690–1.705	+	3.179
orange	none	19.02 ct	1.688–1.690	1.683–1.698	+	3.192
pink	O$_2$-heated	3.30 ct	1.685–1.694	1.690–1.700	+	3.208
pink	O$_2$-heated	0.67 ct	1.675–1.690	1.680–1.696	+	3.122
blue	standard heated	7.75 ct	1.672–1.698	1.680–1.702	+	3.172

Table 3. Color and concentration of trace elements in ppmw of the zoisite samples.

Sample	^{24}Mg	^{47}Ti	^{51}V	^{52}Cr	^{55}Mn	^{56}Fe
1	1913.9	217.9	706.7	29.7	280.9	238.0
1	1996.9	267.4	689.8	29.6	282.4	238.1
1	1914.6	225.9	709.2	27.5	268.5	229.4
2	199.7	59.5	530.0	13.4	197.1	356.9
2	183.3	74.6	507.9	12.1	191.1	355.3
2	192.8	61.6	685.4	14.0	210.2	382.6

Table 3. *Cont.*

Sample	^{24}Mg	^{47}Ti	^{51}V	^{52}Cr	^{55}Mn	^{56}Fe
3	166.2	42.6	543.3	3.4	307.9	2441.6
3	194.2	37.7	484.4	2.9	286.8	2218.8
3	207.7	41.4	485.9	2.6	300.6	2265.0
4	280.4	18.3	125.0	0.6	69.3	36.3
4	225.3	44.6	132.5	0.8	66.1	36.4
4	227.7	6.9	107.1	1.3	77.5	65.3
5	117.9	55.5	246.9	9.4	206.8	310.8
5	114.0	72.7	312.4	9.0	181.5	0.0[1]
5	97.6	73.9	321.2	7.0	187.7	272.1
6	103.3	28.6	1150.5	8.5	427.9	1955.6
6	108.5	34.6	1420.0	11.1	350.1	1464.2
6	93.9	38.3	1218.9	9.8	404.2	1869.6
7	97.5	39.2	737.0	8.3	214.4	1514.1
7	152.0	45.8	743.5	8.4	313.5	1502.3
7	106.8	45.7	795.6	9.6	261.7	1637.1
8	1820.2	151.4	1032.0	85.4	183.0	325.1
8	3082.6	117.6	687.1	53.5	237.2	434.7
8	2679.3	130.1	739.6	61.5	230.5	400.6
8	851.2	171.8	1184.5	105.6	115.8	184.2
8	1555.2	120.3	920.7	83.8	154.4	267.4
8	2964.8	138.5	713.4	56.1	243.8	452.6
8	1206.0	149.5	1010.0	87.5	127.8	226.3
9	461.0	136.1	455.5	25.6	68.6	39.6
9	621.6	118.4	496.5	25.6	83.2	2309.4
9	726.6	128.9	513.3	30.8	105.0	66.4
10	1424.8	148.0	487.4	26.2	187.8	130.6
10	1356.3	154.9	463.3	23.3	176.0	118.1
10	1393.5	145.8	499.3	29.3	185.4	159.4
11	2154.2	191.9	489.2	21.7	272.5	166.2
11	2617.2	2.2	498.3	43.4	283.0	304.1
11	2283.5	200.1	480.8	21.1	293.4	202.7
12	1779.4	241.3	477.2	25.6	249.9	135.1
12	1873.1	253.9	476.7	23.9	260.1	161.3
12	1648.2	213.7	439.1	22.3	229.6	142.9
13	857.1	28.1	64.8	0.0[1]	929.1	885.9
13	959.4	61.8	107.2	77.0	835.2	821.9
13	910.3	85.2	78.9	0.3	844.8	851.1
14	65.6	172.3	124.2	8.2	96.7	75.6
14	95.4	213.7	132.6	9.0	104.4	80.5
14	83.0	208.7	121.2	8.6	95.6	82.7
15	408.6	82.9	58.2	0.1	1067.1	811.0
15	354.6	58.6	51.9	0.1	938.0	681.4
15	388.6	68.8	61.3	0.4	1052.1	783.0
16	630.3	363.5	636.6	37.0	69.8	83.2
16	545.0	337.6	591.7	35.7	60.4	83.8
16	562.3	365.5	674.5	37.0	69.9	91.0

3.2. UV-Vis-NIR Spectroscopy

The UV-Vis spectra of an untreated, natural piece of gem-quality pink zoisite (entry 4 in Table 3) are seen in Figure 3. Upon visual inspection, this piece, like most of the pure pinks, seems to be dichroic, displaying a pink and a yellowish color. The spectrum of the pink orientation shows a distinct band at 530 nm, which has been assigned to Mn(III) [9], and a strong absorption in the UV range below 400 nm. In the yellow direction, the 530 nm band is considerably smaller and an absorption centering at around 420 nm is present.

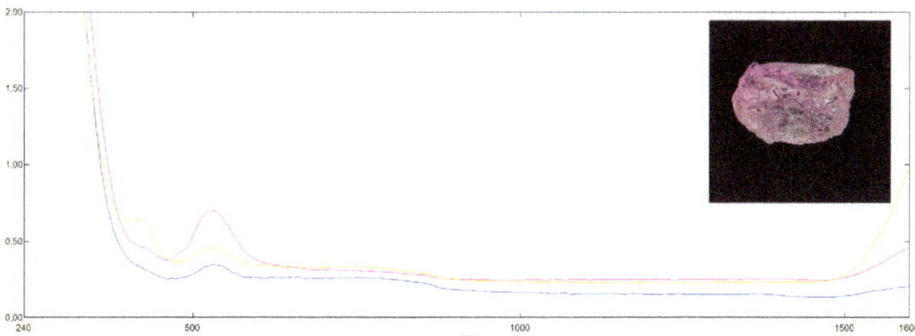

Figure 3. UV-Vis-NIR spectra of a 1.55 ct untreated, naturally pink zoisite (sample 4) along the three axis and a photo of the zoisite.

Orange is also amongst the rarer colors for zoisite as it requires a lack of blue components (vanadium) and a good ratio of pink/red and yellow. Stones that can be cut into orange gems are typically of lower saturation and their trichroic colors are pink or purplish pink, yellow, and a faint blue or purple. Olivier has described orange zoisite to be rich in rare earth elements (REEs) and lists concentrations of 2300–12,000 ppm [12], which is in good agreement with the values of 2800–20,500 ppm (average 10,800 ppm for entrees 8, 9 and 10 in Table 2, data available upon request) found in this study. He attributes the absorption bands to neodymium (Nd^{3+}): 520–535, 575–585, 745–750, and 795–805 nm, and in part titanium (Ti^{3+}): 465 nm.

The gemmy piece described here (entry 2 in Table 2) has a vanadium concentration of 500–600 ppm, 60–75 ppm Ti, 13 ppm Cr, and 200 ppm Mn. Iron is with 360 ppm in the higher range, but maximum values are above 2000 ppm, as mentioned above. The UV-Vis-NIR spectra displayed in Figure 4 show that the major difference to the spectra of a pink zoisite is the dominance of the 445 nm band causing the stone to look yellow in one direction. The contribution of vanadium is seen in the two broad bands centered at 600 and 765 nm with a shoulder at 530 nm [12,17,18]. Figure 5 shows the isolated colors as the gem is rotated in front of a polarized light source, in this case a computer screen. Pink and yellow are rather saturated, the blue color is only faint.

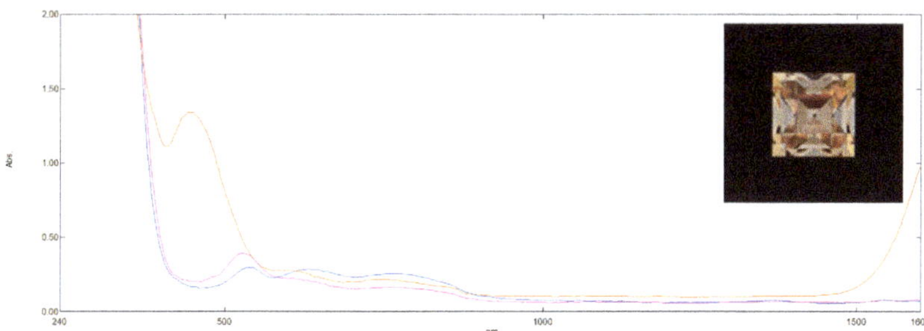

Figure 4. UV-Vis-NIR spectra of a 0.67 ct orange zoisite (sample 2) along the three axes and a photo of the gem.

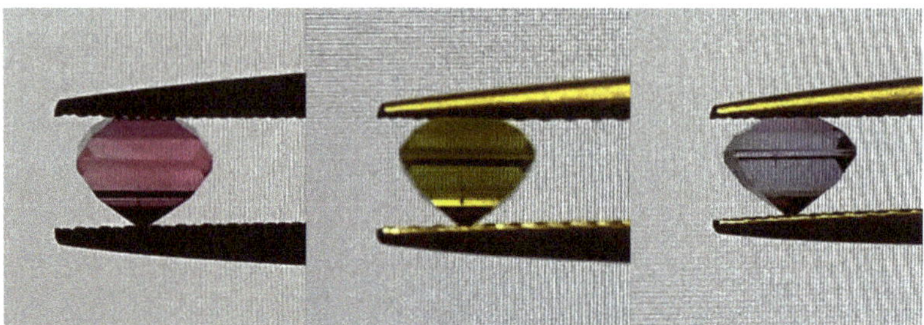

Figure 5. Individual colors of the 0.67 ct orange gem (sample 2) viewed with a polarized light source.

3.3. Heat Treatment of Orange Zoisite

In contrast to conventional heating processes, where the rough or cut stones are embedded in clay, jewelry-casting investment powder or carbon black and heated to temperatures of 500–600 °C under ambient atmosphere, we heated material that has a pinkish color component in an autoclave under a pressure of 30 to 40 bar of pure oxygen. Heating was carried out with 2 °C min^{-1} up to 500 °C, and after keeping this temperature for 30 minutes, the autoclave was cooled down with a maximum temperature rate of again 2 °C min^{-1}. Under these conditions the manganese remained in the 3+ oxidation state, thus the pink color (band at 530 nm) was preserved. The 445 nm band on the other hand was still removed by this treatment, resulting in an overall color change of the gem from orange to a slightly purplish pink (Figure 6). Figure 7 shows again the isolated colors as the treated gem is rotated in front of a polarized light source. Pink remains rather saturated; the blue color is also almost unchanged but the yellow is converted into almost colorless. Depending on the intensity of the blue axis, which also remains unchanged, stones that are treated with this process can show pink to pinkish purple colors.

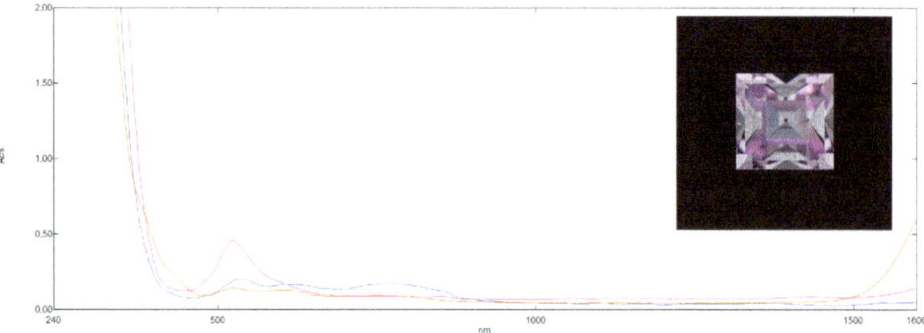

Figure 6. UV-Vis-NIR spectra of a 0.67 ct orange zoisite (sample 2) after heat treatment with oxygen under pressure along the three axes and a photo of the treated gem.

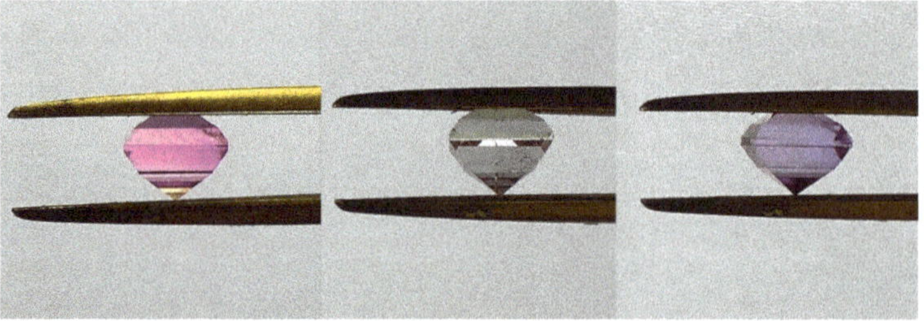

Figure 7. Individual colors of the 0.67 ct orange gem (sample 2) after heat treatment with oxygen under pressure viewed with a polarized light source.

The color change becomes very obvious when the visible spectra are analyzed using the CIELAB or L*a*b color space. In the L*a*b system the a-axis describes green (negative values) to red (positive values) color, the b-axis blue (negative values) to yellow (positive values) color, and the L value more or less describes the lightness. In order to enable a two-dimensionalview, the three-dimensional color space is sliced at a fixed L value, in our case, if not stated otherwise, 50.

In Figure 8, the values of the a and b components are displayed and the change due to heating is indicated by arrows. Whereas the blue and pink points remain relatively constant, the biggest change is seen in the yellow component. The a and b values change from 10 and 70 to almost 0 and 0, while the L value is also lowered from 74 to 44 (thus the starting point is outside the color frame in the large graph; the insert shows its position with the original L value).

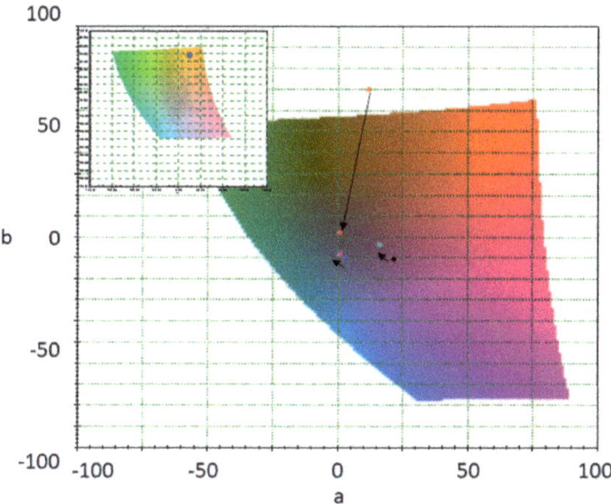

Figure 8. L*a*b coordinates of the orange zoisite before and after heat treatment under oxygen atmosphere. The arrows indicate the change in a and b values, the small insert shows the original starting point of the yellow color component at correct L value of 74.

When heating such zoisites in the conventional way, either embedded in clay, jewelry-casting investment powder, or carbon black to temperatures of 500−600 °C, the yellow as well as the pink color component are lost, and the material turns a light purplish blue color (Figure 9).

Figure 9. From left to right: untreated orange zoisite; bicolor sample treated at 430 and subsequently 460 °C under pressure with oxygen; sample treated at 430 and 460 °C under pressure with oxygen and subsequently at 500 and 550 °C in ambient atmosphere; sample treated directly at 550 °C in ambient atmosphere. All samples are parts of the same crystal with two fragments listed as samples 8 and 9 in Table 1.

To further investigate especially the temperature dependence of this process, several pieces of zoisite that contained a pink color component were subjected to various treatments. All pieces seemed to originate from the same crystal and displayed a greenish orange color, one having a greenish orange rim and a reddish pink core; two of the pieces are listed in Table 1 as samples 8 and 9. A first heating step was done at 430 °C (max oxygen pressure of 40 bar in all experiments), causing outer rims to turn a more green color while the core remained unchanged; a second treatment at 460 °C resulted in the rim to turn light blue with practically no yellow left, while the core still remained unchanged (Figure 10).

The second piece was subjected to the same treatment steps as mentioned above but further heated without pressure in the ambient atmosphere, embedded in charcoal to 500 °C and in a final step to 550 °C. The stone turned a purplish blue of medium saturation-the pink was no longer visible without the use of a polarizer. UV-Vis-NIR spectra show a complete vanishing of the 455 nm band as well as a drop in absorption between 340 and 350 nm leaving an isolated band at 384 nm behind (Figure 11). After heating to 550 °C, a further decrease in the deep UV region could be seen, allowing a band at 294 nm to stand out. The resulting spectrum is very similar to those of standard blue tanzanites.

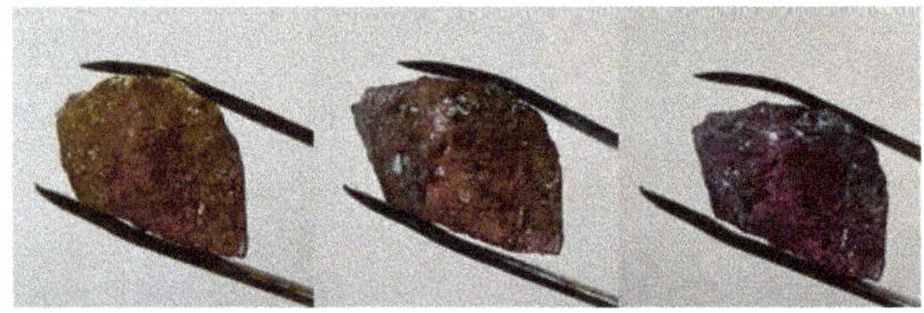

Figure 10. 6.22 g bicolor zoisite with reddish core without treatment and after oxygen treatment at 430 and 460 °C; photos taken with polarized light to select the pink color direction. Parts of the same crystal are listed as samples 8 and 9 in Table 1.

Figures 11 and 12 show the UV-Vis-NIR spectra of the untreated stone and after each treatment step, once with the polarization filter set to isolate the yellow axis (indicated by a maximum absorbance at 1600 nm), and once at a 90° angle to isolate the pink color. The first (yellow direction) spectra all have a band system in common that is dominated by two broad bands, one centered at 770 nm and the other around 630 nm. Those remain unchanged during the treatment and contribute to the bluish color component which is also very prominent in blue tanzanites. The band at 445 nm, which is the main reason for the yellow coloration, is gradually removed by this treatment and correlated more or less with the temperature, independent of atmosphere or pressure. While the absorption in the UV region remains rather unchanged during the heat treatment under oxygen atmosphere at elevated pressures, a transmission window is formed upon conventional heating to 500 and 550 °C in the region of 330-350 nm and thus an absorption band at 380 nm (Figure 11).

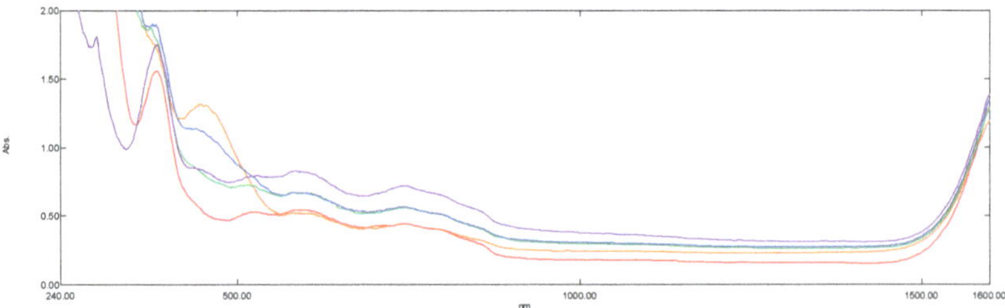

Figure 11. UV Vis spectra of the greenish-orange rough before treatment (orange), after 430 °C in oxygen (blue), 460 °C in oxygen (green), 500 °C in charcoal (red) and 550 °C in charcoal (purple). Polarization parallel to yellow.

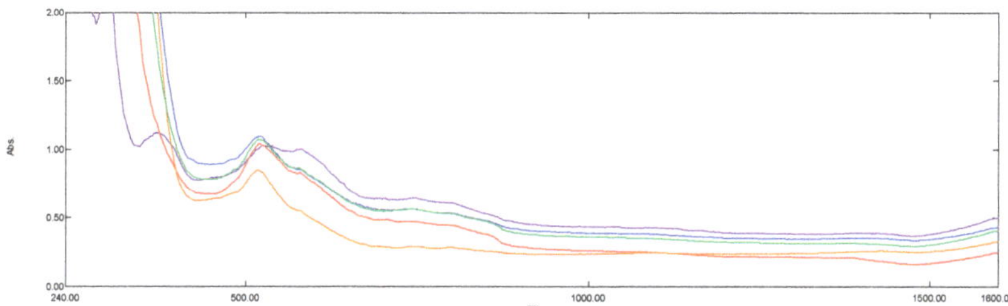

Figure 12. UV Vis spectra of the greenish-orange rough before treatment (orange), after 430 °C in oxygen (blue), 460 °C in oxygen (green), 500 °C in charcoal (red) and 550 °C in charcoal (purple). Polarization parallel to pink.

When looking at the spectra recorded in the pink direction the band at 520 nm can clearly be seen in the untreated sample. During the first 3 treatment steps this band remains rather constant, while after heating to 550 in ambient atmosphere the band seems to be reduced and the side band at 580 nm is strongly increased. As long as the sample is untreated or treated with pressurized oxygen there is a UV absorption continuum starting at roughly 400 nm. When further heating to 500 °C in ambient atmosphere this absorption is shifted towards the UV by 50 nm and upon heating to 550 °C the absorption in the UV region is generally lowered giving rise to two new bands centered at 375 and 300 nm (Figure 12).

A clearer picture can again be obtained using the L*a*b presentation, which shows in the yellow direction an almost linear decrease in the b value from 50 to 0, whereas in the a value, a change of anyway only 10 to 0 is observed (Figure 13a). Looking at the pink direction, things are more complicated. The b value goes down to about 0 during the oxygen treatment, and when further heated to 500 and 550 °C, respectively, the value goes to −10, which indicates a blue instead of a yellow color component. The a value is at first less affected by the various treatment steps, but the final heating to 550 °C in ambient atmosphere results in a distinct lowering of this parameter and thus a loss of pink color as can be seen in Figure 13b.

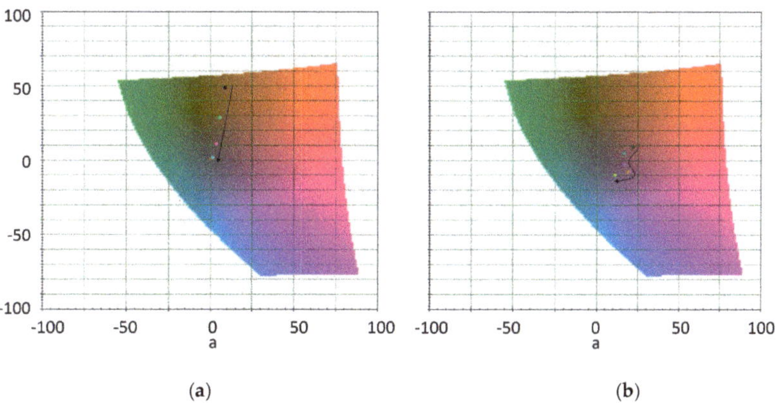

(a)   (b)

Figure 13. Change of color in the (**a**) yellow and (**b**) pink axis in L*a*b coordinates during the treatment steps.

3.4. Detection of the Treatment

The detection of treatments of gemstones is of particular interest as it may have a large influence on their value. In this particular case it is not easily possible to distinguish between treated and untreated gems as the temperature used in the process is rather low which means that typical inclusions, that are scarce in gem-quality zoisite anyway, are not altered or only marginally altered in the process [33]. Liquid hydrogen sulfide or mercaptane inclusions have been reported to be possibly indicative of a heat treatment above 600 °C at ambient pressure but have not been considered in the course of this work [34–36].

This can be compared, for example, to the heat treatment of sapphires [37,38]. While conventional treatment is done at temperatures above 1500 °C, leading to alteration of inclusions or cracks in the host sapphire due to the higher thermal expansion coefficient of the inclusion, more recent treatments at temperatures of around 800 °C (so called low-heat treatments) also do not alter the inclusion scenery and thus pose a problem in terms of identification of the treatment [39].

While most natural pure pink zoisites have rather low levels of color-relevant trace elements with the exception of manganese (70–1000 ppm), those that have been treated to improve their color started out with titanium concentrations that are on the higher side. Unfortunately, some natural pinks do also show titanium levels that have to be considered high. However, those latter stones do also have higher concentrations of vanadium, which masks the yellow color components and gives the material an overall bluish pink to purple tint. When combining this knowledge with interpretation of the UV-Vis-NIR spectra, in some cases the treatment becomes obvious, e.g., material having a high titanium concentration and not showing a 445 nm band should be considered suspicious. Another indication is the reduced absorption in the UV region; however, this effect is most pronounced in conventional heat treatment and at temperatures above 500 °C. Overall, it has to be admitted that the number of treated samples is still not high enough to enable a statistically significant statement on which absolute values are indicative for this novel treatment procedure.

FTIR and Raman spectra also remain unchanged (data not shown), so the best indication for this treatment so far is a two-step combination of UV-Vis-NIR spectroscopy and determination of trace elements:

Step 1: determination of trace elements, especially Ti, V, and Mn.

Step 2: measuring polarized UV-Vis-NIR spectra and checking for the 445 nm band in the direction of the c-axis (indicated by an intense absorption at 1600 nm).

If titanium is present but no 445 nm band, the chances are very high that the zoisite has been treated.

4. Conclusions

It has been shown that zoisites that do have a pink and yellow color component can be treated to remove the yellow whilst maintaining the pink color. If the blue color component is not too dominant, such stones will end up being pink to purplish pink. In order to achieve such a color change, it is necessary to perform the heat treatment in an oxygen atmosphere under elevated pressure, e.g., 30–40 bar. The temperatures can be adjusted depending on the type of material used–best results have been obtained between 430 and 500 °C. The chemical effect of this treatment is that the coloring element manganese remains in its 3+ oxidation state, whereas under conventional heating conditions, it is reduced to Mn^{2+}. Detection of such a treatment is not easy and currently, a combination of trace element distribution together with UV-Vis-NIR spectroscopy seems to be most promising, although more data have to be acquired in order to provide statistically relevant results.

In order to further investigate the reason for color in zoisite techniques such as EPR, luminescence spectroscopy and crystallographic studies might help shed some light onto still controversially discussed topics. The contribution of theoretical predictions must also not be underestimated but what has proven to be most important is the collection of a representative sample library, which allows to determine statistically relevant data.

Funding: This research received no external funding.

Data Availability Statement: Data are contained within the article.

Acknowledgments: The author is grateful to Steve Ulatowski for providing samples and many discussions on the topic.

Conflicts of Interest: The author declares no conflict of interest.

References

1. Zoisite. Available online: https://www.mindat.org/min-4430.html (accessed on 20 October 2022).
2. Von Radics, P. Zois von Edelstein, Siegmund. In *Allgemeine Deutsche Biographie (ADB)*; Duncker & Humblot: Leipzig, Germany, 1900; Volume 45, pp. 403–406.
3. Thulite. Available online: https://www.mindat.org/min-3955.html (accessed on 20 October 2022).
4. Game, P.M. Zoisite-amphibolite with corundum from Tanganyika. *Miner. Mag. J. Miner. Soc.* **1954**, *30*, 458–466. [CrossRef]
5. Bruce, G.A. Zoisite, a new gem discovery. *Rocks Miner.* **1968**, *43*, 332–334. [CrossRef]
6. Hurlbut, C.S. Gem Zoisite from Tanzania. *Am. Miner.* **1969**, *54*, 702–709.
7. Bank, H.; Berdesinski, W.; Nuber, B. Strontiumhaltiger trichroitischer. *Z. Dtsch. Ges. Für Edelsteinkd.* **1967**, *61*, 27–29.
8. Wilson, W.E.; Saul, J.M.; Pardieu, V.; Hughes, R.W. The Merelani Tanzanite Mines. *Miner. Rec.* **2009**, *40*, 347–408.
9. Liddicoat, R.T.; Crowningshield, G.R. More about zoisite. *Lapid. J.* **1968**, *22*, 736–740.
10. Cairncross, B. Connoisseur´s Choice: Tanzanite, Gem Variety of Zoisite, Merelani Hills, Simanjiro District, Manyara Region, Tanzania. *Rocks Miner.* **2019**, *94*, 530–539. [CrossRef]
11. Cairncross, B. The Where of Mineral Names: Tanzanite, A Variety of Zoisite, Merelani Hills, Simanjiro District, Manyara Region, Tanzania. *Rocks Miner.* **2020**, *95*, 48–462. [CrossRef]
12. Olivier, B. The Geology and Petrology of the Merelani Tanzanite Deposits, NE Tanzania. Ph.D. Thesis, University of Stellenbosch, Stellenbosch, South Africa, 2006.
13. Blauwet, D. Alchuri Shigar Valley, Northern Areas, Pakistan. *Miner. Rec.* **2006**, *37*, 513–540.
14. Beaton, D.; Lu, R. Zoisite from Afghanistan. *Gems Gemol.* **2009**, *45*, 70.
15. Ulatowski, S.; 14923 Rattlesnake Rd., Grass Valley, CA 95945, USA. Personal communication, 2022.
16. Smith, C.P. Natural-color tanzanite. *Gems Gemol.* **2011**, *47*, 119–120.
17. Faye, G.H.; Nickel, E.H. On the pleochroism of vanadium-bearing zoisite from Tanzania. *Canad. Miner.* **1971**, *10*, 812–821.
18. Koziarska, B.; Godlewski, M.; Suchocki, A. Optical properties of zoisite. *Phys. Rev. B* **1994**, *50*, 12297–12300. [CrossRef] [PubMed]
19. Javier-Ccallata, H.; Watanabe, S. Crystal field effect on EPR and optical absorption properties of natural green zoisite. *Spectrochim. Acta A Mol. Biomol. Spectr.* **2013**, *104*, 505–511. [CrossRef]
20. Langer, K.; Tillmanns, E.; Kersten, M.; Almen, H.; Arni, R.K. The crystal chemistry of Mn^{3+} in the clino- and orthozoisite structure types, $Ca_2M_3^{3+}[OH \mid O \mid SiO_4 \mid Si_2O_7]$: A structural and spectroscopic study of some natural piemontites and "thulites" and their synthetic equivalents. *Z. Kristallogr.* **2002**, *217*, 563–580. [CrossRef]
21. Zoisite Visible Spectra (400–1000 nm). Available online: http://minerals.gps.caltech.edu/FILES/Visible/zoisite/Index.html (accessed on 3 November 2022).
22. Hainschwang, T.; Notari, F. Zoisite – More than just Tanzanite. *InColor* **2012**, *19*, 44–49.
23. Pink Zoisite var. Thulite from Tanzania Reaching almost the Carat. Available online: https://www.spec4gem.info/reports/8-reports/313-pink-zoisite-var-thulite-from-tanzania-of-almost-a-carat.html (accessed on 13 January 2022).
24. Fritsch, E.; Rossman, G.R. An update on color in gems. Part 1: Introduction and colors caused by dispersed metal ions. *Gems Gemol.* **1987**, *23*, 126–139. [CrossRef]
25. Thulite. Available online: https://en.wikipedia.org/wiki/Thulite (accessed on 20 October 2022).
26. Thongnopkun, P.; Chanwanitsakun, P. Effect of heat treatment on spectroscopic properties of tanzanite. *J. Phys. Conf. Ser.* **2018**, *1144*, 012183. [CrossRef]
27. Pluthmetwisute, T.; Wanthanachaisaeng, B.; Saiyasombat, C.; Sutthirat, C. Cause of Color modification in Tanzanite after Heat Treatment. *Molecules* **2020**, *25*, 3743. [CrossRef]
28. Jochum, K.P.; Weis, U.; Stoll, B.; Kuzmin, D.; Yang, Q.; Raczek, I.; Jacob, D.E.; Stracke, A.; Birbaum, K.; Frick, D.A.; et al. Determination of reference values for NIST SRM 610–617 glasses following ISO guidelines. *Geostand. Geoanal. Res.* **2011**, *35*, 397–429. [CrossRef]
29. Bocchio, R.; Adamo, I.; Bordoni, V.; Caucia, F.; Diella, V. Gem-quality zoisite from Merelani (Northeastern Tanzania): Review and new data. *Period. Mineral* **2012**, *81*, 379–391.
30. Dominy, G.M. *Handbook of Gemmology*; Amazonas Gem Publications: Palma de Mallorca, Spain, 2018; Volume 1, Appendix A.
31. IGS: Specific Gravity Values of Selected Gems. Available online: https://www.gemsociety.org/article/select-gems-ordered-density/ (accessed on 8 November 2022).
32. Moore, P.B. Cell data of orientate and its relation to ardennite and zoisite. *Canad. Miner.* **1964**, *8*, 262–265.
33. Yang, S.; Ye, H.; Liu, Y. The Different Inclusions' Characteristics between Natural and Heat-Treated Tanzanite: Evidence from Raman Spectroscopy. *Crystals* **2021**, *11*, 1302. [CrossRef]
34. Rankin, A.; Taylor, D.; Treloar, P. Mercaptans in Tanzanite from Merelani, Tanzania—A New Type of Stinkstone? *J. Gemmol.* **2014**, *34*, 11–13.

35. Taylor, D.; Rankin, A.H.; Treloar, P.J. Liquid hydrogen sulphide (H2S) fluid inclusions in unheated tanzanites (zoisite) from Merelani, Tanzania: Part 1. Recognition, characterization and gemmological importance. *J. Gemmol.* **2013**, *33*, 149–159. [CrossRef]
36. Rankin, A.H.; Taylor, D.; Treloar, P.J. Liquid hydrogen sulphide (H2S) fluid inclusions in unheated tanzanites (zoisite) from Merelani, Tanzania: Part 2. Influence on gem integrity during and after heat treatment. *J. Gemmol.* **2013**, *33*, 161–169. [CrossRef]
37. Crowningshield, R.; Nassau, K. The heat and diffusion treatment of natural and synthetic sapphires. *J. Gemmol.* **1981**, *17*, 528–541. [CrossRef]
38. Themelis, T. *The Heat Treatment of Ruby and Sapphire*; Ted Themelis: Bangkok, Thailand, 2018; Volume 1.
39. Saeseaw, S.; Khowpong, C.; Vertriest, W. Low-temperature heat treatment of pink sapphires from Ilakaka, Madagascar. *Gems Gemol.* **2020**, *56*, 448–457. [CrossRef]

Article

Dehydration of Diaspore and Goethite during Low-Temperature Heating as Criterion to Separate Unheated from Heated Rubies and Sapphires

Michael S. Krzemnicki [1,2,*], **Pierre Lefèvre** [1], **Wei Zhou** [1], **Judith Braun** [1] and **Georg Spiekermann** [3]

[1] Swiss Gemmological Institute SSEF, Aeschengraben 26, 4051 Basel, Switzerland; pierre.lefevre@ssef.ch (P.L.); wei.zhou@ssef.ch (W.Z.); judith.braun@ssef.ch (J.B.)

[2] Department of Environmental Sciences, University Basel, 4056 Basel, Switzerland

[3] Experimental Mineral Physics, Institute for Geochemistry and Petrology, Federal Institute of Technology, 8092 Zurich, Switzerland; georg.spiekermann@erdw.ethz.ch

[*] Correspondence: michael.krzemnicki@ssef.ch

Abstract: Gem-quality rubies and sapphires are often commercially heat treated at about 800 °C or higher to enhance their color and clarity, and hence quality. For this study, selected corundum samples containing diaspore and goethite inclusions were heated step-by-step to a maximum of 1000 °C with the aim of monitoring the dehydration and phase transformation of these oxyhydroxides to corundum and hematite during heating. Based on our experiments and in agreement with the literature, the dehydration of diaspore in corundum occurs between 525 and 550 °C, whereas goethite transforms to hematite between 300 and 325 °C. As both diaspore and goethite may be present as inclusions in rubies, sapphires, and other corundum varieties (e.g., pink sapphires, padparadscha), these dehydration reactions and phase transformations can be considered important criteria to separate unheated from heated stones, specifically in cases in which other methods (e.g., microscopy, FTIR) are unsuccessful.

Keywords: ruby; sapphire; low-temperature heating; diaspore; goethite; Raman

Citation: Krzemnicki, M.S.; Lefèvre, P.; Zhou, W.; Braun, J.; Spiekermann, G. Dehydration of Diaspore and Goethite during Low-Temperature Heating as Criterion to Separate Unheated from Heated Rubies and Sapphires. *Minerals* **2023**, *13*, 1557. https://doi.org/10.3390/min13121557

Academic Editors: Stefanos Karampelas, Emmanuel Fritsch and Frederick Lin Sutherland

Received: 26 September 2023
Revised: 1 December 2023
Accepted: 16 December 2023
Published: 18 December 2023

1. Introduction

Since historical times, it has been known that the visual appearance of certain gemstones may be enhanced by heat treatment [1] and references therein. For gem-quality corundum Al_2O_3, various heat treatment processes with or without additives have been developed, specifically in the last few decades. They are widely applied in the gem trade (e.g., in cutting/manufacturing and trading hubs in Sri Lanka, Thailand, and India) to enhance the color and additionally—in some cases—the transparency and stability of corundum. By using such heat treatments, it is possible to modify corundum of lower quality into visually beautiful stones of better color and clarity and consequently guarantee a steady supply of gems to the international market.

Heat treatment of ruby and sapphire varieties is usually carried out at temperatures ranging from about 800 to 1800 °C. Historically by using a simple and artisanal blow-pipe (e.g., in Sri Lanka) reaching temperatures of about 1200 °C [2], heating has evolved into a multitude of treatment options, which are often carried out under controlled conditions (e.g., atmosphere) in electric muffle furnaces [2–5]. Heating of corundum at higher temperatures (>1200–1350 °C) may have a considerable impact on its internal features and color. Notably, local discoid expansion cracks may develop around (transformed) solid and fluid inclusions. In addition, tiny rutile (TiO_2) needles commonly present in corundum as so-called "silk" may start to (partially) dissolve at temperatures > 1400 °C. In contrast to this, a heat treatment process at about 800 °C to 1200 °C is generally known in the trade as "low-temperature" heating. It usually results in a slight (but desirable) to distinct (e.g., Fe-rich metamorphic sapphires) shift of color and may leave inclusions visually unaltered [6,7]. As such, the detection of heat treatment in corundum today relies not only on "classic" microscopic observations but also on infrared spectroscopy (FTIR) and Raman micro-spectroscopy.

The presence and arrangement of OH^--Ti^{4+} related bands in the mid-infrared spectral range [8–14] of corundum of metamorphic origin (not applicable for basaltic sapphires) and the peak width (FWHM) of the main Raman peak (at about $1010\ cm^{-1}$) of tiny zircon inclusions in corundum (mainly pink sapphires from Ilakaka, Madagascar) are considered important criteria to separate unheated corundum from heated stones [6,15,16]. However, both mentioned analytical approaches have certain limitations, i.e., hydroxide-related bands in infrared spectra are not always present in heated rubies and sapphires [17,18], and the FWHM of the main Raman peak of zircon inclusions in unheated and heated corundum may show considerable overlap [6,16]. Additionally, the Raman bandwidth of zircon inclusions is strongly dependent on the concentration of radioactive trace elements, crystallinity (metamictisation), and finally, geological and geographic origin where the corundum formed [19–21]. As shown in a recent study [7], further inclusions (e.g., apatite, spinel) may show a broadening of Raman peaks as a consequence of heating, and as such, these inclusions may assist in certain cases in detecting heat treatment of corundum.

It has been known and documented for many decades that the two isostructural oxyhydroxides goethite α-FeO(OH) and diaspore α-AlO(OH) transform upon heating into the anhydrous oxides hematite α-Fe$_2$O$_3$ and corundum α-Al$_2$O$_3$ (coupled with a release of water H_2O) [22–29]. The pseudomorphous character of the dehydration phase transformation is due to the fact that there is no major change in the relative positions of the remaining oxygen atoms of the oxides compared with the originally present oxyhydroxides goethite and diaspore.

Although the temperature at which the phase transformation of these oxyhydroxides into hematite and corundum occurs is to some extent related to the duration of heating (minutes vs. days), pressure, and the size of the crystallites, goethite generally dehydrates to hematite at about $350\ °C$ [26], whereas for diaspore this dehydration occurs at higher temperatures at about $550\ °C$ [30]. XRD studies on the dehydration of goethite and diaspore suggest the formation of intermediate structural states during their phase transformation to hematite and corundum [26,27,30–34].

Raman spectroscopy has proven to be a very useful analytical method to document these dehydration phase transformations. In contrast to crystal structure analyses by XRD (X-ray powder diffraction), Raman spectroscopy has the advantage of being a non-destructive technique (NDT), requiring quasi-no sample preparation, and with the capability to analyze structural changes in situ during heating experiments [35–38]. However, as the focused laser beam used for Raman analyses may locally affect oxyhydroxides (e.g., goethite) considerably by degradation and dehydration effects [35,39], it is necessary to carefully keep laser power at low levels to avoid artifacts and misinterpretation.

The dehydration of goethite as a marker for a gemstone heat treatment has been known in the gemmological literature for quite some time [40–43]. The focus of this study is to present detailed data of in situ heating experiments not only for the goethite–hematite phase transformation but also for the dehydration of diaspore. Using real cases, this study also demonstrates the potential of this analytical approach to separate unheated from heated rubies and sapphires, even in cases in which other methods (e.g., FTIR, microscopy) were unsuccessful.

2. Materials and Methods

To study the dehydration of diaspore and goethite as inclusions in ruby and sapphires, we selected three unheated rough ruby samples and one faceted sapphire. The rubies were cut and polished into small flat discs perpendicular to the optic axis, containing either diaspore or goethite as inclusions. Two of these samples (rubies 120553_B and 85933_C3) originated from the economically important ruby deposit in the Montepuez area in northern Mozambique and were supplied by Gemfields Ltd., London, UK. The two other samples (ruby 126993_6 and sapphire 106424_21) reportedly originated from the Mogok area in northern Myanmar, known since historical times as a source of exceptional rubies and other gemstones and were supplied by a Burmese gemstone trader in Bangkok (SilkenEast Ltd.; Bangkok, Thailand). Their initial treatment status (unheated) and reported origin were analytically confirmed prior to the experiments. In addition, a small crystal fragment of gem-quality diaspore from the Muğla Province in western Turkey was included in our heating experiments (see Table 1).

Table 1. The samples heated and analyzed for this study, their identification, shape, color), geographic origin, and heating experiment parameters (system, maximum temperature, color after heating).

Sample		ID	Weight (ct)	Shape	Colour	Origin	Heating	Max T °C	Colour after Heating
97003		Diaspore	0.40	flat fragment	colourless	Muğla Prov., Turkey	Electric furnace	800	whitish
126993_6		Ruby with diaspore	0.19	polished slab	red	Mogok, Myanmar	Heating stage	700	no change
106424_21		Sapphire with diaspore	1.03	faceted	blue	Mogok, Myanmar	Heating stage	700	no change
85933_C3		Ruby with goethite	0.52	polished slab	red	Montepuez, Mozambique	Heating stage	400	no change
120553_B		Ruby with goethite	1.31	polished slab	red	Montepuez, Mozambique	Electric furnace	1000	no change

Two different experimental setups were chosen for this study (see Table 2): (a) a Linkam TS-1500 heating stage, which was fixed directly to the Raman sample stage and which allows Raman analyses on exactly the same analytical spot during heating experiments, and (b) an electrical muffle furnace (Nabertherm LHT 18) similar to those commonly used for commercial heat treatment of gemstones. In both cases, heating was performed step-by-step, reaching a maximum temperature of 1000 °C in one experiment. Raman spectra were recorded after each heating step only after cooling to room temperature (about 25 °C) again.

Table 2. Summary of the heating experiment setups used for this study.

Samples 126993_6 (ruby with diaspore), 106424_21 (sapphire with diaspore), and 85933_C3 (ruby with goethite) were heated in air to different temperatures using a heating stage (Linkam TS-1500) fixed to the Raman sample stage.

- Step-wise heating up to a max. temperature 700 °C
- On each temperature step, the temperature was kept stable for 4 min
- Ramping up of temperature 80 °C per minute
- Raman analyses at room-temperature after each heating step

Samples 97003 (diaspore) and 120553_B (ruby with goethite) were heated in air to different temperatures using an electric muffle furnace (Nabertherm LHT 18).

- Step-wise heating up to a max. temperature 1000 °C
- On each temperature step, the temperature was kept stable for 1 h
- Ramping up of temperature 50 °C per minute
- Raman analyses at room-temperature after each heating step

Although it would be possible to analyze a sample at peak temperature in situ with the Linkam heating stage, the very strong Cr-related luminescence of the ruby samples at elevated temperatures when using a green laser (514 or 532 nm) made it necessary to cool the samples after each heating step. Both setups come with a controller unit, which allows for time-controlled heating (ramping up) and cooling (ramping down) of the sample. The temperature ramping rate was pre-selected at 80 °C/min for the Linkam stage and 50 °C/min for the electric furnace. The temperature was kept stable at each heating step for a defined time (4 min with the Linkam stage and 1 h with the electric furnace) and, after each heating step, subsequently cooled to room temperature for Raman analyses.

We used a Renishaw confocal Raman microscope (InVia^{TM}) with a Peltier-cooled CCD detector (1024 $\times$ 256 pixels), resulting in a maximum spectral resolution of about 1.5 cm^{-1}. The system is coupled with a Leica stereo microscope and a set of objectives (10$\times$ to 50$\times$). The Raman spectra were collected using an argon-ion laser emitting at 514.5 nm. The laser beam was focused on the samples and inclusions using a 50$\times$ long-range objective, resulting in a spot size of ca. 2 µm. The laser power for analyzing the ruby surface and diaspore was set at a standard of 35 mW (100% laser emission in our setup, measured on the sample surface). For goethite analyses, the laser power was kept below 3 mW at the sample (10% laser) to avoid laser-induced degradation and dehydration effects during analysis [33,37]. The Raman spectra of all our samples are accumulated from 5 to 7 spectral scans at 15 s exposure time on the same analytical spot. All spectra were baseline corrected (Wire^{TM} software) using a built-in cubic-spline integration baseline function.

Additionally, all samples were analyzed using FTIR spectroscopy (Nicolet iS50 equipped with a DLaTGS detector using a diffuse reflectance accessory with 128 scans and 4 cm^{-1} resolution in the range from 400–6000 cm^{-1}) before and after the heating experiments. The two Mozambique rubies (samples 85933_C3 and 120553_B) and the ruby from Myanmar initially showed very broad goethite and diaspore bands in their FTIR spectra, respectively, which disappeared after heating. The sapphire (sample 106424_21) showed no diaspore bands before heating. Importantly, the samples showed no features that would have allowed a straightforward heat treatment detection after the experiments (no 3232 cm^{-1} peak), except for the sample 123993_C (ruby with goethite), in which a small peak at 3232 cm^{-1} peak developed as a result of the heating experiment.

3. Experiments and Results

3.1. Heating of Diaspore in Ruby and Sapphires

For the first experiment, a flat polished ruby from the Mogok valley (Myanmar, sample 126993_6) containing a cluster of colorless diaspore inclusions (>2 mm) at just the surface (Figure 1) was heated at increasing temperature steps using the Linkam heating stage. For Raman analyses a defined position within the diaspore was chosen. Starting at 200 °C, the sample was subsequently heated in 25 °C steps until 600 °C, with a final heating step at 700 °C. At each step, the temperature was kept stable for 4 min. Raman spectra of a defined position within the diaspore inclusion were registered only after cooling each time back to room temperature. The aim of this dehydration experiment was to monitor the temperature range of the phase transformation from diaspore to corundum as closely as possible.

Figure 1. Diaspore aggregate in ruby (sample 126993_6) before (**left**) and after (**right**) heating to 500 °C. Image width 2.2 mm.

Up to 500 °C, Raman spectra reveal only peaks related to diaspore with the main peak at 447 cm^{-1} and adjacent side bands. At 525 °C, a diaspore spectrum is still visible, but for the first time, with a small peak at 416 cm^{-1} indicating the beginning of the phase transformation to corundum (see Table 3). At 550 °C, however, the Raman spectra only show peaks corresponding to corundum, but no diaspore peaks anymore, thus indicating that the dehydration and phase transformation from diaspore to corundum has occurred (Figure 2). As a consequence of this phase transformation, the colorless diaspore cluster has become whitish (see Figure 1). Further heating up to 700 °C does not have much further effect on the Raman spectrum. The absence of the 643 cm^{-1} peak in the heated ruby sample compared with the reference corundum is an orientation effect well-known in anisotropic crystal structures.

Table 3. Position (Raman shift cm^{-1}) of the main Raman peaks of diaspore, corundum, goethite, and hematite. All Raman bands are referenced by the RRUFF database (https://rruff.info/, accessed 25 September 2023). For the interpretation of the Raman-related vibrations in corundum, diaspore, goethite, and hematite, see the literature [44–47].

Diaspore	Corundum	Goethite	Hematite
157			119 *
		247	223
331		301	291
	378		
447	**416**	**394–400**	**411**
500		480	500
		553	615
665	643	683	665
792	750		
1192		1320	1320

* not detected with our analytical setup; **in bold** the strongest Raman peak of each mineral phase.

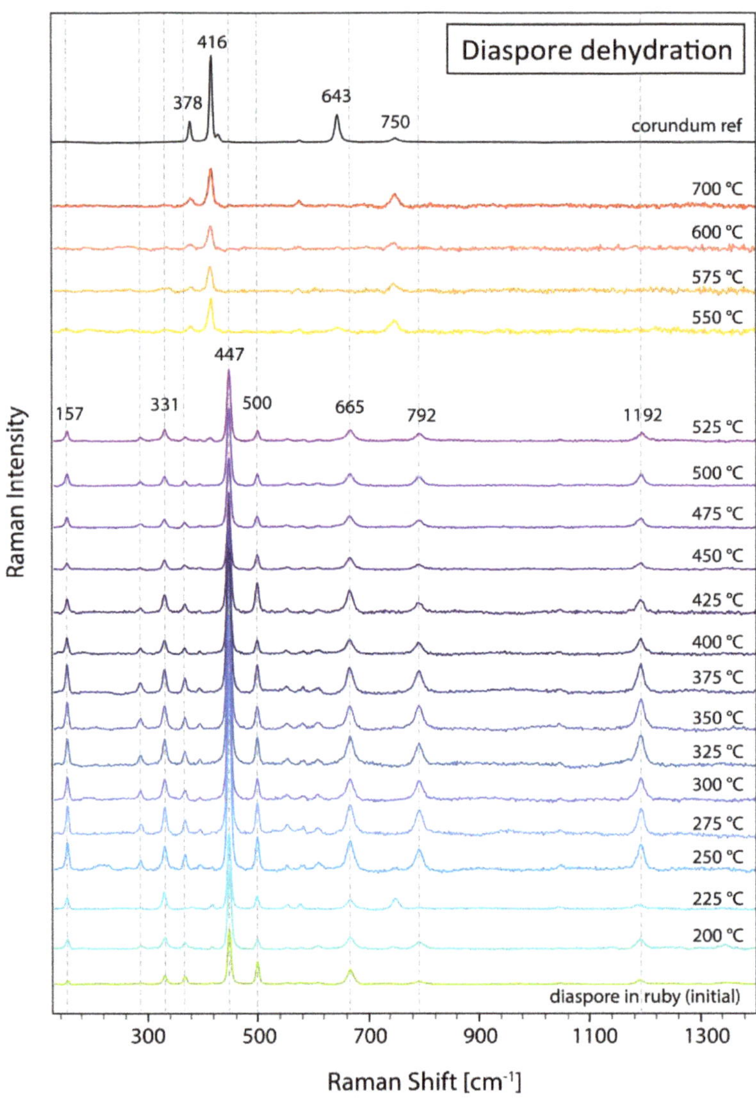

Figure 2. Raman spectra of diaspore in the ruby sample (126993_6) heated up to a maximum temperature of 700 °C in the heating stage. The dehydration of diaspore to corundum occurs at about 525 °C. For comparison, a corundum reference spectrum is indicated at the top. The dotted vertical lines indicate the main diaspore Raman peaks. The spectra all have been baseline subtracted and vertically displaced for clarity.

To verify that the phase transformation of diaspore to corundum is not influenced by the size of the inclusion, we selected a sapphire (106424_21) from Myanmar containing a tiny diaspore needle (few microns) within a fluid inclusion of a healed fissure (Figure 3a,b). Using the Linkam heating stage, this tiny inclusion was analyzed in five steps, starting from room temperature up to a maximum temperature of 700 °C (Figure 4).

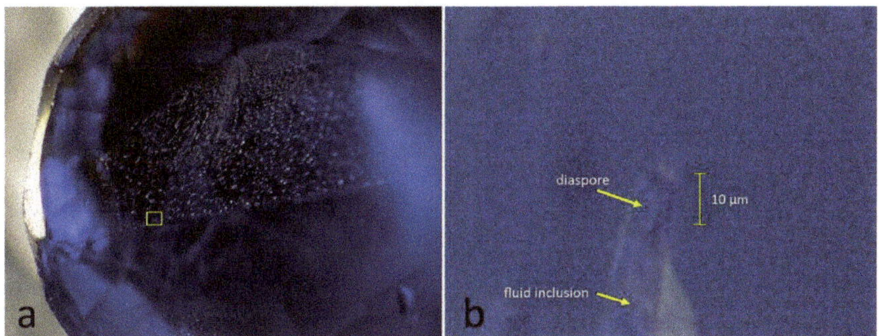

Figure 3. (**a**) Sapphire (sample 106424_21) exhibiting a naturally healed fissure. The yellow rectangle indicates the location of the micrograph on the right. Image width approx. 3 mm. (**b**) Tiny fluid tube within this healed fissure containing a diaspore needle of approx. 10 µm before the heating experiment. Image width 80 µm.

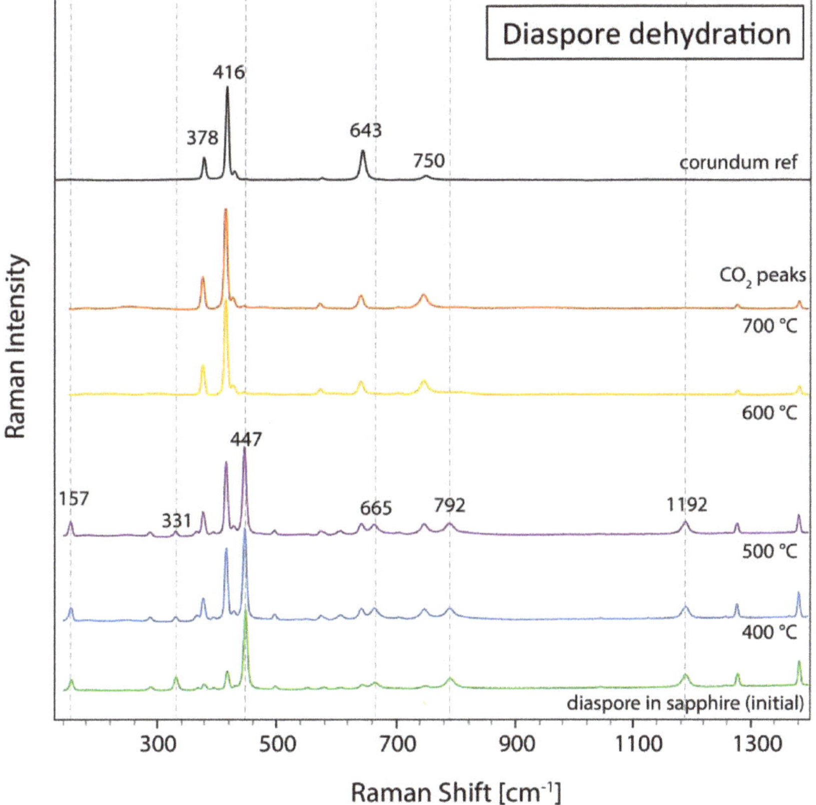

Figure 4. Raman spectra of the tiny diaspore inclusion (about 10 µm) present in a fluid inclusion in sapphire (sample 106424_21). The spectra were registered on the same spot during heating with the heating stage up to a maximum temperature of 700 °C. The dehydration of diaspore to corundum occurs above 500 °C. The dotted vertical lines indicate the main diaspore Raman peaks. The spectra all have been baseline subtracted and vertically displaced for clarity.

The dehydration of the tiny diaspore inclusion (about 10 μm) occurred in a similar temperature range as the larger diaspore in ruby. At 600 °C and beyond, the diaspore has completely transformed into corundum. It is important to know that the presence of corundum peaks at various heights in the spectra below 500 °C in Figure 4 is related to the sapphire matrix, in which this fluid inclusion with the tiny diaspore is located. These peaks do not represent a gradually ongoing phase transformation process, as the diaspore remains stable up to 500 °C.

In addition to these two diaspore inclusions in ruby and sapphire, a transparent diaspore crystal fragment (sample 97003) was heated step-by-step to a maximum of 800 °C in an electric furnace to confirm the dehydration and phase transformation of diaspore to corundum during a so-called "low-temperature" heating treatment as commercially applied on ruby and other corundum varieties. For this experiment, the setup (electric furnace) and conditions were chosen in a manner similar to those used in the gem trade. Visually, this experiment transformed the colorless diaspore into a whitish corundum characterized by numerous tiny micro-fissures, which were induced by the heating and concurrent expulsion of water during the dehydration (Figure 5).

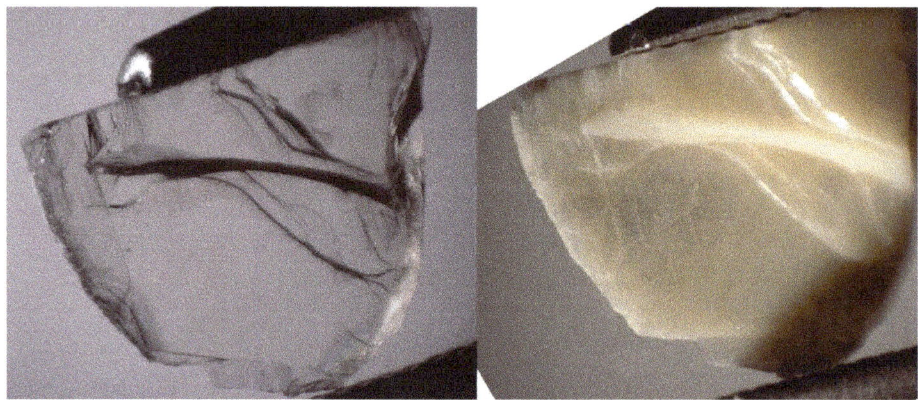

Figure 5. The colorless diaspore fragment (sample 97003) before heating (**left**), transformed to whitish corundum with tiny micro-fissures after dehydration (**right**). Image width 10 mm.

In total five Raman spectra were taken (before heating and after each heating step), always in the same crystallographic orientation and approximately the same position under the Raman microscope (Figure 6). As expected, the diaspore dehydration had fully succeeded after heating at 600 °C for 1 h, revealing now only Raman bands related to Al-O vibrational modes of corundum. By increasing the temperature further to 800 °C, the corundum spectrum became even more pronounced with a distinctly better peak/noise ratio.

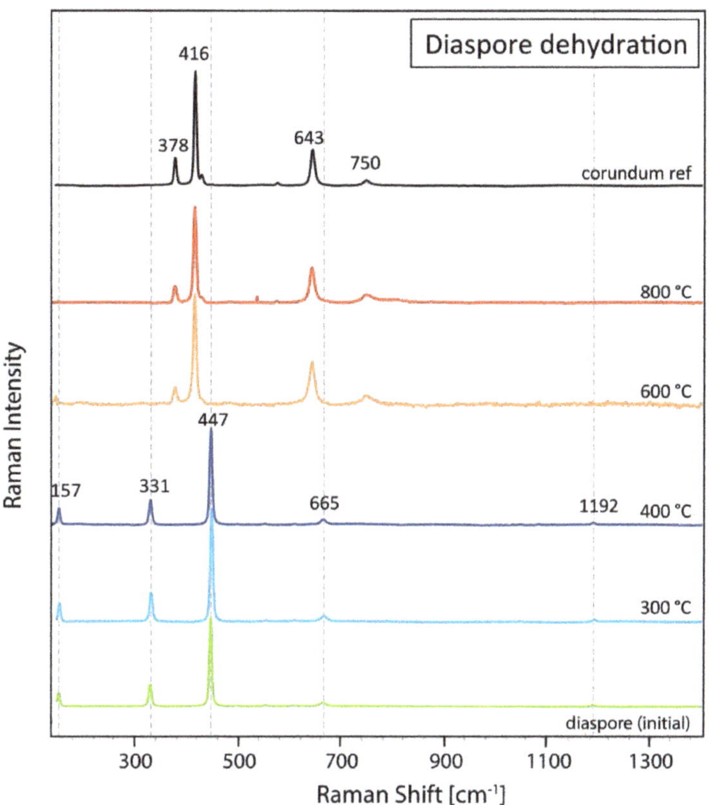

Figure 6. Raman spectra of diaspore crystal fragment (sample 97003) heated up to a maximum temperature of 800 °C in the electric furnace. The dotted vertical lines indicate the main diaspore Raman peaks. The spectra all have been baseline subtracted and vertically displaced for clarity purposes.

3.2. Heating of Goethite in Ruby

Similar to the heating experiment for diaspore dehydration, a flat polished ruby sample (85933_C3) containing orangey–brown goethite in fissures was heated repeatedly at increasing temperature steps using the Linkam heating stage. For Raman analyses a defined position within a goethite-bearing fissure was chosen. Starting at 150 °C, the sample was subsequently heated in 25 °C steps until 400 °C. At each step, the temperature was kept stable for 4 min. Raman spectra of a defined position within the goethite-containing fissure were registered only after cooling each time back to room temperature. The aim of this dehydration experiment was to monitor the temperature range of the goethite to hematite phase transformation when occurring as a thin film in a tiny fissure within the corundum as closely as possible (Figure 7).

Up to 300 °C, the goethite spectrum predominates, with small peak interferences with corundum peaks from the ruby host adjacent to the goethite-bearing fissure (Figure 8). At 325 °C, the Raman bands shift towards hematite positions. Namely, the main broad goethite band at about 400 cm^{-1} is transformed to a rather sharp peak (ideally 411 cm^{-1}), coincidentally superposed by the main corundum peak at 416 cm^{-1}, thus resulting in an inclined shoulder to the main corundum peak rather than a singular hematite peak as illustrated in the hematite reference spectrum. In addition, the smaller side peaks of goethite at 247, 301, and 683 cm^{-1} shift towards the hematite positions at 223, 291, and 665 cm^{-1}. In addition, a distinct broad band at about 1320 cm^{-1} appears and develops further in size with increasing heating temperature. The weak and broad band at about 1320 cm^{-1} in the goethite spectra may be the result of small amounts of hematite in the unheated sample, but more likely is a result of local laser-induced degradation as it appeared during the accumulation of the five spectral scans for each Raman analysis. Only after heating above 325 °C does the phase transformation occur, resulting in the shift of bands and a strong band at about 1320 cm^{-1}, which is characteristic of hematite.

Figure 7. Ruby from Mozambique (sample 85933_C3) with orange–brown goethite in a fissure. Image width 4 mm.

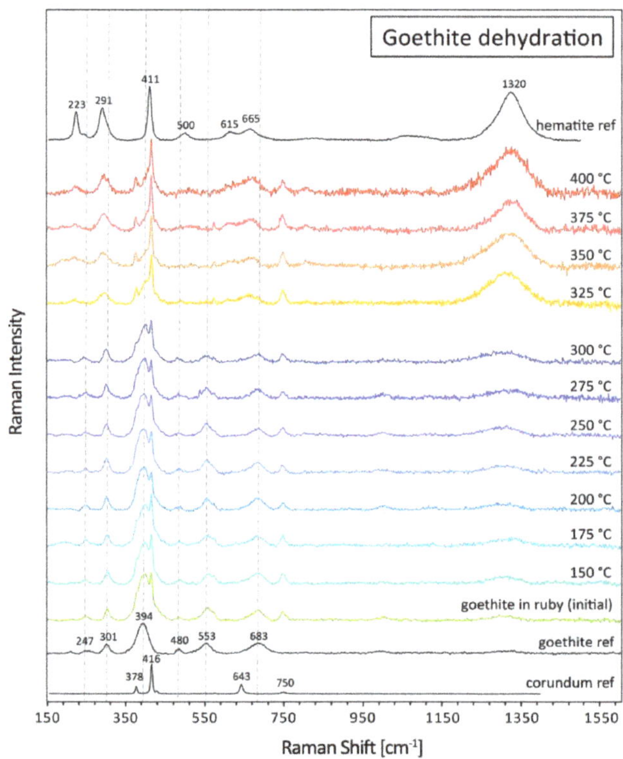

Figure 8. Raman spectra of goethite in a fissure in ruby (sample 85933_C3) heated with the heating stage up to a maximum temperature of 400 °C. The dehydration of goethite to hematite occurs at about 325 °C. The dotted vertical lines indicate the main goethite Raman peaks. The spectra all have been baseline subtracted and vertically displaced for the sake of clarity.

To confirm that the dehydration and phase transformation of goethite to hematite occurs during heat treatment of rubies commonly adopted in the trade, we heated a second ruby from Mozambique (sample 120553_B) containing goethite in an electric furnace up to a maximum temperature of 1000 °C, and repeatedly took Raman spectra always at room temperature after each heating step (Figure 9).

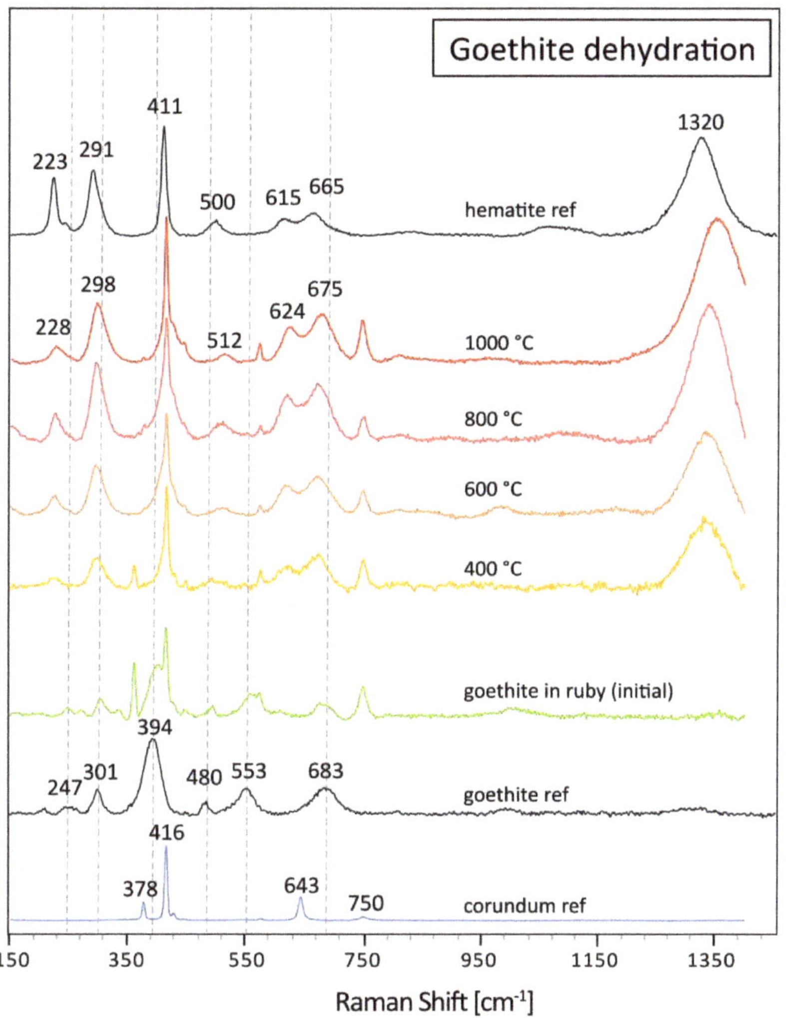

Figure 9. Raman spectra of goethite in a fissure in ruby (sample 120553_B) heated up to a maximum temperature of 1000 °C in an electric furnace. The dehydration of goethite to hematite occurs at about 325 °C. The dotted vertical lines indicate the main goethite Raman peaks. The spectra all have been baseline subtracted and vertically stacked for the sake of clarity.

Similar to the experiment with the Linkam stage, the dehydration of goethite to hematite is completed after heating to 400 °C for 1 h in the electric furnace. Further heating steps just increased the hematite peak intensities in the sample but with no further changes in peak positions. After heating in the electric furnace at a higher temperature, the Fe-staining in the fissures changes from an originally brownish–orange (goethite) to a brownish–red (hematite) color.

4. Discussion

The dehydration of the oxyhydroxides diaspore and goethite was observed in all five samples within the same temperature ranges (325 °C for goethite to hematite and 525 °C for diaspore to corundum, see Figure 10), regardless of the size of the inclusions and the experimental setup, i.e., a Linkam heating stage coupled to a Raman system for in situ analyses or an electric muffle furnace for heating at higher temperatures using similar setup and conditions as used in the gem trade for heat treatments. In both our setups, the accumulated heating time at which the phase transformations occurred was in total between 0.5 and 3 h, a rather short but realistic heat treatment duration when applied commercially on gem-quality rubies and other corundum varieties. Our findings are well in accordance with the extensive literature about goethite and diaspore dehydration using similar heating parameters [26,27,30,35,37,48,49].

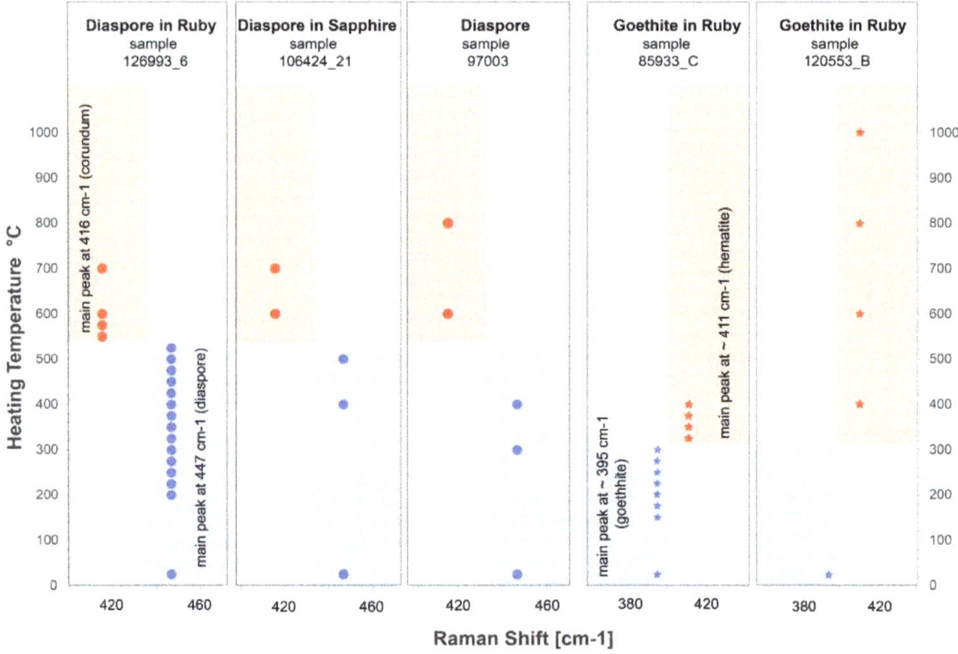

Figure 10. Comparison of the dehydration of diaspore to corundum and goethite to hematite in our five samples and experiments. Regardless of the size of the inclusions and the experimental setup (heating stage or electric muffle furnace), all diaspores and goethite each transformed in the same temperature range (about 525 °C for diaspore; about 325 °C for goethite).

Gem-quality rubies and other corundum varieties often contain microscopic inclusions, among them specifically diaspore and goethite. Diaspore is commonly found in corundum as a retrograde hydrothermal alteration product, e.g., as whitish acicular needles in ruby fissures or as tiny prismatic solids in fluid inclusions and negative inclusions [50]. Goethite is often present in corundum in fissures and (tiny) hollow channels as a pedogenic weathering product [43], introduced by Fe-enriched meteoric waters circulating within the gravels of secondary gem deposits. Another option is to find goethite as a solid inclusion due to the weathering and hydration of a former Fe-oxide inclusion in a gemstone. Although it is, in principle, possible to introduce goethite deliberately into fissures and cavities of gemstones, even after a heat treatment process, we have found no indication so far that such a filling process was applied, nor were our experiments successful in penetrating goethite powder deep into very fine fissures and hollow channels in corundum. For this, we diluted goethite in water and immersed two samples in the water for two hours before drying the samples in the air. In cavities at the surface, some Fe-hydroxide could be observed, but no Raman signal could be measured (only strong luminescence). In the small fissures and fine hollow channels, no goethite could be detected in the samples after our experiments.

To illustrate the direct application of the results of our study to separate unheated from heated corundum, we add two real cases: a ruby from Mozambique with a diaspore inclusion (in a negative crystal) and a sapphire from Sri Lanka with brownish goethite in a fissure. In both cases, it was readily possible to analyze these inclusions and to identify them as diaspore and goethite, respectively (Figure 11). Consequently, the presence of these two oxyhydroxides positively confirms that these ruby and sapphire samples were not heat treated. The two inclusion phases would have dehydrated during heating as commercially applied on corundum (usually >800 °C).

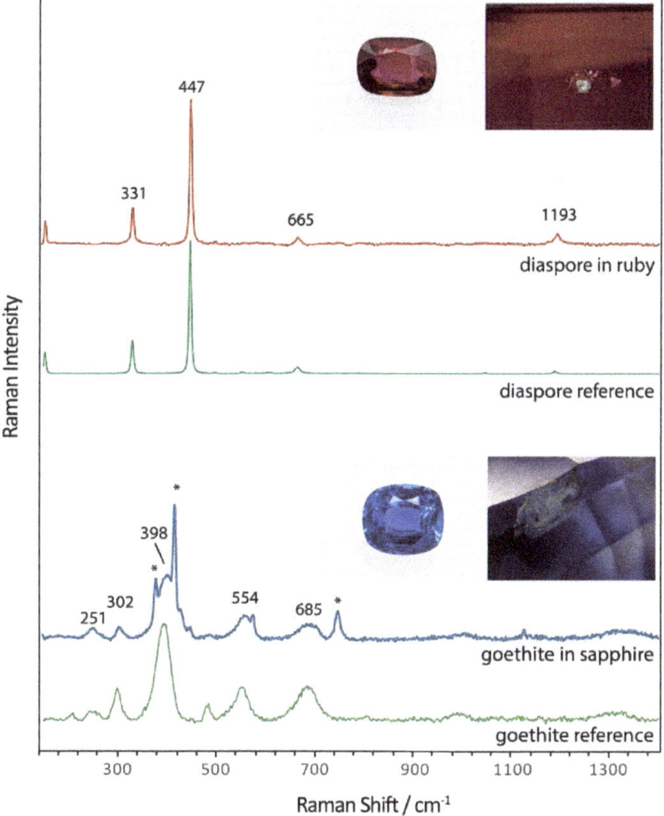

Figure 11. Raman spectra of goethite in sapphire fissure and diaspore in a ruby inclusion. The presence of these two hydroxides positively confirms that this sapphire and ruby of gem quality are unheated. The spectra all have been baseline subtracted and vertically displaced for clarity purposes. The peaks marked with asterisk * are related to corundum.

These two unheated examples are of specific interest, as they both show peaks (mainly at 3309 and at about 3230 cm^{-1}) in their infrared spectra (Figure 12), which are similar to those encountered in heated corundum [8–12,18]. Due to this, these examples may be wrongly interpreted as heated. However, it is important to note that the broad peak at 3229 cm^{-1} in the ruby sample is not related to the 3232 cm^{-1} peak often observed after heat treatment. Similarly, the 3234 cm^{-1} peak analyzed in the sapphire sample is, in fact, forming a series with the peaks at 3395 and 3380 cm^{-1} and is unrelated to heating. This shows how relevant this study is for separating unheated from heated corundum, as the detection of diaspore and/or goethite inclusions may help to avoid misinterpretation of an unheated corundum as heated. A more detailed study about FTIR spectra of corundum regarding heat treatment detection is in preparation.

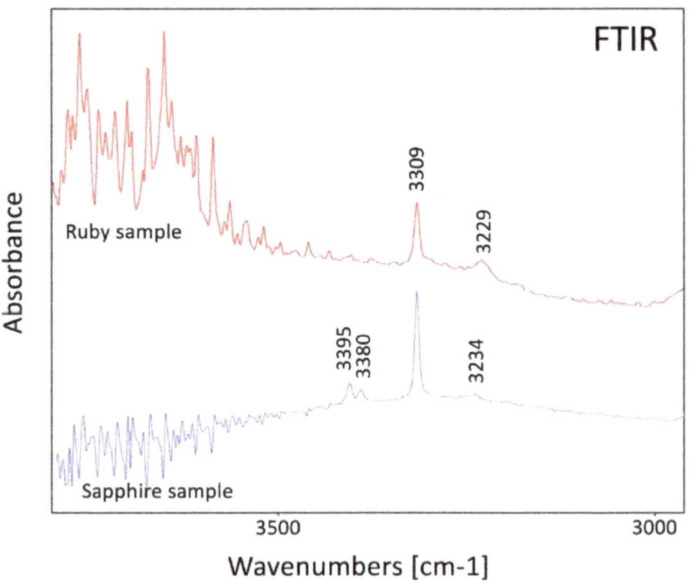

Figure 12. Infrared spectra of the ruby and the sapphire sample are presented in Figure 11.

The results of our experiments confirm that phase transformations such as the described dehydration of diaspore to corundum and goethite to hematite occur at temperatures (about 325 °C for goethite and 525 °C for diaspore) much lower than any successful heat treatment on corundum (generally well above 800 °C) described in the gemmological literature [1,3–5,7,13,15,18,51,52] and known to the authors. Furthermore, these dehydration reactions cannot be suppressed or shifted towards higher temperatures by any heat treatment commonly applied on gemstones and proceed in a rather small temperature range and thus result in a quasi-sudden shift of the Raman spectrum from goethite to hematite and diaspore to corundum as soon as the dehydration reaction occurs.

We have not seen any influence of the size of the inclusion on dehydration temperature, and to our knowledge, such an effect has never been described in the literature. In our experiments, regardless of whether the oxyhydroxide inclusion was only a few microns (sapphire 106424_21) or several mm (ruby 126993_6) in size, the dehydration could be observed in the same temperature range well below commercially applied heat treatment of corundum. This is, in fact, also the advantage of this method, in our opinion, as the dehydration of diaspore (and goethite) is unrelated to the size or position of these inclusions within corundum.

Interestingly, a long duration of heating at low temperatures will only shift the dehydration to lower temperatures [26]. By using the Linkam heating stage, we actually used short heating times (4 min at maximum temperatures) to check if the dehydration is also occurring after very short heat treatment. This was proven; see, for example, the sapphire (sample 106424_21) with only a tiny diaspore in a fluid inclusion.

It is thus safe to say that both described dehydration phase transformations and, specifically, the presence of diaspore and goethite can be used in many cases as criteria to separate unheated from heated rubies or other corundum varieties. This is even valid for many other gemstones that are commercially heated, as long as the heating temperature is higher than the dehydration temperatures of goethite (about 325 °C) and diaspore (about 525 °C).

However, it is important to note that the absence of diaspore or goethite does not necessarily imply that ruby or other corundum variety (or any other gemstone) has been heat-treated. For example, it is possible to find naturally formed hematite in fissures and at the surface of corundum, specifically when the gemstone in its rough form is still partially coated with Fe-rich soil. In such cases, however, careful analyses of several Fe-rich pedogenic inclusions in the same gemstone will often reveal that hematite is present together with Fe-hydroxides, thus still proof that this stone was not commercially heat treated to enhance its visual appearance.

5. Conclusions

Diaspore and goethite are often found as inclusions in ruby and other corundum varieties and are readily identifiable using Raman micro-spectroscopy. For this Raman study, we have carried out heating experiments on gem-quality rubies (corundum) containing tiny inclusions of goethite and diaspore. For these experiments, we used a heating stage for in situ Raman analyses of these inclusions at different temperatures and an electric muffle furnace similar to those commercially used for gemstone treatment, e.g., in Sri Lanka and Thailand.

Our experiments reveal that regardless of the size of the inclusion, the dehydration of goethite to hematite occurs at about 325 °C, whereas the phase transformation of diaspore to corundum occurs at about 525 °C. These temperatures are well in accordance with the literature about the thermal transformation of both oxyhydroxides and their anhydrous oxides. Compared with this, commercial heat treatment of corundum is known to be carried out at temperatures above 800 °C, thus much higher than the described dehydration reactions. Consequently, experiments reveal that the presence of one of these two oxyhydroxides is a very safe criterion to positively confirm that a ruby or other corundum variety has not been heat treated to enhance its visual appearance. This is even true in cases in which other analytical methods, such as FTIR spectroscopy or Raman analyses on the width of the main Raman peak of zircon inclusions ($FWHM_{1010}$) within corundum are not conclusive for the separation of unheated and heated rubies and other corundum varieties.

Author Contributions: All authors were involved in designing the experiments. M.S.K., P.L. and W.Z. collected and prepared the samples, including gemological characterization. M.S.K., G.S. and J.B. conducted heating experiments. M.S.K. performed data analysis and prepared the manuscript. All authors contributed to the figures and the manuscript. M.S.K., P.L. and W.Z. conceived of and supervised the project. All authors have read and agreed to the published version of the manuscript.

Funding: This research received no external funding.

Data Availability Statement: The data underlying the results in this paper are not publicly available at this time but may be obtained from the corresponding author upon request.

Acknowledgments: The authors would like to thank Sean Gilbertson and Elena Basaglia from Gemfields Ltd. for supplying unheated ruby samples from Mozambique. The Burmese ruby sample for this study was kindly donated by Miemie Htin Tut, Silkeneast Ltd., Bangkok. The authors would like to further thank SSEF staff, namely Walter A. Balmer, Alexander Klumb, Michael Rytz, Laurent Cartier, Hao Wang, Markus Wälle, and Carole Lachavanne for fruitful discussions and comments.

Conflicts of Interest: The authors declare no conflict of interest.

References

1. Hughes, R.W.; Manorotkul, W.; Hughes, E.B. *Ruby & Sapphire: A Gemologist's Guide*; RWH Publishing/Lotus Publishing: Bangkok, Thailand, 2017; p. 886.
2. Notari, F.; Hainschwang, T.; Caplan, C.; Ho, K. The heat treatment of corundum at moderate temperature. *InColor Mag.* **2019**, *42*, 76–85.
3. Nassau, K. Heat treating ruby and sapphires: Technical aspects. *Gems Gemol.* **1981**, *17*, 121–131. [CrossRef]
4. Themelis, T. *The Heat Treatment of Ruby and Sapphire*; Publisher Gemlab Inc.: Wheatland, PA, USA, 1992.

5. GIT. A brief history of gem corundum heat treatment in Thailand. *InColor Mag.* **2019**, 68–74, Spring issue.
6. Karampelas, S.; Hennebois, U.; Mevellec, J.-Y.; Pardieu, V.; Delaunay, A.; Fritsch, E. Pink to purple sapphires from Ilakaka, Madagascar: Insights to separate unheated from heated samples. *Minerals* **2023**, *13*, 704. [CrossRef]
7. Hughes, E.B.; Vertriest, W. A Canary in the ruby mine: Low-temperature heat treatment experiments on ruby. *Gems Gemol.* **2022**, *58*, 400–423. [CrossRef]
8. Smith, C.P. A contribution to understanding the infrared spectra of rubies from Mong Hsu, Myanmar. *J. Gemmol.* **1995**, *24*, 321–335. [CrossRef]
9. Smith, C.P. Infrared spectra of gem corundum. *Gems Gemol.* **2006**, *42*, 92–93.
10. Beran, A.; Rossman, G.R. OH in naturally occurring corundum. *Eur. J. Mineral.* **2006**, *18*, 441–447. [CrossRef]
11. Krzemnicki, M.S. New Research by SSEF Studies Methods for Detecting Low-Temperature Heated Rubies from Mozambique. SSEF Press Release, 2018. Available online: https://www.ssef.ch/detecting-low-temperature-heated-rubies-from-mozambique/ (accessed on 1 July 2023).
12. Saeseaw, S.; Khowpong, C.; Vertriest, W. Low-temperature heat treatment of pink sapphires from Ilakaka, Madagascar. *Gems Gemol.* **2020**, *56*, 448–457. [CrossRef]
13. Balmer, W.; Leelawatanasuk, T.; Atichat, W.; Wathanakul, P.; Somboon, C. Update on FTIR characteristics of heated and unheated yellow sapphire. In Proceedings of the GIT Conference, Bangkok, Thailand, 6–9 December 2006; Volume 6–7, p. 91.
14. Balmer, W.A. Petrology, Geochemistry, and Gemological Characteritics of Marble-Hosted Ruby Deposits of the Morogoro Region, Tanzania. Unpublished. Ph.D. Thesis, Depart. of Geology, Faculty of Science, Chulalongkorn University, Bangkok, Thailand, 2011; p. 185p.
15. Wang, W.; Scarratt, K.; Emmett, J.L.; Breeding, C.M.; Douthit, T.R. The effects of heat treatment on zircon inclusions in Madagascar sapphires. *Gems Gemol.* **2006**, *42*, 134–150. [CrossRef]
16. Krzemnicki, M.S.; Lefèvre, P.; Zhou, W.; Wang, H.A.O. Zircon inclusions in unheated pink sapphires from Ilakaka, Madagascar: A Raman spectroscopic study. In Proceedings of the International Gemmological Conference, Online, 20–21 November 2021.
17. Saeseaw, S.; Kongsomart, B.; Atikarnsakul, U.; Khowpong, C.; Vertriest, W.; Soonthorntantikul, W. Update on "Low-Temperature" Heat Treatment of Mozambican Ruby: A Focus on Inclusions and Ftir Spectroscopy. GIA Lab Report. 2018. Available online: https://www.gia.edu/doc/low_HT_Moz_report.pdf (accessed on 1 July 2023).
18. Pardieu, V.; Saeseaw, S.; Detroyat, S.; Raynaud, V.; Sangsawong, S.; Bhusrisom, T.; Engniwat, S.; Muyal, J. GIA Lab Reports on Low-Temperature Heat Treatment of Mozambique Ruby. GIA Lab Report, 2015. Available online: https://www.gia.edu/gia-news-research-low-temperature-heat-treatment-mozambique-ruby (accessed on 25 September 2023).
19. Nasdala, L.; Irmer, G.; Wolf, D. The degree of metamictization in zircon: A Raman spectroscopic study. *Eur. J. Mineral.* **1995**, *7*, 471–478. [CrossRef]
20. Nasdala, L.; Wenzel, M.; Vavra, G.; Irmer, G.; Wenzel, T.; Kober, B. Metamictisation of natural zircon: Accumulation versus thermal annealing of radioactivity-induced damage. *Contrib. Mineral. Petrol.* **2001**, *141*, 125–144. [CrossRef]
21. Xu, W.; Krzemnicki, M.S. Raman spectroscopic investigation of zircon in gem-quality sapphire: Application in origin determination. *J. Raman Spectrosc.* **2021**, *52*, 1011–1021. [CrossRef]
22. Deflandre, M.M. La structure cristalline du diaspore. *Bull. Minéralogie* **1932**, *55*, 140–165. [CrossRef]
23. Goldsztaub, M.S. Etude de quelques derives de l'oxyde ferrique (FeOOH, FeO_2Na, FeOCl); determination de leurs structures. *Bull. De Minéralogie* **1935**, *58*, 6–76.
24. Rooksby, H.P. *X-ray Identification and Crystal Structure of Clay Minerals*; Brindley, G.W., Ed.; Mineralogical Society: London, UK, 1951; p. 250.
25. Ervin, G., Jr. Structural interpretation of the diaspore–corundum and boehmite–a-Al_2O_3 transitions. *Acta Crystallogr.* **1952**, *5*, 103–108. [CrossRef]
26. De Faria, L.J.; Gay, P. Disordered structural states in the dehydration of goethite and diaspore. *Mineral. Mag.* **1962**, *33*, 37–41.
27. De Faria, L.J. Dehydration of goethite and diaspore. *Z. Krist.* **1963**, *119*, 176–203. [CrossRef]
28. Ruan, H.D.; Frost, R.L.; Kloprogge, J.T.; Duong, L. Infrared spectroscopy of goethite dehydroxylation: III. FT-IR microscopy of in situ study of the thermal transformation of goethite to hematite. *Spectrochim. Acta Part A Mol. Biomol. Spectrosc.* **2002**, *58*, 967–981. [CrossRef]
29. Kloprogge, J.T.; Ruan, H.D.; Frost, R.L. Thermal decomposition of bauxite minerals: Infrared emission spectroscopy of gibbsite, boehmite and diaspore. *J. Mater. Sci.* **2002**, *37*, 1121–1129. [CrossRef]
30. Iwai, S.-I.; Yamamoto, H.; Morikawa, H.; Isobel, M. Topotactic thermal-transformation of diaspore to corundum. *Mineral. J.* **1973**, *7*, 137–158. [CrossRef]
31. Carim, A.H.; Rohrer, G.S.; Dando, N.R.; Tzeng, S.-Y.; Rohrer, C.L.; Perrotta, A.J. Conversion of diaspore to corundum: A new alpha-alumina transformation sequence. *J. Am. Ceram. Soc.* **1997**, *80*, 2677–2680. [CrossRef]
32. Pomiès, M.P.; Morin, G.; Vigneuad, C. XRD study of the goethite-hematite transformation: Application to the identification of heated prehistoric pigments. *Eur. J. Solid State Inorg. Chem.* **1998**, *35*, 9–25. [CrossRef]
33. Grevel, K.D.; Burchard, M.; Fasshauer, D.W. Pressure-volume-temperature behavior of diaspore and corundum: An in-situ X-ray diffraction study comparing different pressure media. *J. Geophys. Res. Solid Earth* **2000**, *105*, 27877. [CrossRef]
34. Löffler, L.; Mader, W. Transformation mechanism of the dehydration of diaspore. *J. Am. Ceram. Soc.* **2003**, *86*, 534–540. [CrossRef]

35. De Faria, D.L.A.; Venâncio Silva, S.; de Oliveira, M.T. Raman microspectroscopy of some iron oxides and oxyhydroxides. *J. Raman Spectrosc.* **1997**, *28*, 873–878. [CrossRef]
36. De Faria, D.L.A.; Lopes, F.N. Heated goethite and natural hematite: Can Raman spectroscopy be used to differentiate them? *Vib. Spectrosc.* **2007**, *45*, 117–121. [CrossRef]
37. Gialanella, S.; Girardi, F.; Ischia, G.; Lonardelli, I.; Mattarelli, M.; Montagna, M. On the goethite to hematite phase transformation. *J. Therm. Anal. Calorim.* **2010**, *102*, 867–873. [CrossRef]
38. Canımoglu, A.; Garcia-Guinea, J.; Correcher, V.; Karabulut, Y.; Tuncer, Y.; Can, N. Luminescent, structural, and thermal properties of the unusual "Anatolian" diaspore (zultanite) from Turkey. *Spectrosc. Lett.* **2014**, *47*, 292–300. [CrossRef]
39. Hanesch, M. Raman spectroscopy of iron oxides and (oxy)hydroxides at low laser power and possible applications in environmental magnetic studies. *Geophys. J. Int.* **2009**, *177*, 941–948. [CrossRef]
40. Koivula, J.I. Goethite inclusion alteration during the heat conversion of amethyst to citrine. *Aust. Gemmol.* **1987**, *16*, 271–272.
41. Kammerling, R.C.; Koivula, J.I. Thermal alteration of Inclusions in "rutilated" topaz. *Gems Gemol.* **1989**, *25*, 165–167. [CrossRef]
42. Koivula, J.I. Useful visual clue indicating corundum heat treatment. *Gems Gemol.* **2013**, *49*, 160–161. [CrossRef]
43. Sripoonjan, T.; Wanthanachaisaeng, B.; Leelawatanasuk, T. Phase transformation of epigenetic iron staining: Indication of low-temperature heat treatment in Mozambique ruby. *J. Gemmol.* **2016**, *35*, 156–161. [CrossRef]
44. Porto, S.P.S.; Krishnan, R.S. Raman Effect of Corundum. *J. Chem. Phys.* **1967**, *47*, 1009–1012. [CrossRef]
45. Delattre, S.; Balan, E.; Lazzeri, M.; Blanchard, M.; Guillaumet, M.; Beyssac, O.; Haussühl, E.; Winkler, B.; Salie, E.K.H.; Calas, G. Experimental and theoretical study of the vibrational properties of diaspore (α-AlOOH). *Phys. Chem. Miner.* **2012**, *39*, 93–102. [CrossRef]
46. Abrashev, M.V.; Ivanov, V.G.; Stefanov, B.S.; Todorov, N.D.; Rosell, J.; Skumrye, V. Raman spectroscopy of alpha-FeOOH (goethite) near antiferromagnetic to paramagnetic phase transition. *J. Appl. Phys.* **2020**, *127*, 205108. [CrossRef]
47. Beattie, I.R.; Gibson, T.R. The single crystal Raman spectra of nearly opaque materials. Iron (III) oxide and chromium (III) oxide. *J. Chem. Soc.* **1970**, *A 6*, 980–986. [CrossRef]
48. Frost, R.L.; Kloprogge, J.T.; Russell, S.C.; Szetu, J. Dehydroxylation and the vibrational spectroscopy of aluminium (oxo)hydroxides using infrared emission spectroscopy. Part III: Diaspore. *Appl. Spectrosc.* **1999**, *53*, 829–835. [CrossRef]
49. Liu, H.; Chen, T.; Zou, X.; Qing, C.; Frost, R. Thermal treatment of natural goethite: Thermal transformation and physical properties. *Thermochim. Acta* **2013**, *568*, 115–121. [CrossRef]
50. Schmetzer, K.; Medenbach, O. Examination of three-phase inclusions in colorless, yellow, and blue sapphires from Sri Lanka. *Gems Gemol.* **1988**, *24*, 107–111. [CrossRef]
51. Abraham, J.S.D. Heat treating corundum: The Bangkok operation. *Gems Gemol.* **1982**, *18*, 79–82. [CrossRef]
52. Emmett, J.L.; Douthit, T.R. Heat treating the sapphires of Rock Creek, Montana. *Gems Gemol.* **1993**, *29*, 250–272. [CrossRef]

Article

Pink to Purple Sapphires from Ilakaka, Madagascar: Insights to Separate Unheated from Heated Samples

Stefanos Karampelas [1,*], Ugo Hennebois [1], Jean-Yves Mevellec [2], Vincent Pardieu [3], Aurélien Delaunay [1] and Emmanuel Fritsch [2]

[1] Laboratoire Français de Gemmologie (LFG), 30 rue de la Victoire, 75009 Paris, France; a.delaunay@lfg.paris (A.D.)

[2] Institut des Matériaux de Nantes Jean Rouxel, IMN, Nantes Université, CNRS, 44000 Nantes, France; emmanuel.fritsch@cnrs-imn.fr (E.F.)

[3] VP Consulting, Manama, Bahrain; vince@fieldgemology.org

[*] Correspondence: s.karampelas@lfg.paris

Abstract: The present study is focused on the analysis of zircon inclusions found in pink to purple sapphires from Ilakaka (Madagascar) with an optical microscope, Fourier-transform infrared (FTIR), and micro-Raman spectroscopy in order to update previous knowledge and find insights to separate heated from unheated samples. In total, 157 zircon inclusions in 15 unheated samples and 74 zircon inclusions in 6 heated samples are analysed using micro-Raman spectroscopy with standardised parameters. The full width at half maximum (FWHM) of the main Raman band due to anti-symmetric stretching vibration v_3 of the SiO_4 tetrahedron in the zircon structure has been carefully measured. In the unheated samples, it ranges from 6.26 to 21.73 cm^{-1} with an average of 10.74 cm^{-1}, a median of 10.04 cm^{-1}, and a standard deviation of 2.84 cm^{-1}. On the other hand, it is lower in the heated samples, ranging from 4.83 to 14.97 cm^{-1} with an average of 7.23 cm^{-1}, median of 7.06 cm^{-1}, and standard deviation of 1.63 cm^{-1}. In our unheated samples, the FWHM was rarely below 7 cm^{-1}. In our heated samples, the FWHM was rarely above 12 cm^{-1} but mostly below 8 cm^{-1}, with a variation restricted to less than 3 cm^{-1} in the same sample. The present work will hopefully further contribute to more accurately identifying the low-temperature heat treatment of pink sapphires from Ilakaka, Madagascar.

Keywords: pink sapphire; Ilakaka (Madagascar); zircon inclusions; "low" temperature heat treatment; micro-Raman spectroscopy

Citation: Karampelas, S.; Hennebois, U.; Mevellec, J.-Y.; Pardieu, V.; Delaunay, A.; Fritsch, E. Pink to Purple Sapphires from Ilakaka, Madagascar: Insights to Separate Unheated from Heated Samples. *Minerals* **2023**, *13*, 704. https://doi.org/10.3390/min13050704

Academic Editor: Evgeny Galuskin

Received: 28 April 2023
Revised: 17 May 2023
Accepted: 19 May 2023
Published: 22 May 2023

1. Introduction

Pink sapphire is corundum (Al_2O_3) with pink as the main colour. It is principally coloured by Cr^{3+}, with minor influences from the Fe^{2+}-Ti^{4+} charge transfer or V^{3+} (both giving a purplish component), or Fe^{3+} and colour centres (giving a brownish component) [1–3]. Separation of rubies and pink sapphires is performed visually using master stones (Figure 1) or sometimes colour charts [4].

Pink sapphires of gem quality can be found in various mines around the globe, for example, in Myanmar (Burma), Vietnam, Sri Lanka (Ceylon), Mozambique, and Madagascar [5–7]. Most gem-quality pink sapphires on the market today come from Ilakaka, Madagascar, from a secondary gem deposit discovered at the end of the 1990s [8,9]. Faceted stones from Ilakaka rarely exceed 10 ct. Many of those present the characteristic of numerous rounded zircons measuring from several to more than 100 μm in dimension (Figure 2a,b). They may form clusters or "nests" (Figure 2c), are less often shaped as elongated prisms (Figure 2d), and are less frequently accompanied by coloured monazites (see again Figure 2a).

Figure 1. Four synthetic-coloured corundum used as master stones in LFG to separate rubies from pink sapphires. The two samples on the left are synthetic rubies, and the two on the right are pink sapphires. The left samples are about 14.10 × 12.06 × 7.19 mm and weigh about 2.54 ct each. Photo: Aurélien Delaunay; © LFG.

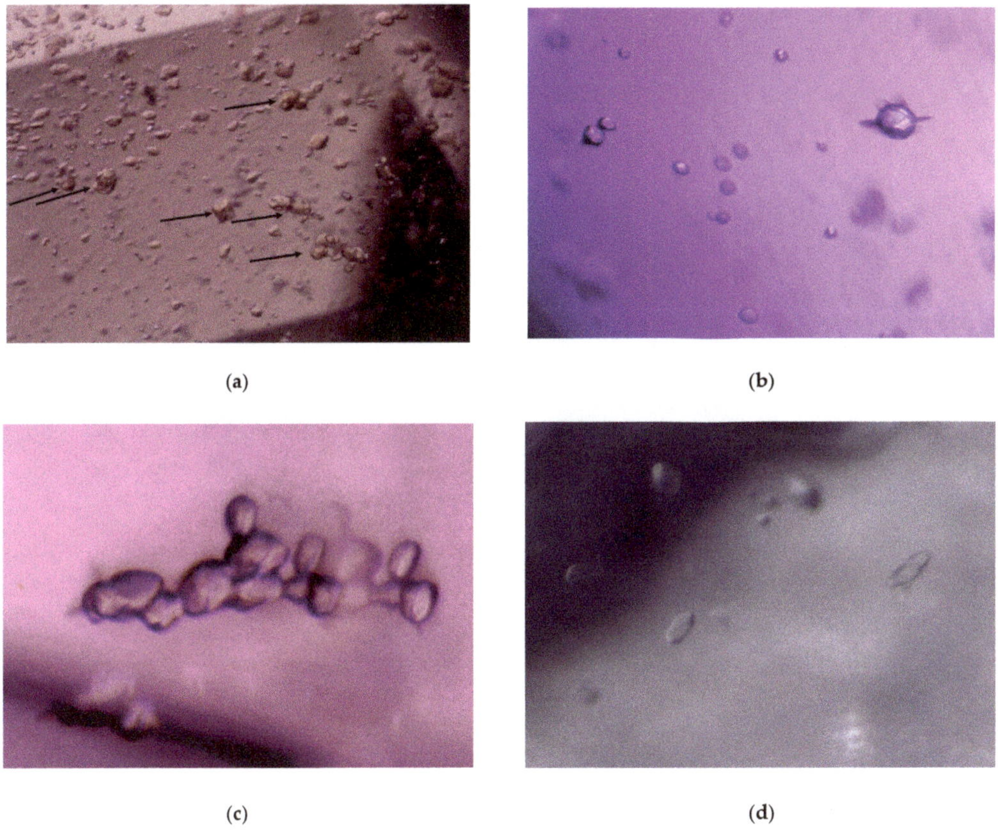

(a)

(b)

(c)

(d)

Figure 2. Inclusion scene observed frequently in unheated pink sapphires from Ilakaka, Madagascar. The photomicrographs are not colour calibrated, and their colours might not be accurate. (a) Numerous transparent colourless zircons of rounded shapes and brown coloured monazites indicated by black arrows, field of view (FOV): 2 mm; (b) Zircons of rounded shapes with tension halos and fractures in the host pink sapphire, FOV: 3 mm; (c) Clustered transparent rounded colourless zircons, FOV: 2 mm; (d) Elongated prismatic (euhedral) colourless zircon inclusions, FOV: 1 mm. Microphotos by Ugo Hennebois; © LFG.

Heat treatment under various conditions has been used to improve their colour by decreasing or removing their violet-to-purple (and sometimes brown) component and enhancing their pink appearance [9–16]. This treatment to improve corundum's colour has been known for more than 1500 years [17]. Heat treatment at temperatures above

1200 °C visibly alters the appearance of some inclusions in corundum. It is considered that the boundary between heat treatment at "low" and "high" temperatures is from 1200 to 1350 °C [13,18–20]. Unheated gems fetch higher prices than heated ones; thus, separating unheated from heated gems is one of the main issues of gemological laboratories today [9–22].

In pink sapphires, heat treatment at "high" temperatures can be identified by a variety of criteria and technologies. First, simple observation with a binocular microscope reveals the alteration of the zircon, such as a "frosty" appearance of slightly melted zircon inclusions (Figure 3a) and altered fissures (Figure 3b) [10,21]. Under a short-wave ultraviolet (SWUV) lamp, some heated sapphires (including pink ones) present a chalky blue luminescence due to the high-temperature internal diffusion of titanate groups in the mineral [23]. FTIR spectroscopy reveals the presence of absorption bands at around 3309 and 3232 cm^{-1}, with sometimes a band at 3185 cm^{-1} [12,13,16]. These infrared bands are due to different complexes of titanium associated with hydroxyl defects in corundum [24] and present strong polarisation [25]. The band at 3232 cm^{-1} is especially considered proof of heat treatment. On the contrary, FTIR bands from 2900 to 3700 cm^{-1}, linked with hydrous minerals inclusions, indicate that the sample is unheated [25], as these inclusions would lose their structural water during heat treatment [26].

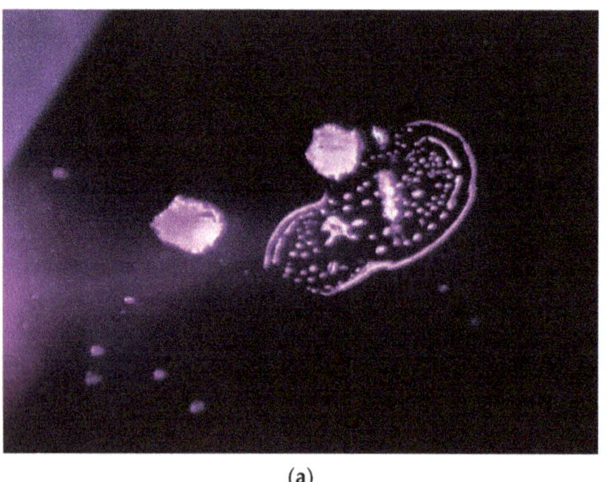

(a)

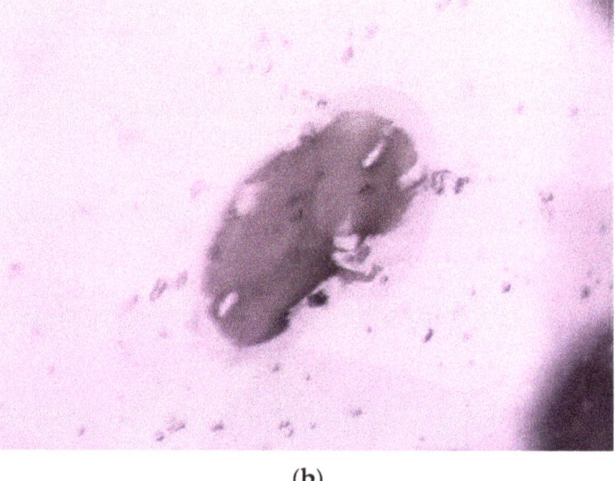

(b)

Figure 3. Inclusion scene observed in "high" temperature heated pink sapphires from Ilakaka, Madagascar. The photomicrographs are not colour calibrated, and their colours might not be accurate. (a) Altered zircon inclusions with "frosty" appearances and melt aureoles, FOV: 1.5 mm; (b) Fissure traces of alteration and numerous zircons, FOV: 3 mm. Microphotos by Ugo Hennebois; © LFG.

One newer criterion is related to micro-Raman spectroscopy of zircon inclusions, specifically in pink sapphires from Ilalaka, Madagascar. Many of those zircons are disordered because of self-irradiation (radiation-damaged, also known as metamictisation) with broadened bands compared to perfectly crystallised references. Thus, heating reduces disorder in those zircons and, consequently, the FWHM of all bands [9,10,13]. The easiest band on which to measure this effect is the strongest one, at around 1010 cm^{-1}. This feature will be called from here on the main Raman band. It is the anti-symmetric stretching vibration ν_3 of the SiO$_4$ in the zircon structure. In well-crystalised zircons, the FWHM of this feature is below 5 cm^{-1}, whereas in highly radiation-damaged zircons, it may reach more than 30 cm^{-1} [27]. The degree of radiation damage in the zircon inclusions of gem-quality corundum, as well as their age determined by isotopic methods, may be used in some cases for corundum's origin determination [28–32].

Unlike its high-temperature counterpart, heat treatment of pink sapphires at relatively "low" temperatures, below 1200 °C and sometimes 800 °C, is challenging to identify [11–13,16,20]. One rare possibility concerned coloured monazite inclusions in pink sapphires from Ilakaka, Madagascar. They keep their colour after being heated up to 600 °C; thus, the presence of colourless monazite inclusions can be used as one of the rare microscopic indications of low-temperature heat treatment [13]. On the other hand, the presence of coloured monazite inclusions cannot be used as an indication that the sample is unheated [13,33]. As the treatment only subtly affects the inclusions, one turns to spectroscopic methods. To this day, none of the series of FTIR bands indicating heat treatment at 3309 and 3232 cm^{-1}, sometimes along with a band at 3185 cm^{-1}, have been observed in an unheated pink sapphire [12,13]. Yet, the absence of these bands does not indicate that the sample is unheated.

In view of this lack of adequate criteria to identify unheated pink sapphires from Ilakaka, measuring with precision the full width at half maximum (FWHM) of the main Raman band around 1010 cm^{-1} in zircon inclusions could be potentially very useful for gemological laboratories. Values reported in the literature vary considerably (see Table 1). In unheated samples, it spans from 10.1 to 13.5 cm^{-1} (average of 11.5 cm^{-1} [10]) to 8.8–13.8 cm^{-1} [13], from 7.5 to 17.6 cm^{-1} (median value below 10 cm^{-1} [9]), and from 7.1 to 21.7 cm^{-1} with a median value of 11.3 cm^{-1} and an average of 11.6 cm^{-1} [14,15]. By comparison, the FWHM reported in heat-treated pink sapphires from Ilakaka averages 8.7 cm^{-1} for samples heated to 1400 °C [10] and from 6.6 to 12.7 cm^{-1} for samples heated to 1000 °C [13]. The importance of the spatial and spectral resolutions, as well as of the instrument used, was recently revealed [14,15]. Additionally, samples' reliability is very important [34]; one of the authors (VP) observed that most of the samples were heated before being faceted in the mining area. In the present study, updated Raman data measured to minimise the influence of spectroscopic and instrumental parameters in zircon inclusions in pink to purple sapphires from Ilakaka, themselves carefully selected, are presented to gauge the usefulness of this approach.

Table 1. FWHM, median, and mean values, when mentioned (na*: not available), of the main Raman band at around 1010 cm^{-1} of zircon inclusions found in (unheated and heated) pink sapphires from Ilakaka, Madagascar, with the reference from which they are extracted. Spatial and spectral resolutions were mentioned only in articles [13,14], and these are similar to those used in the present study.

Treatment	FWHM (cm^{-1})	Median Value (cm^{-1})	Average Value (cm^{-1})	Reference
Unheated	10.1–13.5	na*	11.5	[9]
Unheated	8.8–13.8	na*	na*	[12]
Unheated	7.5–17.6	<10	na*	[8]
Unheated	7.1–21.7	11.3	11.6	[13,14]
Heated to 1000 °C	6.6–12.7	na*	na*	[12]
Heated to 1400 °C	na*	na*	8.7	[9]

2. Materials and Methods

We studied zircon inclusions of fifteen unheated pink sapphires from Ilakaka, Madagascar (Table 2). All the samples are reliable as they were collected by one of the authors (VP) in 2005, where gravels were washed; i.e., they represent B-Type samples according to a recently published degree of confidence [34]. All samples were obtained rough (not faceted), and one side was polished so their inclusions are easier to observe via optical microscope.

Table 2. Characteristics of the fifteen studied unheated pink sapphires from Ilakaka, Madagascar (1 ct = 0.2 g).

Sample Number	Mass (ct)	Dimensions (mm)	Colour	Photo
SK-007	0.373	4.43 × 4.15 × 2.02	Light pink	
SK-008	0.412	5.09 × 4.64 × 1.86	Light pink	
SK-009	0.525	6.44 × 4.13 × 2.04	Pink	
SK-010	0.566	6.83 × 5.31 × 1.63	Pink	
SK-011	0.346	5.12 × 4.07 × 2.00	Purplish-pink	
SK-012	0.328	5.13 × 3.50 × 2.16	Light pink	
SK-013	0.307	3.96 × 3.03 × 2.04	Light pink	
SK-014	0.337	4.71 × 3.75 × 2.19	Pink	
SK-015	0.599	5.01 × 4.49 × 2.51	Pinkish-purple	
SK-016	0.460	4.76 × 3.59 × 2.29	Purple-pink	
SK-017	0.346	4.28 × 3.23 × 2.44	Pinkish-purple	
SK-018	0.319	4.15 × 3.49 × 2.06	Light pink	
SK-023	0.255	4.98 × 3.77 × 1.65	Pink	
SK-024	0.550	5.37 × 4.16 × 2.62	Pink	
SK-025	0.357	4.35 × 3.70 × 2.29	Pink	

Six faceted heated pink sapphires from Ilakaka, Madagascar, from the LFG collection (Z-Type [34]) were also studied (Table 3). Two of the samples (LFG101 and LFG102) presented indications of heating under the microscope (Figure 4a,b) and might be considered as heated at "high" temperature. The four other samples (LFG103, LFG 104, LFG105, and LFG106) presented FTIR absorption bands at 3232 and 3309 cm^{-1}, observed to this day only in pink sapphires from Ilakaka, Madagascar, if they have been heated [12,13]. One of these four samples (LFG103) also presented indications of heating under the microscope (Figure 4c) and might thus be considered as heated at "high" temperature, but the other three samples presented no indications of heating under the microscope and might be considered as heated at "low" temperature (Figure 5a–c).

Table 3. Characteristics of the six heated pink sapphires from Ilakaka, Madagascar.

Sample Number	Mass (ct)	Dimensions (mm)	Colour	Photo
LFG101	1.797	9.02 × 6.22 × 4.28	Purplish pink	
LFG102	1.893	8.94 × 6.52 × 4.02	Purplish pink	
LFG103	2.038	7.78 × 6.30 × 4.86	Purplish pink	
LFG104	3.022	9.57 × 7.74 × 4.55	Purplish pink	
LFG105	3.214	10.27 × 8.39 × 4.57	Purplish pink	
LFG106	3.027	9.22 × 7.93 × 4.87	Purplish pink	

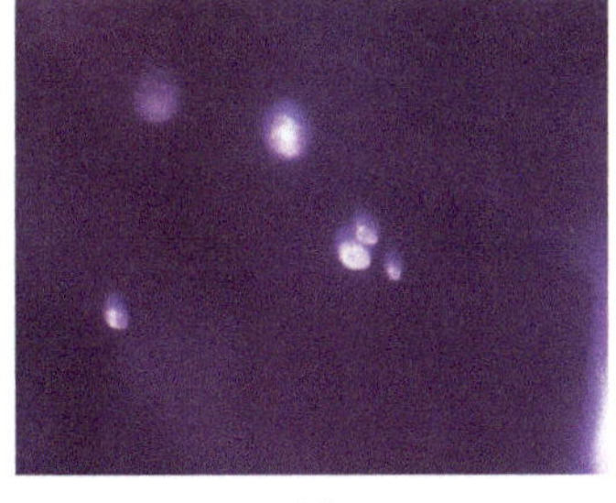
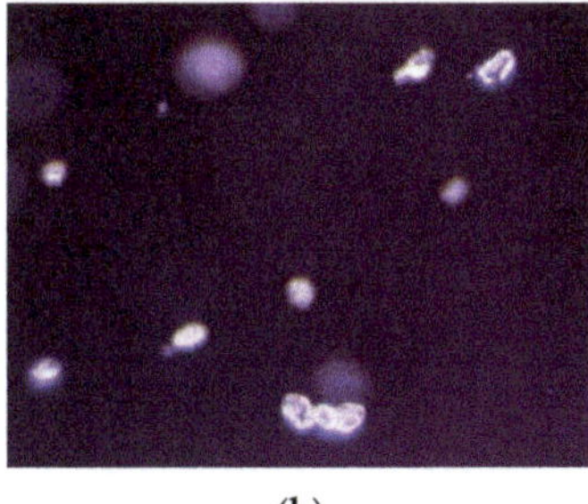
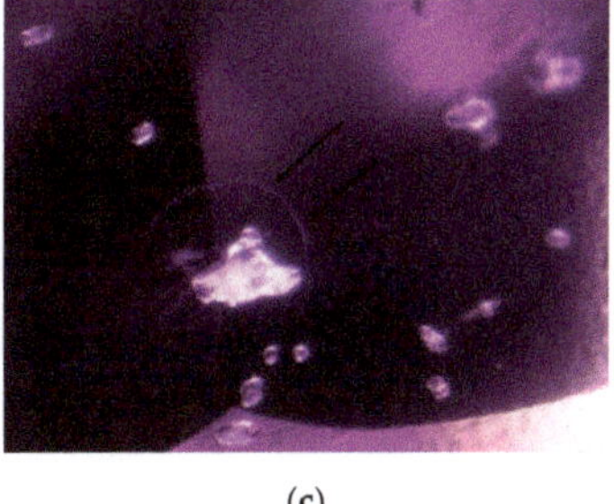

Figure 4. Inclusion scene observed in three of our heated pink sapphires from Ilakaka, Madagascar. Indications of heating under microscope are observed in these samples. The photomicrographs are not colour calibrated, and their colours might not be accurate. (**a**) Altered zircons with turbid/frosty appearance in the sample LFG101, FOV: 2 mm; (**b**) altered zircons with turbid/frosty appearance in the sample LFG102, FOV: 1.5 mm; (**c**) altered zircons with Toll-like structures by black arrows in the sample LFG103, FOV: 2 mm. Microphotos by Ugo Hennebois; © LFG.

All samples were examined under a Zeiss Stemi 508 binocular microscope (Oberkochen, Germany). FTIR spectra (400–8000 cm^{-1}) were obtained using the Nicolet iS5 spectrometer (Thermo Fischer Scientific, Waltham, MA, USA) with a 4 cm^{-1} resolution and averaging 200 scans. Raman spectra were acquired using identical instrumentation and conditions to those detailed in previous publications [15,16]; Raman spectra were acquired on a Renishaw inVia spectrometer (Renishaw plc, Wotton-under-Edge, Gloucestershire, UK) coupled with an optical microscope (always using 50× long-working-distance objective lens) with a 514 nm laser excitation (diode-pumped solid-state laser of about 10 mW laser power on the sample), a confocal mode (20 μm entrance slit), a grating of 1800 grooves/mm, and about 1.5 cm^{-1} spectral resolution. The acquired spectra are from 600 to 1200 cm^{-1}, with 5 accumulation and 20 s exposure time. The FWHM of the main Raman band was calculated using the instrument software, choosing a fit with a Lorentzian function after baseline correction. A diamond was used for the calibration of Raman spectrometer using the 1331.80 cm^{-1} Raman band. When the sample geometry allowed it, several zircon inclusions were measured (up to twenty). However, in one sample (SK-016), no zircon inclusions could be analysed. When zircon inclusions are near the surface, and due to relaxation effects, the exact position of the Raman bands may be affected [32]. Whenever possible, zircon inclusions in the sample, and not at the proximity of the surface, were analysed. This was, however, not possible for one sample (e.g., sample SK-018), as the two inclusions accessible are situated near the surface. In some large enough (i.e., ca. 100 microns) zircon inclusions, more than one spot was analysed in order to check for possible zoning.

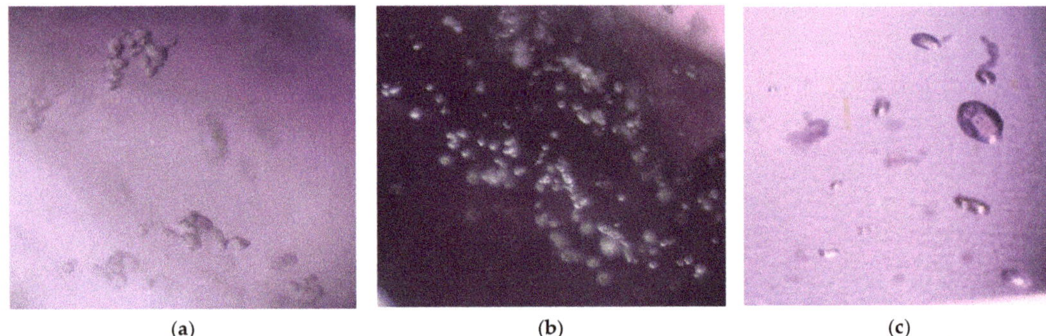

(a)   (b)   (c)

Figure 5. Inclusion scene observed in three studied heated pink sapphires from Ilakaka, Madagascar. No indications of heating under microscope are observed in these samples. The photomicrographs are not colour calibrated, and their colours might not be accurate. (**a**) Clustered transparent rounded colourless zircons of various sizes in the sample LFG104, FOV: 1 mm; (**b**) transparent colourless zircons of different shapes and sizes in the sample LFG105, FOV: 2.5 mm; (**c**) transparent colourless prismatic (euhedral) colourless zircon with other zircons of different shape in the sample LFG106, FOV: 0.5 mm. Microphotos by Ugo Hennebois; © LFG.

3. Results and Discussion

Table 4 lists the ranges in position and FWHM of the main Raman band of zircon measured on 157 zircon inclusions found in 15 unheated pink to purple sapphires from Ilakaka (Madagascar). In Figure 6a (black traces), the main Raman zircon band is presented on an expanded scale from 980 to 1050 cm^{-1} so that differences in width can be better appreciated in the spectra of two different zircon inclusions in two unheated pink to purple sapphires. As previously observed [9,10,13–16], a large variation of the FWHM and position of the zircon inclusion's ν_3 Raman band can be observed (Table 4) from sample to sample, and even within the same unheated sample [14]. The exact position of the maximum of the main Raman band ranged from 1003.76 cm^{-1} to 1021.61 cm^{-1} with an average of 1015.70 cm^{-1}, median of 1016.26 cm^{-1}, and standard deviation of 2.82 cm^{-1}. The FWHM of this band ranged from 6.26 to 21.73 cm^{-1} with an average of 10.75 cm^{-1}, median value of 10.04 cm^{-1}, and standard deviation of 2.84 cm^{-1}. In accordance with the previous studies [10,21], zircon inclusions in pink to purple sapphires from Ilakaka, Madagascar, are radiation-damaged.

Table 4. Range of position and FWHM of Raman main band observed in zircon inclusions of unheated pink to purple sapphires from Ilakaka, Madagascar.

Sample	Range of Peak Position (cm^{-1})	Range of FWHM (cm^{-1})	Number of Analysed Zircon Inclusions	Total Number of Raman Analyses
SK-007	1013.86–1019.02	7.87–17.70	20	28
SK-008	1017.56–1020.65	15.07–18.42	5	7
SK-009	1018.81–1019.73	9.47–21.73	7	7
SK-010	1005.10–1017.21	7.06–13.14	10	10
SK-011	1011.87–1019.01	9.60–13.39	10	10
SK-012	1011.13–1015.32	8.11–12.09	5	5
SK-013	1014.12–1017.22	7.11–11.65	8	8
SK-014	1011.20–1016.63	7.87–13.80	10	10
SK-015	1012.74–1021.11	7.55–14.21	7	7
SK-016	-	-	-	-
SK-017	1012.37–1018.22	7.25–16.15	13	17
SK-018	1003.76–1005.25	14.48–15.85	2	2
SK-023	1014.14–1021.61	6.26–15.39	20	26
SK-024	1014.55–1018.21	6.83–13	20	26
SK-025	1010.00–1016.14	6.81–10.31	20	26

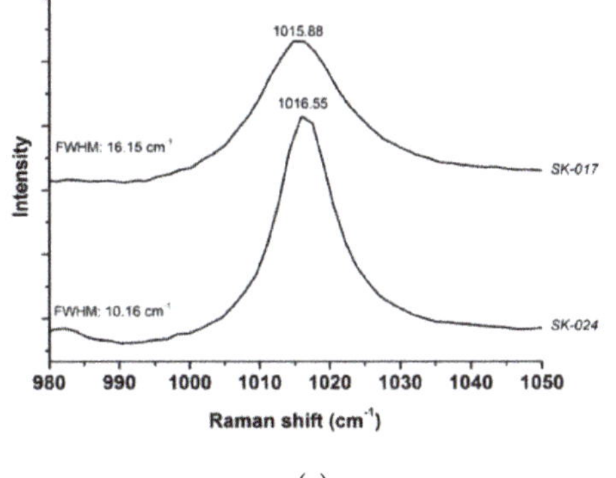

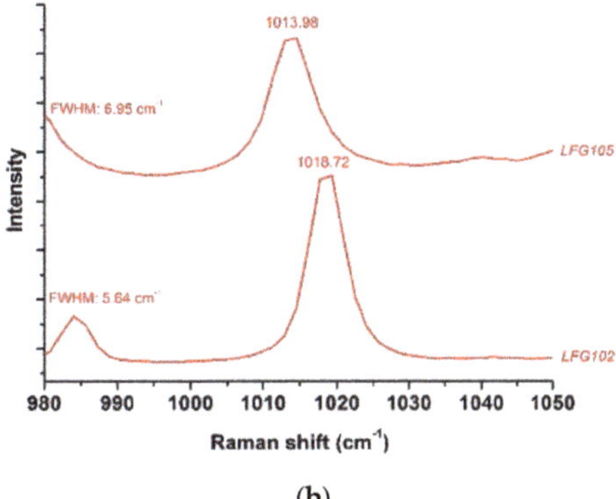

Figure 6. FWHM and exact position of the main zircon band measured in two zircon inclusions in two pink to purple sapphires from Ilakaka, Madagascar: (**a**) unheated (black traces) and (**b**) heated (red traces). Spectra are shifted vertically for clarity.

In Table 5, the ranges in position and FWHM of 74 analysed zircons found in six heated pink to purple sapphires from Ilakaka (Madagascar) are listed. In Figure 6b (red traces), examples of Raman spectra are presented. The exact position of the maximum of the main Raman band ranged from 1007.29 cm^{-1} to 1020.23 cm^{-1} with an average of 1014.38 cm^{-1}, median of 1013.78 cm^{-1}, and standard deviation of 2.96 cm^{-1}. The FWHM of this band ranged from 4.83 to 14.97 cm^{-1} with an average of 7.23 cm^{-1}, median value of 7.06 cm^{-1}, and standard deviation of 1.63 cm^{-1}. These values, taken as a whole, are relatively lower than those observed in unheated samples (see again Table 4). This is because heat treatment leads to structural reconstitution and better crystalline order in zircon inclusions, thus leading to a decrease in the FWHM. It confirms what was observed previously in pink sapphires from Madagascar [10,13] as well as other radiation-damaged zircons [21,35–38].

Table 5. Range of position and FWHM of Raman main band observed in zircon inclusions of heated pink to purple sapphires from Ilakaka, Madagascar.

Sample	Range of Peak Position (cm^{-1})	Range of FWHM (cm^{-1})	Number of Analysed Zircon Inclusions	Total Number of Raman Analyses
LFG101	1018.34–1020.23	5.47–5.91	6	6
LFG102	1017.95–1019.57	4.83–6.86	10	10
LFG103	1007.29–1013.24	6.56–9.32	11	11
LFG104	1009.75–1013.82	6.36–7.79	10	10
LFG105	1010.58–1014.12	6.19–11.38	17	22
LFG106	1018.34–1020.23	5.16–14.97	20	34

In Figure 7, the position and FWHM of the main band obtained on zircon inclusions in all unheated (black squares) and heated (red circles) pink sapphires from Ilakaka, Madagascar, from this study are shown. Unheated and heated samples presented comparable main Raman band positions, with the vast majority of these situated above 1010 cm^{-1}. It has been presented that this band shifts towards the highest wavenumbers due to compressive strain; thus, as in previous studies, the band positions indicate pressure around the zircon inclusion [10,21,31,32,39,40]. It has been suggested that in some cases (e.g., for diamonds' inclusions), compressive strain (also referred to as "fossilised pressure" or "overpressure") on the inclusion can be up to 2–3 GPa (see [21,40–42] and references therein). Free-standing radiation-damaged zircons present a significantly lower Raman peak maximum position (see isobar line in Figure 7). The inclusions situated by samples' surfaces present lower peak maximum positions, sometimes below 1010 cm^{-1}, possibly due to pressure relaxation [31]. Suggested calculations of the compressive stress of zircon inclusions using Raman spectroscopy [42] indicate a large variation from below 1 GPa to above 4 GPa. It has been suggested that calculated stress accuracy might be affected due to partial non-hydrostatic stress or inhomogeneous damage within the analysed sample volume (see [42] and references therein). Moreover, compressive strain affects the zircon inclusions' main Raman position change during heat treatment; during a heat treatment up to 1000 °C, the band shifts towards lower wavenumbers, and on the other hand, when it is heated above 1000 °C it shifts towards higher wavenumbers [21,37,42].

Table 6 presents the frequency of a given FWHM range by a step of 1 cm^{-1} observed for the main Raman band in zircon inclusions and the corresponding statistics for both non-heated and heated samples. Unheated samples presented zircon inclusions with a broadened main Raman band, having a FWHM sometimes over 20 cm^{-1} (see Figure 7 and sample SK-009 in Table 4). Heated samples presented zircon inclusions with a sharper main Raman band, and the FWHM may sometimes be below 6 cm^{-1} (see also Table 5, samples LFG101, LFG102, LFG103, and LFG 104). A FWHM of <7 cm^{-1} for the main Raman zircon band was observed in 34 out of 74 inclusions of heated samples. On the other hand, for the unheated samples, no zircon inclusions with a FWHM < 6 cm^{-1} were observed in the studied samples, and only 3 out of 157 had a low FWHM between 6 and 7 cm^{-1}. The majority (68 out of 74, i.e., ca. 92%) of the studied zircon inclusions in the heated samples present a FWHM ranging below 9 cm^{-1}. Moreover, a FWHM > 12 cm^{-1} was observed in only 1 of 74 studied zircon inclusions in heated samples, but in 47 out of 157 in unheated samples. One might think that zircon inclusions with a relatively high FWHM (e.g., above 17 cm^{-1}) before heating would present a relatively high FWHM after heat treatment at low temperatures because of partial structural reconstitution of the inclusion. This is possibly the reason why we measured a FWHM of 14.97 cm^{-1} in one sample, a FWHM from 9 to 12 cm^{-1} (5 out of 74), and otherwise a FWHM below 9 cm^{-1} in the heated samples (see again Tables 5 and 6).

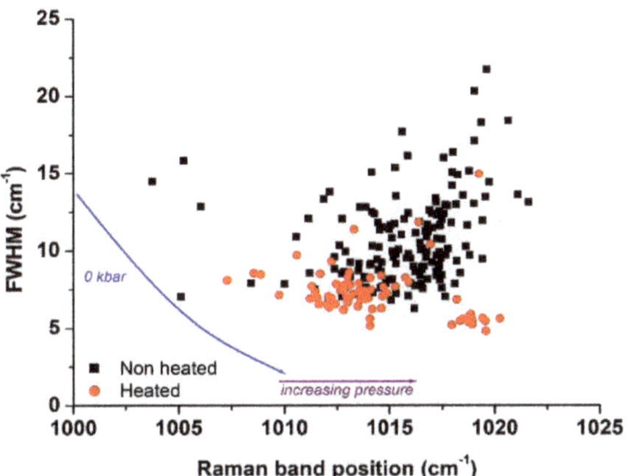

Figure 7. FWHM and exact position of the main zircon band measured in zircon inclusions of unheated (black squares) and heated (red circles) samples from Ilakaka, Madagascar. Due to compressive strain, the exact position of the main Raman band of the zircon inclusions is shifted towards higher wavenumbers compared to those expected for the free-standing radiation-damaged zircons (0 kbar isobar blue line) [21].

Table 6. Frequency of observation of a specific range of FWHM observed in 157 zircon inclusions in 15 unheated, and 74 zircon inclusions in 6 heated, pink to purple sapphires from Ilakaka, Madagascar.

Range of FWHM (cm^{-1})	Non-Heated	Heated
<6	0/157	17/74 (ca. 23%)
6–7	3/157 (ca. 1.9%)	17/74 (ca. 23%)
7–8	21/157 (ca. 13.4%)	25/74 (ca. 34%)
8–9	25/157 (ca. 15.9%)	9/74 (ca. 12%)
9–10	28/157 (ca. 17.8%)	2/74 (ca. 2.7%)
10–11	15/157 (ca. 9.6%)	1/74 (ca. 1.3%)
11–12	18/157 (ca. 11.5%)	2/74 (ca. 2.7%)
12–13	19/157 (ca. 12.1%)	0/74
13–14	10/157 (ca. 6.4%)	0/74
14–15	4/157 (ca. 2.5%)	1/74 (ca. 1.3%)
15–16	5/157 (ca. 3.1%)	0/74
16–17	3/157 (ca. 1.9%)	0/74
17–18	2/157 (ca. 1.3%)	0/74
18–19	2/157 (ca. 1.3%)	0/74
>20	2/157 (ca. 1.3%)	0/74

Heat treatment of pink to purple sapphires at temperatures above 1400 °C can partially (or entirely) decompose zircon inclusions to baddeleyite (ZrO_2) and a SiO_2-rich phase; these can be identified via micro-Raman spectroscopy [10,21]. Only one inclusion in the heated sample LFG102 presented such characteristics reflected in the Raman spectrum. Moreover, photoluminescence bands linked to REE are observed in Raman spectra of zircon; using a 514 nm laser excitation, these bands can affect the signal of the main zircon Raman band [10,21,37]. These bands were observed after heat treatment accompanied by an FHWM below 7 cm^{-1} for the main zircon Raman band [10]; however, in our case, these were also observed in a zircon inclusion in an unheated sample (SK-025).

When several zircon inclusions can be measured in the same sample, the variation of the FWHM and position of the zircon inclusion main Raman band can be checked. In several cases, variations of the FWHM are large in unheated samples and less important in heated ones [16]. Again, this is because heat treatment leads to structural reconstitution of zircon inclusions and a decrease in the FWHM and its ranges. Comparing the minimum and the maximum value of the FWHM observed in Tables 4 and 5, a more important variation is observed in our unheated samples than in heated gems. For instance, four heated (LFG101, LFG102, LFG103, and LFG 104) samples out of the six present a variation between the maximum and the minimum FWHM of less than 3 cm^{-1}. For the studied unheated samples, the difference can be over 10 cm^{-1} (see SK-009 in Table 4), larger than those previously reported for a single sample [9,14]. However, heated samples LFG105 and LFG 106 also present relatively large variations of the FWHM, and unheated samples SK-012, SK-013, and SK-025 present a smaller variation. In our sampling, a range of variation of less than 3 cm^{-1} for the FWHM combined with a FWHM below 8 cm^{-1} is only observed for the heated samples. This provides a criterion applicable to our limited population of Ilakaka gems.

When the inclusion was large enough, several micro-Raman analyses were acquired in a single zircon inclusion. This was performed for 16 zircons in six unheated samples and 7 zircons in two heated samples. The FWHM in one unheated zircon (SK-017) varied over 7.88 cm^{-1}, from 8.27 to 16.15 cm^{-1} [14]. In other unheated samples, the FWHM varied less than 4.5 cm^{-1}, and in some zircon inclusions, the FWHM did not vary much (<1 cm^{-1}). In the heated samples, two zircon inclusions in the sample LFG106 present a variation of slightly above 4 cm^{-1}, with the other measurements having a variation being less than 2.5 cm^{-1} and sometimes below 1 cm^{-1}. Differences in the FWHM in the same zircon inclusion reveal heterogeneous radiation damage [42]. In cathodoluminescence (CL) images, zoning was observed for zircon inclusions in sapphires from the same region [28,29]. The fact that relatively large variations of the FWHM in the same zircon inclusion could also be observed in heated samples further confirms that low-temperature heating leads only to partial structural reconstruction of zircons.

Zircon inclusions can be found in several gem-quality sapphires of various colours and from different mining areas [7,21,28–32,43]. The characteristics presented here are for pink to purple sapphires from Ilakaka, Madagascar. Preliminary studies on unheated similar coloured sapphires from Myanmar (Burma) and Sri Lanka (Ceylon) showed that these present zircon inclusions with narrower FWHMs. Additionally, rubies and sapphires from Madagascar might present different Raman spectroscopic characteristics; data on such stones need to be acquired so they can be used for similar applications.

If another instrument is used instead of the one described for the present study, the criteria should be used with caution as, apart from the spectral and spatial resolution, different parameters linked to the instrument could play a role in the exact shape of the bands, and thus, on the measured FWHM. In order to compare studies using another instrument with the results presented here, instrument profile functions should also be taken into consideration [44].

4. Conclusions

Determining whether a pink to purple sapphire from Ilakaka, Madagascar, is heat-treated at low temperatures or not is vital to gemological laboratories around the world. This is because this locality has produced most of the pink to purple sapphires in the market today. Under the microscope, samples heated at low temperatures are challenging to identify. Thus, identification of such treatment can be accurately performed only using spectroscopy. FTIR spectroscopy is in some cases useful as the presence of the series of FTIR bands at 3309 and 3232 cm^{-1}, sometimes along with a band at 3185 cm^{-1}, in pink to purple sapphires, indicates heat treatment [12,13]. Frequently, those pink to purple sapphires present numerous zircon inclusions of various shapes, often over 100 μm in maximum dimension. These are radiation damaged, and even low-temperature heat treatment may lead to structural reconstitution of zircon inclusions. Micro-Raman spectroscopy can also give valuable clues to the identification with careful measurement of the peak position and FWHM of the main Raman band, that is, the anti-symmetric stretching vibration v_3 of the SiO_4 at about 1010 cm^{-1} [9,10,13–16]. In this study, unheated samples and heated samples of reliable geographic and thermal history present a large variation of the FWHM and position of this band, always using the same spectroscopic parameters. For unheated samples, the FWHM was rarely observed below 7 cm^{-1}, and for heated samples, the FWHM was rarely above 12 cm^{-1} but mostly below 8 cm^{-1}.

Analyses by micro-Raman spectroscopy of various zircon inclusions within one single sample provide useful indications as to the band position and FWHM. Only in heated samples does the FWHM have a value below 8 cm^{-1}, with a variation range limited to 3 cm^{-1}, indicating relatively good crystallinity. This might be used as an additional criterion for the identification of heated pink to purple sapphires from Ilakaka, Madagascar. The criteria proposed here apply strictly to zircon that can be easily analysed by micro-Raman in pink to purple sapphires from the Ilakaka mining area in Madagascar. These criteria cannot work in the case where no such inclusions are present, or if they are present but difficult to access with the Raman spectrometer. Zircon inclusions might be found in pink to purple sapphires from other countries, such as Myanmar (Burma) and Sri Lanka (Ceylon). Additionally, other coloured sapphires from Madagascar containing zircon inclusions might not present similar spectroscopic characteristics.

Further studies on reliable samples will improve the statistics presented here. Additionally, results on the same samples before and after heat treatment, and also by performing line scans or even better maps using micro-Raman spectroscopy, will help understand the exact effect of low-temperature heat treatment on the zircon inclusions of pink to purple sapphires from Ilakaka, Madagascar. Moreover, as zircon inclusions of different shapes are documented in the studied samples, it might be useful to look for possible relations between the FWHM and the inclusions' shape (e.g., check whether sharper bands are linked with specific zircon shapes). Similar extended studies with micro-Raman spectroscopy, for example, on monazite inclusions as well as epigenetic iron staining (e.g., goethite, which changes form in relatively low temperature), might give additional clues for more accurate identification of this challenging treatment.

Author Contributions: S.K. and U.H. formulated the paper, selected the samples, designed the experiments, performed data reduction, participated in the data interpretation, drew the figures, and wrote the manuscript. U.H., S.K., J.-Y.M. and E.F. performed the experiments. V.P. found most of the studied samples and edited the manuscript. J.-Y.M., A.D. and E.F. participated in the data interpretation and edited the manuscript. All authors have read and agreed to the published version of the manuscript.

Funding: This research received no external funding.

Data Availability Statement: Not applicable.

Acknowledgments: The authors wish to thank all the colleagues at LFG, Paris, and IMN, Nantes, for their help during this study.

Conflicts of Interest: The authors declare no conflict of interest.

References

1. Fritsch, E.; Rossman, G.R. An update on color in gems. Part I. Introduction and colors caused by dispersed metal ions. *Gems Gemol.* **1987**, *23*, 126–139. [CrossRef]
2. Emmett, J.L.; Dubinsky, E.V.; Hughes, R.W.; Scarratt, K. Color, Spectra & Luminescence. In *Ruby & Sapphire: A Gemmologist's Guide*; Hughes, R., Ed.; RWH Publishing: Bangkok, Thailand, 2017; pp. 90–148.
3. Dubinsky, E.V.; Stone-Sundberg, J.; Emmett, J.L. A quantitative description of the causes of color in corundum. *Gems Gemol.* **2020**, *56*, 2–28. [CrossRef]
4. Krzemnicki, M.; Cartier, L.E.; Lefèvre, P.; Zhou, W. Colour varieties of gems—Where to set the boundary? *InColor* **2020**, *45*, 92–95.

5. Giuliani, G.; Ohnenstetter, D.; Fallick, A.E.; Groat, L.; Fagan, A.J. The geology and genesis of gem corundum deposits. In *Geology of Gem Deposits*, 2nd ed.; Mineralogical Association of Canada Short Course Series; Groat, L.A., Ed.; Mineralogical Association of Canada: Tucson, AZ, USA, 2014; Volume 44, pp. 113–134, ISBN 9780921294375.

6. Giuliani, G.; Groat, L. Geology of corundum and emerald deposits: A review. *Gems Gemol.* **2019**, *55*, 464–489. [CrossRef]

7. Palke, A.C.; Saeseaw, S.; Renfro, N.D.; Sun, Z.; McClure, S.F. Geographic origin determination of ruby. *Gems Gemol.* **2019**, *55*, 580–612. [CrossRef]

8. Milisenda, C.C.; Henn, U.; Henn, J. New gemstone occurrences in the south-west of Madagascar. *J. Gemmol.* **2001**, *27*, 385–394. [CrossRef]

9. Krzemnicki, M.S.; Lefèvre, P.; Zhou, W.; Wang, H.A.O. Zircon inclusions in unheated pink sapphires from Ilakaka, Madagascar: A Raman spectroscopic study. In Proceedings of the International Gemmological Conference, Online, 20–21 November 2021.

10. Wang, W.; Scarratt, K.; Emmett, J.L.; Breeding, C.M.; Douthit, T.R. The effects of heat treatment on zircon inclusions in Madagascar sapphires. *Gems Gemol.* **2006**, *42*, 134–150. [CrossRef]

11. Krzemnicki, M.S. How to get the "blues" out of the pink: Detection of low-temperature heating of pink sapphires. *Facette* **2010**, *17*, 12.

12. Saeseaw, S.; Khowpong, C. The effect of low-temperature heat treatment on pink sapphire. *Gems Gemol.* **2019**, *56*, 290–291.

13. Saeseaw, S.; Khowpong, C.; Vertriest, W. Low-temperature heat treatment of pink sapphires from Ilakaka, Madagascar. *Gems Gemol.* **2020**, *56*, 448–457. [CrossRef]

14. Karampelas, S.; Hennebois, U.; Delaunay, A.; Pardieu, V.; Mevellec, J.-Y.; Fritsch, E. Raman spectroscopy of zircon inclusions in unheated pink sapphires from Ilakaka, Madagascar: Opening new perspectives. *J. Gemmol.* **2022**, *38*, 16–18. [CrossRef]

15. Karampelas, S.; Hennebois, U.; Mevellec, J.-Y.; Pardieu, V.; Delaunay, A.; Fritsch, E. Identification of heated pink sapphires from Ilakaka (Madagascar). In Proceedings of the 7th International Gem & Jewelry Conference (GIT2021), Chanthaburi, Thailand, 2–3 February 2022; p. 212.

16. Karampelas, S.; Hennebois, U.; Delaunay, A. Détection du traitement thermique à basse température des corindons. *Rev. Gemmol. A.F.G.* **2022**, *217*, 4–5.

17. Notari, F.; Hainschwang, T.; Caplan, C.; Ho, K. The heat treatment of corundum at moderate temperature. *InColor* **2019**, *42*, 14–23.

18. Emmett, J.L.; Hughes, R.W.; Douthit, T.R. Treatments. In *Ruby & Sapphire: A Gemmologist's Guide*; Hughes, R., Ed.; RWH Publishing: Bangkok, Thailand, 2017; pp. 196–248.

19. McClure, S.F.; Smith, C.P.; Wang, W.; Hall, M. Identification and durability of lead glass-filled rubies. *Gems Gemol.* **2006**, *42*, 22–34. [CrossRef]

20. Hughes, E.B.; Vertriest, W. A canary in the ruby mine: Low-temperature heat treatment experiments on Burmese rubies. *Gems Gemol.* **2022**, *58*, 400–423. [CrossRef]

21. Wanthanachaisaeng, B. The Influence of Heat Treatment on the Phase Relations in Mineral Growth Systems. Ph.D. Thesis, Johannes Gutenberg University of Mainz, Mainz, Germany, 2007; 79p.

22. Sripoonjan, T.; Wanthanachaisaeng, B.; Leelawatanasuk, T. Phase Transformation of Epigenetic Iron Staining: Indication of Low-Temperature Heat Treatment in Mozambique Ruby. *J. Gemmol.* **2016**, *35*, 156–161. [CrossRef]

23. Vigier, M.; Fritsch, E.; Cavignac, T.; Latouche, C.; Jobic, S. Shortwave UV Blue Luminescence of Some Minerals and Gems due to Titanate Groups. *Minerals* **2023**, *13*, 104. [CrossRef]

24. Balan, E. Theoretical infrared spectra of OH defects in corundum (α-Al$_2$O$_3$). *Eur. J. Mineral.* **2020**, *32*, 457–467. [CrossRef]

25. Beran, A.; Rossman, G.R. OH in natural occurring corundum. *Eur. J. Mineral.* **2006**, *18*, 441–447. [CrossRef]

26. Smith, C.P. A contribution to understanding the infrared spectra of rubies from Mong Hsu, Myanmar. *J. Gemmol.* **1995**, *24*, 321–335. [CrossRef]

27. Nasdala, L.; Irmer, G.; Wolf, D. The degree of metamictization in zircon: A Raman spectroscopic study. *Eur. J. Mineral.* **1995**, *7*, 471–478. [CrossRef]

28. Elmaleh, E. Blue Sapphires from Madagascar, Sri Lanka, Tanzania and Burma (Myanmar): Gemological, Chemical, Spectroscopic Characterization and Dating of Zircon Inclusions. Master's Thesis, University of Geneva, Geneva, Switzerland, 2015; 425p.

29. Elmaleh, E.; Karampelas, S.; Schmidt, S.T.; Galster, F. Zircon inclusions in blue sapphire. In Proceedings of the 34th International Gemmological Conference, Vilnius, Lithuania, 26–30 August 2015; pp. 51–52.

30. Link, K. Age determinations of zircon inclusions in faceted sapphires. *J. Gemmol.* **2015**, *34*, 692–700. [CrossRef]

31. Elmaleh, E.; Schmidt, S.T.; Karampelas, S.; Link, K.; Kiefert, L.; Süssenberger, A.; Paul, A. U-Pb Ages of Zircon Inclusions in Sapphires from Ratnapura and Balangoda (Sri Lanka) and Implications for Geographic Origin. *Gems Gemol.* **2019**, *55*, 18–28. [CrossRef]

32. Xu, W.; Krzemnicki, M.S. Raman spectroscopic investigation of zircon in gem-quality sapphire: Application in origin determination. *J. Raman Spectrosc.* **2021**, *52*, 1011–1021. [CrossRef]

33. Hennebois, U.; Delaunay, A.; Karampelas, S. Heated purplish pink sapphire from Ilakaka (Madagascar) with colored monazite inclusions. *Gems Gemol.* **2023**, *59*, 159–160.

34. Vertriest, W.; Palke, A.C.; Renfro, N.D. Field gemology: Building a research collection and understanding the development of gem deposits. *Gems Gemol.* **2019**, *55*, 490–511. [CrossRef]

35. Zhang, M.; Salje, E.K.H.; Capitani, G.C.; Leroux, H.; Clark, A.M.; Schlüter, J.; Ewing, R.C. Annealing of O-decay damage in zircon: A Raman spectroscopy study. *J. Phys. Condens. Matter* **2000**, *12*, 3131–3148. [CrossRef]

36. Nasdala, L.; Lengauer, C.; Hanchar, J.; Kronz, A.; Wirth, R.; Blanc, P.; Kennedy, A.; Seydoux-Guillaume, A.M. Annealing radiation damage and the recovery of cathodoluminescence. *Chem. Geol.* **2002**, *191*, 121–140. [CrossRef]
37. Váczi, T.; Nasdala, L.; Wirth, R.; Mehofer, M.; Libowitzky, E.; Häger, T. On the breakdown of zircon upon "dry" thermal annealing. *Mineral. Petrol.* **2009**, *97*, 129–138. [CrossRef]
38. Marsellos, A.E.; Garver, J.I. Radiation damage and uranium concentration in zircon as assessed by Raman spectroscopy and neutron irradiation. *Am. Mineral.* **2010**, *95*, 1192–1201. [CrossRef]
39. Knittle, E.; Williams, Q. High-pressure Raman spectroscopy of $ZrSiO_4$: Observation of the zircon to scheelite transition at 300 K. *Am. Mineral.* **1993**, *78*, 245–252.
40. Nasdala, L.; Miletich, R.; Ruschel, K.; Váczi, T. Raman study of radiation-damaged zircon under hydrostatic compression. *Phys. Chem. Miner.* **2008**, *35*, 597–602. [CrossRef]
41. Noguchi, N.; Abduriyim, A.; Shimizu, I.; Kamegata, N.; Odake, S.; Kagi, H. Imaging of internal stress around a mineral inclusion in a sapphire crystal: Application of micro-Raman and photoluminescence spectroscopy. *J. Raman Spectrosc.* **2013**, *44*, 147–154. [CrossRef]
42. Zeug, M.; Rodríguez Vargas, A.I.; Nasdala, L. Spectroscopic study of inclusions in gem corundum from Mercaderes, Cauca, Colombia. *Phys. Chem. Miner.* **2017**, *44*, 221–233. [CrossRef]
43. Palke, A.C.; Saeseaw, S.; Renfro, N.D.; Sun, Z.; McClure, S.F. Geographic origin determination of blue sapphire. *Gems Gemol.* **2019**, *55*, 536–579. [CrossRef]
44. Váczi, T. A New, Simple Approximation for the Deconvolution of Instrumental Broadening in Spectroscopic Band Profiles. *Appl. Spectrosc.* **2014**, *68*, 1274–1278. [CrossRef]

Article

A Research of Emeralds from Panjshir Valley, Afghanistan

Quanli Chen [1,2], Peijin Bao [1], Yan Li [1], Andy H. Shen [1], Ran Gao [1], Yulin Bai [3], Xue Gong [3] and Xianyu Liu [4,*]

1   Gemological Institute, China University of Geosciences, Wuhan 430074, China
2   School of Jewelry, West Yunnan University of Applied Sciences, Baoshan 679100, China
3   College of Earth Sciences, Guilin University of Technology, Guilin 541006, China
4   School of Jewelry, Shanghai Jianqiao University, Shanghai 201306, China
*   Correspondence: liuxianyu@gench.edu.cn; Tel.: +86-177-2118-5433

Abstract: In recent years, emeralds from the Panjshir Valley in Afghanistan have taken a large share of the market, with high-quality emeralds comparable to Colombian emeralds. In order to meet the market demand for tracing the origin of emeralds, 20 emeralds from the region were tested using conventional gemology, laser Raman spectroscopy, Fourier infrared spectroscopy, ultraviolet-visible-near-infrared spectroscopy, and laser ablation plasma-mass spectrometry. The results show that the contents of the samples are mainly serrated three-phase inclusions, which are similar to those of Colombian emeralds. There are multiple solid inclusions and two liquids in the serrated voids. The main coloring elements of the sample are chromium and vanadium. The alkali metal content is moderate, among which rubidium (average content: 25.72 ppm) and cesium (average content: 33.15 ppm) content is lower. The near-infrared spectrum reveals that the absorption characteristic was dominated by type I water. A chemical composition analysis indicates that the chemical composition of Panjshir emeralds is similar to that of the emeralds of Davdar Township in China and Coscuez in Colombia, but they could be distinguished by an Na-Sc and Rb-Ga diagram.

Keywords: emerald; Panjshir Valley; gemological characteristics; origin traceability

Citation: Chen, Q.; Bao, P.; Li, Y.; Shen, A.H.; Gao, R.; Bai, Y.; Gong, X.; Liu, X. A Research of Emeralds from Panjshir Valley, Afghanistan. *Minerals* **2023**, *13*, 63. https://doi.org/10.3390/min13010063

Academic Editors: Stefanos Karampelas and Emmanuel Fritsch

Received: 15 November 2022
Revised: 27 December 2022
Accepted: 27 December 2022
Published: 30 December 2022

1. Introduction

Emerald, known as the "king of green gems", is a kind in the beryl family, and its ideal chemical formula is $Be_3Al_2SiO_{18}$. Emerald is a hexagonal crystal system, which is a cyclic silicate mineral. Its structure is formed by stacking hexagonal rings composed of silicon tetrahedrons, which leads to the existence of a channel parallel to the c-axis in the structure, in which many ions and molecules can exist. Emeralds are produced in five continents of the world. At present, emeralds with high quality and of a large quantity in the market mainly come from Colombia, Zambia, Afghanistan, and Pakistan.

Historically, Afghanistan has been known for its lapis lazuli deposits, but since the discovery of gem-grade emeralds in 1970 [1], a large number of high-quality emeralds have emerged (Figure 1). In 2015, Christie's Auction House sold a ring inlaid with 10.11 ct emerald, cutting emerald at the price of HKD17560000. Emeralds are produced in the Panjshir Valley, which is about 110 km northeast of Kabul. In 1977 and 2015, there were five emerald mining areas in the Panjshir Valley—Darkhenj, Mikeni, Yaknow, Buzmal and Darun, which are to the east of Panjshir River (Figure 2) [2].

Previous reports on the emeralds in the Panjshir Valley of Afghanistan are mostly about the geological deposits, mining environment, and mineralogical properties [3–6] but seldom on gemological characteristics, spectral characteristics, and color origin. Emerald color is associated with the presence of trace chemical elements, and chemical element and inclusion characteristics are linked to their geographical origin. These characteristics can expand the existing emerald origin database and have a great impact on the origin traceability of emeralds.

(a)   (b)

Figure 1. High-quality emeralds. In the left photo (**a**), the weight from left to right and from top to bottom is 9.85, 8.18, 7.61, 2.39, and 2.58 ct, respectively. In the right picture (**b**), the weight of emerald is 3.27 ct. Photos by Peijin Bao.

Figure 2. Location of the Panjshir Valley and mining areas related to emeralds. Picture: Ran Gao, Peijin Bao, modified from [3].

Afghan emeralds are becoming more and more popular in the Chinese market and favored by Chinese consumers. Therefore, it is necessary to trace the origin of emeralds in this area. We have carried out a detailed study on the gemological characteristics of emeralds in this area; supplemented it with more-detailed inclusion characteristics, chemical composition data, and spectral information; and preliminarily summarized a set of tracing processes for emeralds in this area. In this study, the history, geology, and exploitation of emeralds in this area are introduced. The gemological properties of emeralds are determined by conventional gemological instruments, the inclusions are identified by Raman spectrometer, and the spectral characteristics of emeralds are determined by infrared spectrometer. The chemical composition and color origin of emeralds with color zonation are studied by using an ultraviolet-visible-near-infrared spectrometer and a laser denudation plasma-mass spectrometer.

2. History

Most people believe that in ancient Greece and Rome, the real emeralds came from Egypt [7]. However, in the first century CE, Pliny's Natural History [3] and Theophrastus's Stone Theory [8] both referred to "Smaragdus," a Latin term that in ancient times referred to emeralds and other green gemstones. It came from the Bactria of what is now Iran and Afghanistan. This suggests that emeralds were known to have been produced in Afghanistan as early as Roman times.

Nevertheless, only Marco Polo mentioned silver ore, Balas rubies, and lapis lazuli from Bactria in his travelogues in 1265 CE. He wrote that "this gem called Balas ruby comes from this province (Bactria)", "And also, there are veins in this mountain, from which a large amount of silver, copper and lead can be obtained" (Biography of Marco Polo). It is believed that Bactrian emeralds might be those from Pakistan and Afghanistan [9]. Nevertheless, we do not know whether there was an emerald trade in Afghanistan at that time. Using oxygen isotope analysis, Giuliani and other gemologists discovered that the antique emeralds in the Nizam treasure from Hyderabad (India) may have come from the Panjshir Valley in Afghanistan [9]. This means that the Panjshir Valley produced high-quality emeralds at least as far back as the 18th century [10].

Over the next hundred years, the Panjshir Valley gems were rarely reported. At the beginning of the 20th century, geologists from Britain, France, the United States, and other countries reported on the geological conditions of Afghanistan [11,12]. At the start of the 1970s, emeralds were discovered in the Buzmal mine, east of Dest-e-Rewat village in the Panjshir Valley. It has been systematically investigated by Russian and Afghan geologists [13,14]. Although the local political situation in Afghanistan affected the geological work, the United Nations Development Programme also published the name and location of the emerald deposit in the report of 1977 [2].

Agnew first discussed emerald deposits in Afghanistan in 1982 [3]. Over the past decade, scholars ventured into Afghanistan one after another at the risk of war to discover the secrets of emeralds and report extensively on their mineral deposits and gemological characteristics [3–5,10,15]. In 1990, Ward predicted that emeralds from Afghanistan could have a huge impact on the future gem market [3]. At the 2016 Toussaint show, Arthur Groom, Jr., the dealer (distributor) of Eternal Natural Emerald, believed that Afghanistan emeralds were one of his biggest sources of goods [16]. The Panjshir emerald, which entered the market in 2017, is comparable in quality to the Colombian emerald [10]. Today, Afghan emeralds have occupied a large market share, and it is necessary to study them because of their similarities to Colombian emeralds.

3. Geological Setting

The emerald mine in Panjshir Valley, Afghanistan, is located about 113 km northeast of Kabul [5], on a hillside southeast of Panjshir, at an elevation of about 3135 to 4270 m (Figure 2) [3]. The Panjshir Valley is a major fault zone between two crustal plates, the Paleo-Asian plate, and the Neo-Merian plate, which Rossovskiy and Konovalenko [17] suggest may be related to Himalayan orogenic events.

The Panjshir emerald deposit is located in the southeastern part of the Panjshir fracture zone, the veins cut through host rocks consisting of metamorphosed limestones, calcareous slates, phyllites, and micaceous schists of the Silurian-Devonian age [5]. The deposits have been affected by intense fracturing, fluid circulation, and hydrothermal alteration, resulting in intense albitization and muscovite-tourmaline replacements [18]. Emerald-bearing rocks throughout the Panjshir Valley are pervasively cut by hydrothermal veins. Emeralds fill veins along fissures; the vein body is mainly composed of quartz and albite; and the exposed thickness can be up to 15 cm. Emeralds are often associated with pyrite but are also found in phlogopite, tourmaline, and carbonate [18].

Fluid inclusions in the emeralds are highly saline, which suggests that the Panjshir deposits, like those in Colombia, are linked to hydrothermal fluids [19]. Giuliani et al. (2019) proposed a new classification of emerald deposits [18], where the emerald occurrences and deposits are classified into two main types: (Type I) tectonic magmatic-related and (Type II) tectonic metamorphic-related. Meanwhile, the Panjshir deposit was classified as Type IIC, associated with metamorphic hydrothermal rocks. Emerald formation is due to the reaction of Be hydrothermal fluids flowing along the veins and the enclosed host rocks with the Cr-rich rocks. The numerous intrusive rocks, including quartz prophyries, of eastern Panjshir would be good sources for the beryllium-bearing hydrothermal fluid. Some of the highest-quality material is found in veinlet networks that cut metasomatically altered gabbro and metadolomite; ultramafic rocks are ideal sources of chromium [3].

4. Materials and Methods

The 20 original emerald samples in this study were provided by Mr. Xiao, a gem dealer (Figure 3). All crystals are hexagonal columns, about 1 cm in length and weigh from 0.15 to 0.38 ct (Figure 3). All the crystals show traces of oil, which fluoresces blue-white under long-wavelength ultraviolet light (Figure 4). Sample A-1, where obvious color banding can be observed, was selected and made into double-sided parallel-polished slices (1 mm thick) with vertical optical axis to study the cause of color banding (Figure 5).

The refractive index and birefringence were determined by a gem refractometer (FGR-003A, FABLE, Shenzhen, China). The dichroism of gemstones was observed under dichroic mirror, and colors were photographed by using a polarizing microscope (FID-1, FABLE, Shenzhen, China). The reaction of the gem was tested under a Charles filter (FCF-25, FABLE, Shenzhen, China). The specific gravity was measured by using hydrostatic weighing method. The inclusions were observed and photographed using the M205A microscope system, with bright field illumination, optical fiber illumination, the ZEISS AXIO Imager, and M2m polarizing microscope, using a maximum magnification of 256x. Since the crystals are too small, UV fluorescence was measured with an ultraviolet lamp pen, and their reactions under a Charles filter were observed.

Figure 3. Emerald samples from Panjshir Valley, Afghanistan. Most of the crystals are hexagonal columnar and 1 cm high, and the weight range is 0.15 to 0.38 ct. Photo by Peijin Bao.

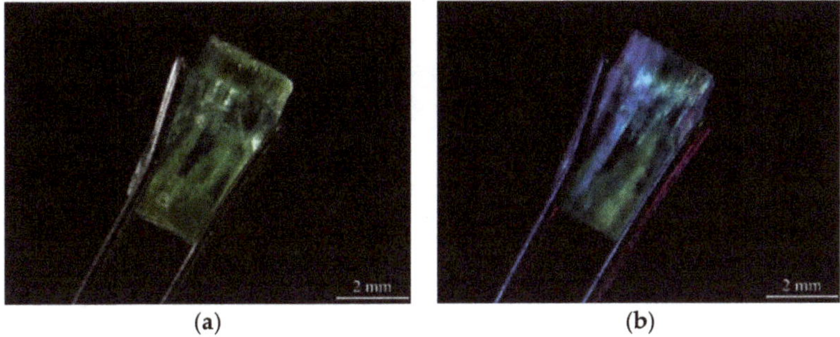

(a)   (b)

Figure 4. Left (**a**): oil-filled sample A-2; right (**b**): sample A-2 fluoresces blue-white under long-wavelength ultraviolet light. Photos by Peijin Bao.

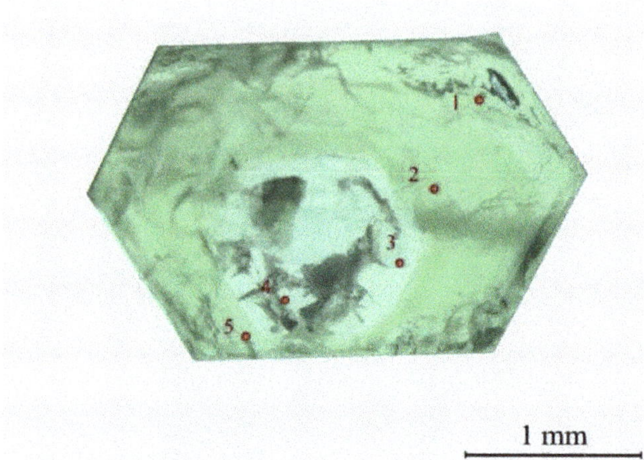

Figure 5. The Panjshir emeralds with color bands show a light blue-green core and green edges in a direction perpendicular to the c-axis. LA-ICP-MS and UV-Vis-NIR measurements were performed at five locations. Photo by Peijin Bao.

The contents of trace elements in the samples were determined by the laser ablation inductively coupled plasma-mass spectrometer (LA-ICP-MS), which is composed of the GeolasPro laser denudation system and Agilent 7700 inductively coupled plasma test system (Agilent, Singapore). The excimer laser was COMPexPro 102 ARF 193 nm, and the optical system was MicroLas. In the process of denudation, helium was used as carrier gas, argon was used as compensation gas to adjust the sensitivity, and the denudation system was equipped with signal smoothing device [20]. The laser beam spot and frequency were 44 μm and 5 HZ, respectively. BHVO-2G, BCR-2G, and BIR-1G were used to calibrate the element concentration without internal standard. Each time-resolved analysis data of LA-ICP-MS includes approximately 20–30 blank signals and 50 sample signals. The offline analysis and processing of data (including the selection of samples and blank signals, instrument sensitivity drift correction, and element content calculation) is completed by ICPMSDataCal software (version 12.2, developed by State Key Laboratory of Geological Processes and Mineral Resources, China University of Geosciences, Wuhan, China). Al was used as the normalized element, and NIST 610 glass was used for time drift correction. The standard for LA-ICP-MS measurements is that the calibration values of the monitored reference materials agree within the error range within the recommended values. Quality control deviations: Major elements within 5% uncertainty and trace elements within 10% uncertainty. Three sites were tested in the clean zoning of sample A-2 and A-20, five sites were tested in the color band region of sample A-1, and UV-Vis-NIR spectra were also tested.

FTIR spectra of transmission were acquired from 5000–2000 cm^{-1} and 9000–4000 cm^{-1} using a Vertex-80 spectrometer (BRUKER OPTICS, Billerica, MA, USA) with 4 cm^{-1} resolution and 32 scanning frequency. The information on the water vibration and other molecular vibrations of the emerald channels was preliminarily obtained. At the same time, the spectra of transmission of 2 emeralds from Davdar Township, Xinjiang, China, and 5 emeralds from Swat, Pakistan, were acquired from 9000–4000 cm^{-1} for comparative analysis.

UV-Vis-NIR spectra were acquired using the Jasco MSV-5200 micro spectrometer equipped with a microscope and Glan-Taylor polarization system in the 250–900 nm spectral range with a spectral bandwidth of 0.5 nm and a scan rate of 1000 nm/min. The normal and extraordinary spectra of sample A-1 at each point were measured.

Raman spectra were acquired using the LabRAM HR Evolution with 325 nm and 532 nm lasers for testing inclusions and channel water types. Channel water types were identified from the 3800–3500 cm^{-1} range with a 325 nm laser, 1800/mm grating, 5.5 cm^{-1} resolution, and a 15× lens. Inclusions were excited at 2000–100 cm^{-1} range with a 532 nm laser, 600 mm grating, 6.4 cm^{-1} resolution, and a 50× lens. The scanning time was 20 s. The Raman shift was calibrated with monocrystalline silicon (at 521 cm^{-1}).

5. Results

5.1. Gemological Properties

All 20 samples are blue-green in color, some crystals having distinct color bands, and the light hexagonal core and dark periphery are shown on the vertical c-axis (Figure 6). Most of the samples have fractures parallel to the c-axis, which were filled with colorless oil to mask the more obvious fractures. Most samples are transparent, with good hexagonal column habit, and some have the phenomenon of continuous growth.

Figure 6. Vertical c-axis sample A-1 shows light hexagonal core and dark periphery. Photo by Peijin Bao.

The refractive index range was 1.575–1.580 under extraordinary-ray and 1.580–1.584 under ordinary-ray with a birefringence between 0.005 and 0.009. The variation range of specific gravity was 2.53–2.76. As the samples are less than 0.5 ct, there may be some errors in density measurement from using the hydrostatic weighing method. Some of the samples (A-2, A-4~A-6, A-12, and A-16) show a very faint red under the Charles filter. No fluorescence was observed under long-wave (365 nm) or short-wave (254 nm) ultraviolet light. Under the polarizing mirror, the samples can be seen as yellow-green in the vertical (perpendicular) c-axis and blue-green in the parallel c-axis. The gemological properties of emeralds from Panjshir Valley are summarized in Table 1.

Table 1. Gemological properties of emeralds from Panjshir Valley, Afghanistan.

Color	Medium bluish green to green
Clarity	Slightly to heavily included
Refractive indices	no = 1.580–1.584; ne = 1.575–1.580
Birefringence	0.005–0.009
Specific gravity	2.47–2.89
Pleochroism	Medium yellowish green (o-ray) and greenish blue (e-ray)
Fluorescence	Inert to long-wave and short-wave UV radiation
Chelsea filter	Slightly red or no reaction
Internal features	• Obvious colorless oil filling

Table 1. *Cont.*

• Obvious color bands
• Yellow matter seeping along the cracks
• Three-phase inclusions contained liquid, gas, and many solids; displayed tubular, needle-like, or fusiform; and were sometimes serrated and irregular

5.2. Gemological Properties

Three-phase inclusions were the most common inclusions in emerald from this area, which contained three states: solid, liquid, and gas. The inclusions are tubular, needle-like, or fusiform, sometimes serrated and irregular (Figure 7b–d). The volume is usually tens to hundreds of microns long, and the width varies from a few microns to tens of microns. The bubbles are usually less than a quarter of the total volume of the inclusion and are circular, elliptical, or extruded because they are confined within the envelope (Figure 7a–d).

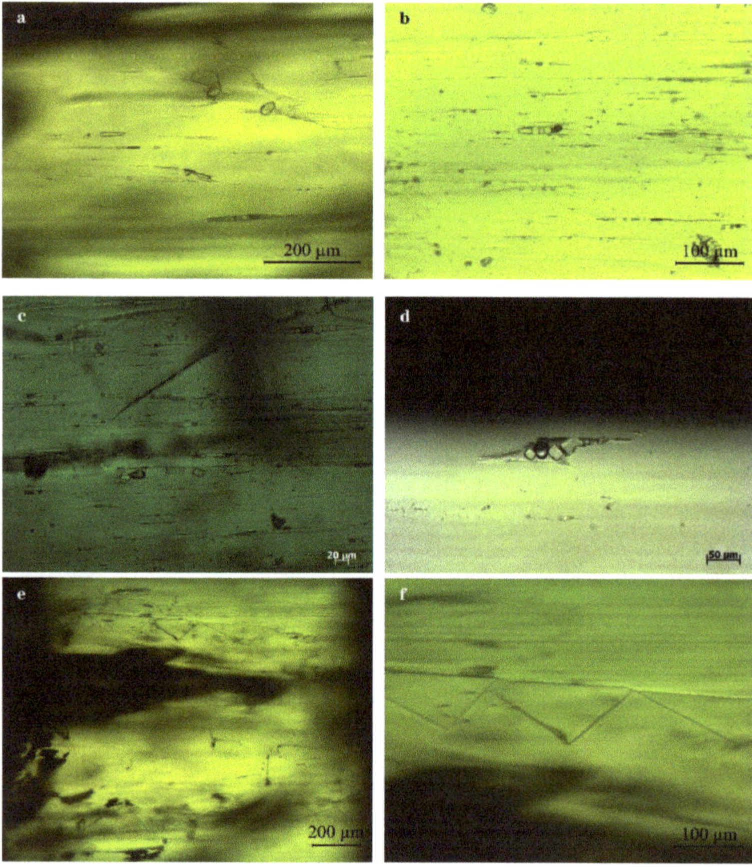

Figure 7. Three-phase inclusions of various shapes in Panjshir emeralds. (**a**) Tubular three-phase inclusions; (**b,c**) needle-like three-phase inclusions containing bubbles and more than four crystals; (**d**) a serrated three-phase inclusion containing a circular bubble and a plurality of cubic crystals; (**e,f**) triangular growth characteristics. Photos by Peijin Bao.

The most distinctive features of the three-phase inclusions in this origin emerald are the high number of solid crystals (Figure 7b,d), most of which are more than three and up to seven (Figure 7c). The size of the crystal varies with the shape of the inclusion: in long tubular inclusions, the crystal is the same width as the tube (Figure 7a–d); nevertheless, in serrated inclusions, the crystal is less than one-third the volume of the inclusion (Figure 7a,d). Most of the crystals are colorless lumps, and no results are obtained by Raman testing. According to the presence of cubic crystals, it is suggested that these inclusions may be NaCl [4]. Triangular growth characteristics parallel to the c-axis are found in sample A-1 (Figure 7e,f).

5.3. Chemical Composition Analysis

In Figure 8, the box diagram of trace element data of 20 emeralds from Panjshir Valley in Afghanistan were measured by LA-ICP-MS as part of a trace element analysis (Table 2). Figure 8 shows that the samples are rich in Mg (5154.67–11,610.89 ppm), with an average content of 7071.91 ppm; a small amount of P (47.37–273.85 ppm) has an average content of 157.88 ppm; a small amount of Sc (71.34–804.98 ppm) has an average content of 381.50 ppm; and Ga (15.34–34.95 ppm) has an average content of 27.22 ppm. The content of other trace elements, such as Ti, Mn, Ni, and Zn, are lower than 5 ppm. The alkali metal concentrations from the lowest to the highest are Rb 19.93–50.07 ppm (average 25.72 ppm), Cs 29.01–53.50 ppm (average 33.15 ppm), Li 74.83–131.01 ppm (average 84.14 ppm), K 194.96–1503.69 ppm (average 359.11 ppm), and Na 5586.07–11,219.99 ppm (average 7313.58 ppm). The total concentration of alkali metal elements ranges from 5904.80 to 12,958.26 ppm. The average is 7815.70 ppm.

In the emerald of the Panjshir Valley in Afghanistan, the content of transition metal elements is high, and Cr, V, and Fe may contribute to the color. Among them, Cr = 136.02–3896.84 ppm, with an average value of 1838.44 ppm; V = 275.03–1687.07 ppm, with an average value of 765.53 ppm; and Fe varied from 3173.22–15,030.29 ppm, with an average value of 4458.49 ppm. Cr/V range spans 0.3 to 3.7. Except for A-20, the content of Cr in all samples was higher than that of V.

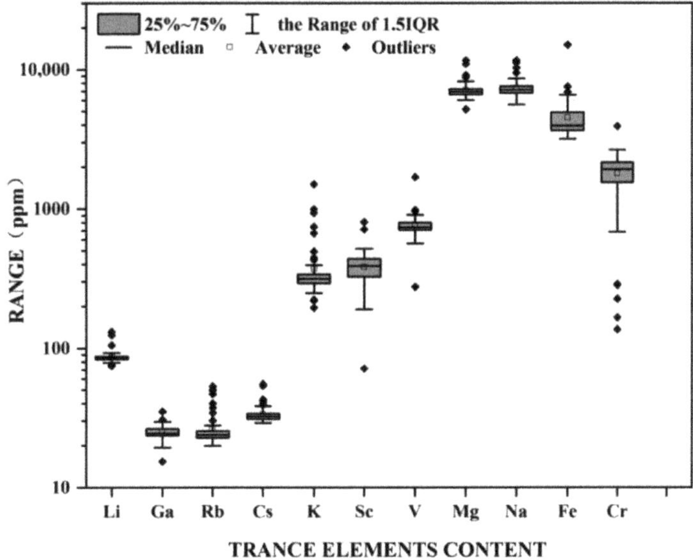

Figure 8. Box chart of trace elements in emeralds from the Panjshir Valley.

Table 2. LA-ICP-MS of samples in ppm.

Samples	Element	Min–Max	Average (SD)	Median	LOD
	Li	74.83~131.01	86.14 (8.84)	85.24	1.32~2.67
	Na	5586.07~11,615.12	7313.58 (1072.42)	7245.95	24.16~67.97
	Mg	5154.67~11,610.89	7071.91 (1010.51)	6991.72	0.70~7.17
	K	194.96~1503.69	159.11 (208.65)	316.78	23.16~37.41
	Sc	71.34~804.98	381.50 (108.15)	387.75	0.30~1.19
Panjshir	Ti	1.52~14.72	5.95 (1.30)	5.73	2.81~3.90
emeralds	V	275.03~1687.07	765.53 (157.13)	764.50	0.17~0.82
20 samples	Cr	136.02~3896.89	1838.44 (655.50)	1866.08	7.06~21.71
64 analyses	Mn	2.91~69.76	13.78 (23.15)	3.11	2.59~6.64
	Fe	3173.22~15,030.29	4458.49 (1704.49)	4337.50	38.05~80.04
	Ni	bdl	bdl	bdl	4.20~12.75
	Zn	1.31~3.82	3.22 (1.78)	3.22	1.22~6.09
	Ga	15.34~34.95	24.70 (2.82)	24.58	0.31~2.05
	Rb	19.93~50.07	25.72 (7.43)	24.08	1.32~3.09
	Cs	29.01~54.94	33.15 (4.70)	32.26	0.74~2.01

SD = standard deviations; LOD = limit of detection; bdl = below detection limit.

5.4. Ultraviolet Visible Near-Infrared Spectroscopy

Figure 9 showed the UV-Vis-NIR spectrum of sample A-1 with high Fe content, showing obvious absorption peaks of Fe^{2+}, Fe^{3+}, and Cr^{3+} (Figure 9). Under ordinary-ray, the spectrum has a wide absorption band of 430 and 604 nm; in extraordinary-ray, 420 and 632 nm wide absorption bands are displayed. George Bosshart (1991) believed that after the Cr and V spectra overlap, the positions of absorption bands will shift to each other's peak positions and the slope of absorption peaks will become steeper. Therefore, the above absorption bands belong to the common absorption of V^{3+} and Cr^{3+} [21]. The 637 and 682 nm absorption peaks displayed under ordinary-ray and 646, 660, and 684 nm absorption peaks displayed under extraordinary-ray belong to the Cr^{3+} absorption [22–24]. In sample A-1, $Cr/V \approx 1$, and at $Cr/V > 1$, the V^{3+} absorption shoulder at 395 nm is not usually shown [3]. The sharp absorption peak at 370 nm displayed under ordinary-ray belongs to Fe^{3+}, while this absorption peak is not displayed under extraordinary-ray. The medium absorption peak of 840 nm in extraordinary-ray and 833 nm in ordinary-ray is attributed to Fe^{2+} [25]. Most of the emeralds in Afghanistan showed obvious Fe absorption, and the absorption peak of Fe^{2+} under o-ray is steeper than that under e-ray, which is consistent with the higher Fe content.

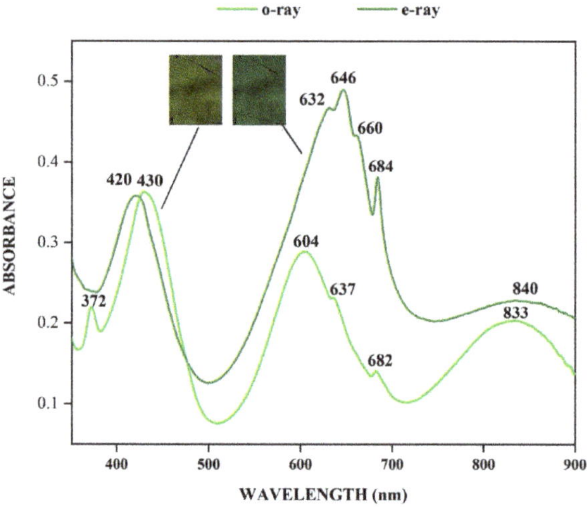

Figure 9. UV-Vis-NIR spectra of sample A-1. It showed obvious Cr^{3+} absorption (646, 660, and 684 nm

under ordinary-ray; 637 and 682 nm under extraordinary-ray), moderate Fe^{2+} absorption (830 nm under ordinary-ray), and moderate Fe^{3+} absorption (372 nm under ordinary-ray).

5.5. Infrared Spectrometry

The hexagonal ring structure of emerald leads to the existence of a structural channel parallel to the c-axis, in which alkali metal ions and water and other macromolecules exist to maintain the charge balance and structural stability of minerals. According to the different occupation directions and coexisting ion types of water molecules in the emerald channel, they can be divided into two types: type I water and type II water. The former axis of symmetry is perpendicular to the c-axis, showing mainly the characteristics of poor alkali, whereas the latter axis of symmetry is parallel to the c-axis, showing mainly the characteristics of rich alkali [24].

Figure 10 shows the IR spectra of Panjshir emeralds in the ranges of 5000–2000 cm^{-1} and 9000–4000 cm^{-1}. The absorption peaks shown in the range of 5000–2000 cm^{-1} are related mainly to the stretching vibration of channel water, the vibration of chemical bonds formed between other alkali metal ions (such as K^+, Na^+, etc.) and water in the channel, and the vibration of CO_2 molecules (Figure 10 left); the peak in the range of 4600–4000 cm^{-1} and the weak peak of 3110 cm^{-1} are attributed to the stretching vibration of M-OH. Meanwhile, the weak Na-H stretching vibration spectrum peak at 3164 cm^{-1} can be seen [22,26]. The spectrum peak in the range of 3900–3500 cm^{-1} belongs to the absorption of water in the emerald channels, and the weak absorption at 3521 cm^{-1} is the asymmetric stretching vibration of type I water. At the same time, owing to its fluorescence, the sample filled with colorless oil is obviously visible under the UV lamp pen, which belongs to the peak of organic matter in the range of 3000–2800 cm^{-1}, and colorless oil peaks of 2812, 2852, 2923, and 2955 cm^{-1} can be found [27]. Spectral peaks of 2733 and 2669 cm^{-1} are related to the OD vibration of hydrogen isotope and oxygen atom. The higher strength of 3238 cm^{-1} is associated with the polymeric ion absorption of $[Fe_2(OH)_4]^{2+}$, whereas 2474 cm^{-1} is associated with Cl^-, and 2356 cm^{-1} is related to CO_2, which is sharp and weak.

In the near-infrared range of 9000–5000 cm^{-1}, the absorption spectrum band of emerald is related mainly to the combined frequency and double-frequency vibration of structural water [28]. The right side of Figure 10 shows the near-infrared spectra of the parallel and vertical c-axis in the direction of optical vibration, separately. The sharp and strong spectral peaks of 7141 cm^{-1} and slightly weaker spectral peaks of 8700, 7274, and 6819 cm^{-1} of type I water can be observed [24,26]. The sharp and strong 5272 cm^{-1} is related to the cofrequency absorption of type I /II water, and the sharp bimodal of 7095, 7074 cm^{-1}, and 5205 cm^{-1} are related to the double-frequency absorption of type II water [26].

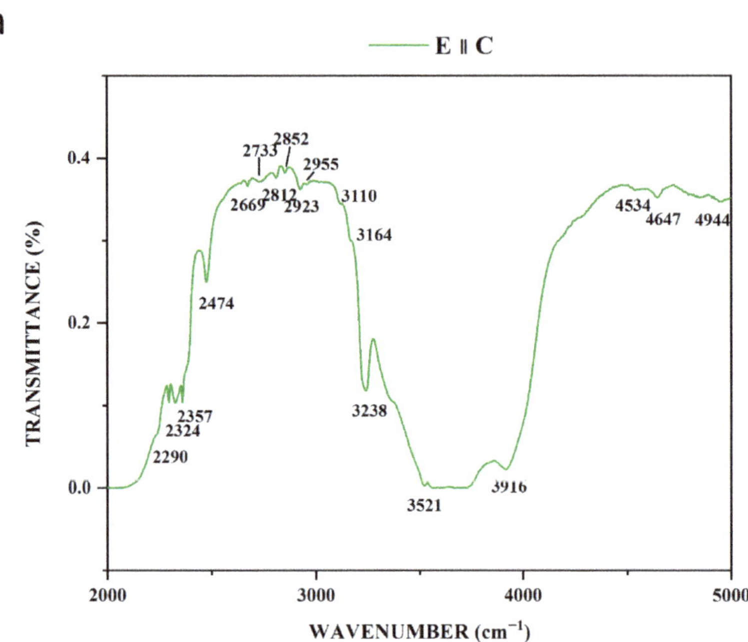

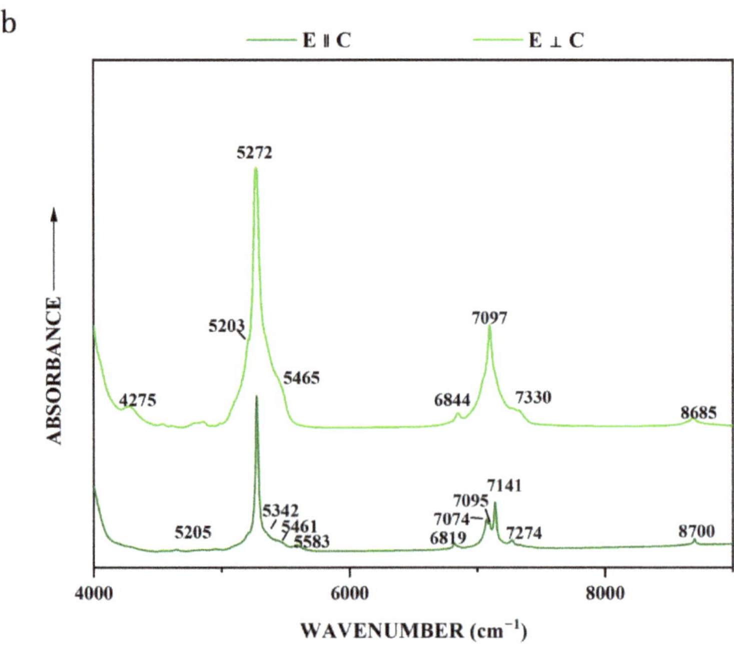

Figure 10. Infrared spectrum of sample A-1. The spectral range is 5000–2000 cm^{-1} (**a**) and 9000–4000 cm^{-1} (**b**). Spectra in (**b**) are vertically offset for clarity.

5.6. Raman Spectroscopy

Figure 11 shows the Raman spectra of Afghanistan emeralds A-1 and A-2 and Pakistan emerald YS-4 in the range of 3500–3800 cm^{-1}. According to Huong and Cui, the Raman peaks at 3598/3604 cm^{-1} were caused by type II water, and the Raman peaks at 3608/3614 cm^{-1} were caused by type I water [22,29–31]. The peak difference in the Raman shift that is used to characterize type I water and type II water in the literature is 6 cm^{-1}, which should be caused by instrument calibration. As can be discovered from Figure 11, the emerald spectrum of Panjshir Valley in Afghanistan is dominated by type I water, which is consistent with the conclusion of the near-infrared region in the infrared spectrum. When compared to the Swat Valley emeralds in Pakistan, the latter showed peaks only in the 3602 cm^{-1} (type II water) band, yet the peaks of type I water are not obvious and could not be separated, because of instruments or conditions. However, the overall migration to type II water indicates that type II water is the dominant emerald in the Swat Valley of Pakistan, which is consistent with the NIR spectra and chemical composition analysis. As shown in Figure 11, the Raman shifts of samples YS-4, A-1, and A-2 are different, which may be caused by different crystal orientations during the test (the direction of the laser beam is parallel to the c-axis of sample YS-4 and perpendicular to the c-axis of samples A-1 and A-2).

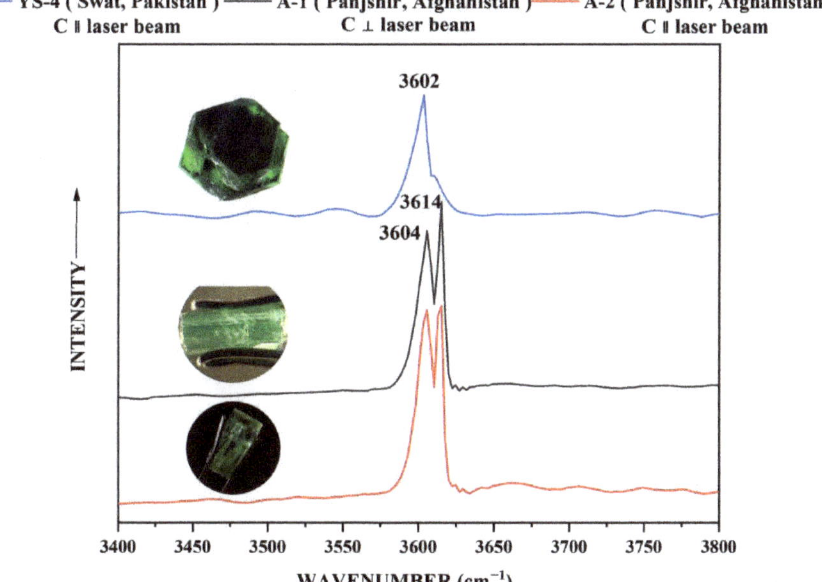

Figure 11. Raman spectra of samples A-1, A-2, and YS-4. Samples from different occurrences show Raman peaks for different types of water: 3602cm^{-1} of YS-4 corresponds to 3598/3604 cm^{-1}, which characterize type II water; 3604 cm^{-1} of A-1 and A-2 corresponds to 3598/3604 cm^{-1}, which characterize type II water; and 3614 cm^{-1} corresponds to 3608/3614 cm^{-1}, which characterize type I water. Spectra are vertically offset for clarity.

6. Discussion

6.1. Chromogenic Cause of Panjshir Emerald

Figure 12 shows the variation of the content of the chromogenic elements Cr, V, and Fe contained in the different color bands in color-zoned sample A-1. From position 1 to 5, the color shades changed in line with the variation of Cr and V contents. In sample A-1, the concentrations of Cr_2O_3 and V_2O_3 are 0.02–0.23 wt% and 0.05–0.15 wt%, respectively. The darker green color at the edges is caused by higher levels of Cr and V. The UV-Vis-NIR spectra corresponding to the points analyzed for chemical composition indicated that the darker the color of the ribbon, the higher the content of Cr and V and the stronger their relative absorption (Figure 13). In addition, the spectrum also shows the absorption of Fe^{3+}, but the variation of Fe content does not coincide with Cr and V, as seen in Figure 13, and the relative intensity of the absorption peak of Fe^{3+} at 370 nm is consistent in Figure 14.

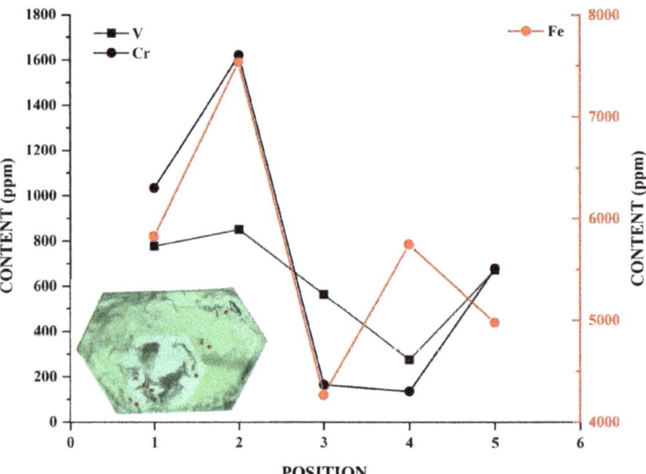

Figure 12. The chemical composition data of sample A-1 shows the change trend of the content of chromogenic ions Cr, V, and Fe with the color band.

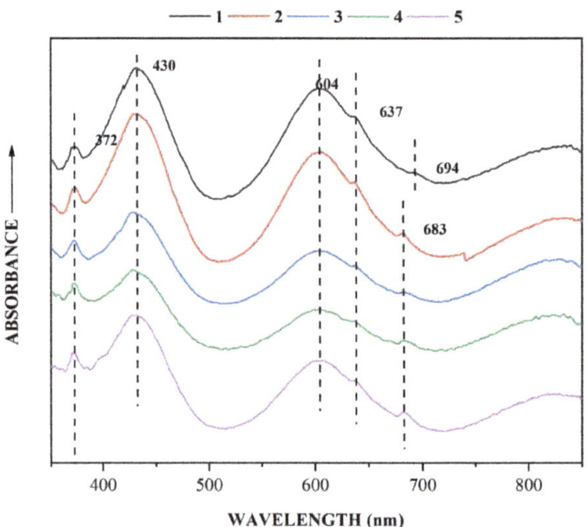

Figure 13. Ultraviolet visible spectrum of sample A-1 shows the changing trend of spectral absorption with color band. Spectra are vertically offset for clarity.

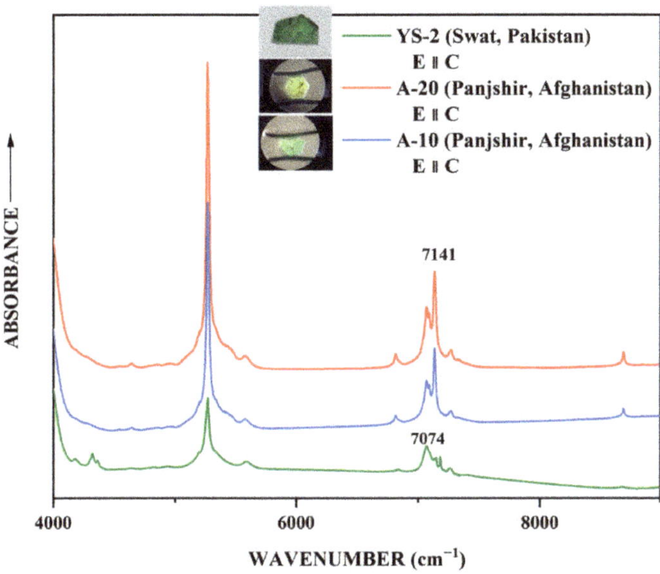

Figure 14. The infrared spectra of Panjshir emeralds and Swat emeralds show the absorption of different types of water. Spectra are vertically offset for clarity.

6.2. The Relationship between the Type of Water and Its Related Alkali Metal Content

The thickness of samples A-20 and A-10 is about the same (about 6 mm), and YS-2 is thinner (about 2 mm). The thickness may have a certain impact on the absorption of the infrared projection spectrum. Figure 14 shows the infrared spectra of Panjshir emeralds A-10 and A-20 in the range of 9000–4000 cm^{-1} and emerald YS-2 from the Swat Valley, Pakistan. The double-peak splitting of 7095–7074 cm^{-1} is obvious in samples A-20 and A-10, but not in YS-2. Although the total alkali metal content in the A-20 and A-10 samples is different (the difference is 2000 ppm), the absorption peak position of the near-infrared spectrum is not different. It is speculated that the content difference is not large, the critical point for the difference of absorption peak is not reached, or the alkali metal does not completely exist in the emerald channels. The absorption peak at 7141 cm^{-1} is related to the stretching vibration caused by type I water molecules, and the peak at 7074 cm^{-1} is attributed to the stretching vibration of type II water molecules. A scatter plot is drawn with alkali metal content as the ordinate and I_{7141}/I_{7074} as the abscissa to explore the relationship between alkali metal Na+K+Rb+Cs content in the infrared spectrum (Li usually exists in the tetrahedron in the way of replacing Be) and the channel's water type (Figure 15). As shown in the Figure 15, the scatter plot of emerald data from three producing areas, namely 20 emerald crystals from the Panjshir Valley in Afghanistan, 2 emerald crystals from Xinjiang in China, and 5 emerald crystals from the Swat Valley in Pakistan, is negatively correlated, indicating that the content of type II water and alkali metal Na+K+Rb+Cs is positively correlated within a certain range and not linearly positively correlated, which is consistent with the assumption that alkali metals do not completely exist in the emerald channels.

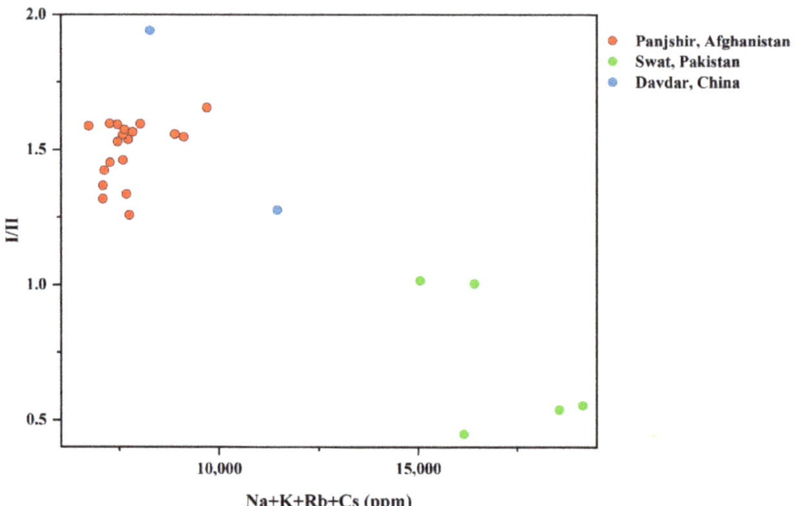

Figure 15. The relationship between the proportion of two water types in the infrared spectra of Panjshir, Swat, and Davdar emeralds and the content of alkali metal Na+K+Rb+Cs. The relative peak height ratio between 7141 and 7074 cm^{-1} is related to the type of water.

According to [26], alkali metal ions enter the emerald structure channels by means of charge balance in an isomorphic substitution, which is mainly $Al^{3+} = Mg^{2+} (Fe^{2+}) + K^+ (Na^+)$. Additionally, the presence of alkali metals affects the O-H vibration of the water molecules in the channels, thus having a distinction between type I and type II water. Therefore, $(Mg^{2+}+Fe^{2+})$ and $(Na^++K^++Rb^++Cs^+)$; $(Mg^{2+}+Fe^{2+})$ and I_{3598}/I_{3608} or I_{3604}/I_{3614}; and $(Na^++K^++Rb^++Cs^+)$ and I_{3598}/I_{3608} or I_{3604}/I_{3614} all have positive correlations within a certain range, as described above, the same as in the IR spectra. This is consistent with the results of Houng et al. (2010). As shown in Figure 16, the sum of alkali metal ions K and Na has a positive correlation with I_{3598}/I_{3608} or I_{3604}/I_{3614}. The total alkali concentration of emeralds from the Panjshir Valley, Afghanistan, is lower than that of the Swat Valley, Pakistan, with a Raman spectral peak ratio around 1 (0.89–1.08); the Raman spectral peak ratio of emeralds from the Swat Valley, Pakistan, is significantly greater than 1 ($\approx$3); and in Colombian emeralds, where the total alkali concentration is even lower, the Raman spectral peak ratio is less than 1. Therefore, as the alkali ion content increases, the proportion of type II water increases. In contrast, the alkali ion content is controlled by the Mg and Fe content of the host rock or mineralized fluid, which is more sensitive to the emerald-formation conditions [26] and can be used for origin tracing. At the same time, the formation of emeralds in a provenance is a dynamic process, its geographic environment will undergo various changes, and the elements contained in the mineralized fluid or host rock will change within a certain range during the dynamic process, which is the reason for the changes in emerald color, Raman spectral peak ratios, alkali metal content, and the ratio of type I and II water in a provenance.

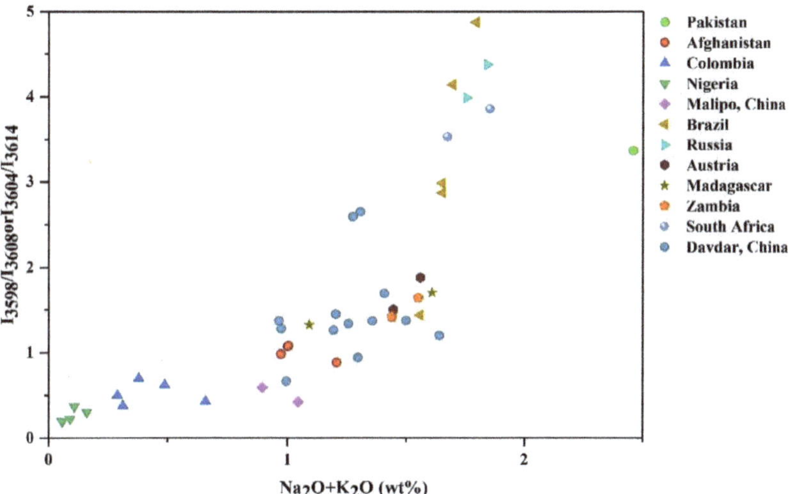

Figure 16. This figure shows the relationship between the content of Na$_2$O+K$_2$O in emeralds from different regions and the intensity ratio of Raman peaks, representing different water types. The spectral peaks of 3598 cm^{-1} (or 3604 cm^{-1}) and 3608 cm^{-1} (or 3614 cm^{-1}) are related to type II water and type I water, respectively. Data source [22,29].

6.3. Origin Traceability Analysis

The identification of the origin of an emerald cannot be separated from the geological deposits of an emerald. Among all the classifications of emerald deposits, we quite agree with the view put forward by Giuliani et al. in 2019: the emerald can be divided into IA, IB, IC and IIA, IIB, and IID, according to the sedimentary type and host rock type [18,32]. According to their research, the emeralds in the Panjshir Valley in Afghanistan belong to class IIC, and the emeralds from Davdar deposit in China also belong to this class, which is the same as our conclusion. The chemical composition and inclusion composition of the emeralds from these two habitats are similar.

The element content of 20 emeralds in the Panjshir Valley, Afghanistan, measured in this study shows that compared with the research of Bowersox et al. in 1991 and 2015 [3,4], the content of elements Mg, Na, Fe, V, and Cr is relatively high, but it is consistent with the research of Saeseaw et al. in 2014 and 2019 [33,34], and the content of alkali metals in other previous research [33,34].

As mentioned above, the proportion of type II water increases with increasing alkali metal ion content in emeralds. Figure 17 shows that the 12 emerald deposits can be divided into two main categories, according to the total K and Na content.

(Type I) K + Na < 12,500 ppm: including Coscuez, Colombia; the Hill Valley and the Panjshir Valley, Afghanistan; Itabira, Brazil; Ural, Russia; Davdar Township, China; and Malipo, China.

(Type II) K + Na > 12,500 ppm: including Kafubu, Zambia; Kagem, Zambia; the Swat Valley, Pakistan; Mananjary, Madagascar; Shakisso, Ethiopia; and Sandawana, Zimbabwe.

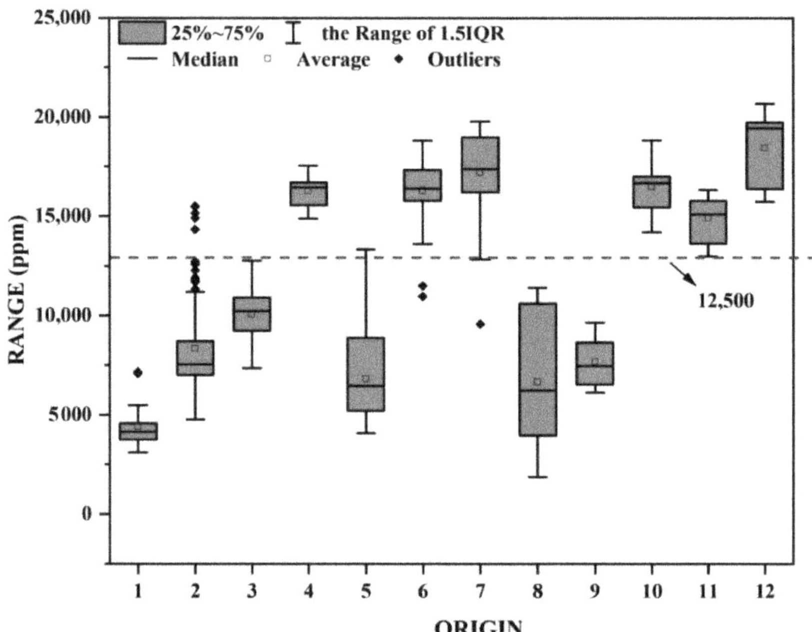

Figure 17. Box diagram of total K and Na contents of emeralds from different occurrences. 1 = Coscuez, Colombia; 2 = Panjshir, Afghanistan; 3 = Itabira, Brazil; 4 = Kafubu, Zambia; 5 = Ural, Russia; 6 = Kagem, Zambia; 7 = Swat, Pakistan; 8 = Davdar, China; 9 = Malipo, China; 10 = Mananjary, Madagascar; 11 = Shakisso, Ethiopia; and 12 = Sandawana, Zimbabwe.

In Type I, all deposits are associated with tectonic metamorphism, except for the Russian emerald deposits, and in Type II, all deposits are also associated with tectonic magma, except for the Pakistani emerald deposits (here cf Giuliani et al. 2019) [18]. Furthermore, as seen in Figure 14, emeralds from Panjshir, Afghanistan, are dominated by type I water, while emeralds from Swat, Pakistan, are dominated by type II water. Is Na + K = 12,500 ppm the cut-off value for the alkali metal content in the NIR spectrum of emeralds dominated by different types of water? This requires more analytical data. Could NIR spectra be used to determine the geographic origin of emeralds as well as UV-Vis spectra? Further research would be required.

Figure 18 shows a binary plot of K versus Na for the major emerald sources in the world. This plot allows the separation of Afghan emeralds from those from Zambia, Pakistan, Zimbabwe, and Ethiopia, because the former samples contain higher levels of sodium or potassium than the Afghan samples. The Russian emeralds can also be distinguished from the Afghan Panjshir emeralds thanks to their lower K content.

Figure 19 shows the distribution of Li and Cs in emeralds from the world's major origins, from which emeralds can be distinguished from most of the world's origins. For example, a part of the Colombian mines, because of their lower Li and Cs content than the Panjshir Valley emeralds' content, while the Swat Valley in Pakistan and the Urals in Russia, Ethiopia, Sandawana, Madagascar, Zambia, and Malipo in China have higher Li and Cs content than that of the Panjshir Valley emeralds [23,33–38].

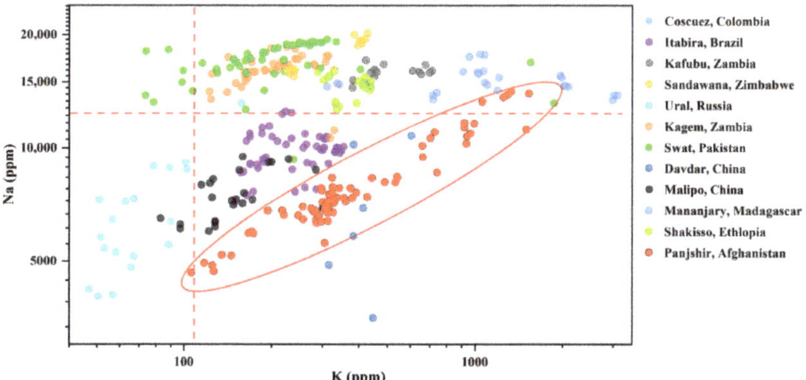

Figure 18. The log binary graph of K and Na shows that it can distinguish Panjshir emeralds from Kagem emeralds in Zambia, Sandawana emeralds in Zimbabwe, Shakisso emeralds in Ethiopia, Kafubu emeralds in Zambia, and Ural emeralds in Russia. Data source: [33–39].

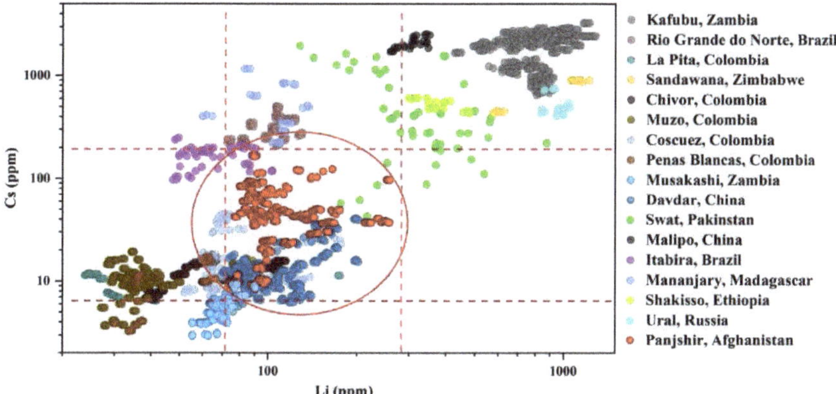

Figure 19. The log binary graph of Li and Cs shows that Panjshir emeralds are similar to Davdar emeralds and Coscuez emeralds in this data. Data source: [33–39].

From the distribution maps of Na and Sc, among the provenances shown in Figure 20, the Panjshir Valley emerald from Afghanistan belongs to the medium-low Na and medium-high Sc, which can be distinguished from the Davdar Township emerald from China, but the sample data from Davdar Township, China, are too small to represent general conclusions. Meanwhile, the data for the Coscuez emerald from Colombia and the Panjshir emerald from Afghanistan overlap in this binary plot [34].

According to Figure 21, Rb and Ga can distinguish between the Colombian Coscuez and Afghan Panjshir Valley emeralds. In conclusion, it is possible to distinguish the emeralds of the Panjshir Valley in Afghanistan from those of most origins in the world by the content of the alkali metals Li, Na, K, Rb, and Cs, as well as Sc and Ga.

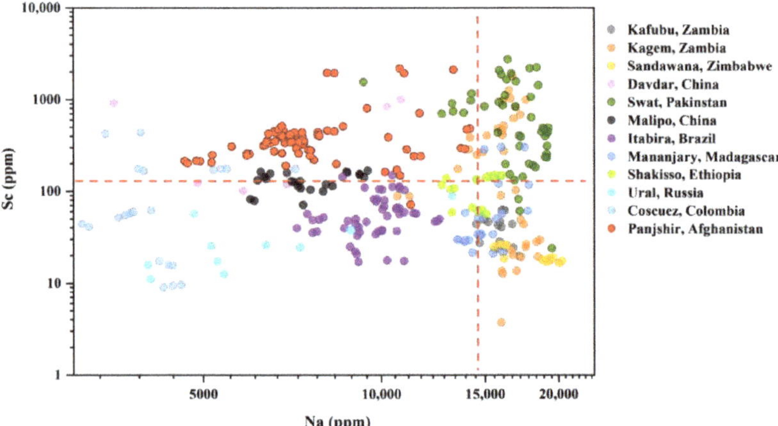

Figure 20. The log binary graph of Na and Sc shows that Panjshir emeralds are similar to Coscuez emeralds and Malipo emeralds in this data. Data source: [23,33].

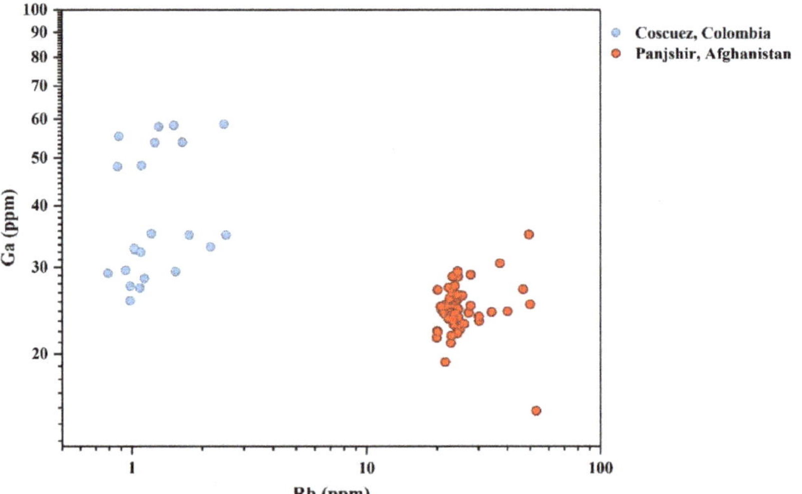

Figure 21. The log binary graph of Rb and Ga shows that Panjshir emerald and Coscuez emerald are not similar in this data and can be distinguished. Data source: [33].

Krzemnicki et al. proposed a new type of emerald from Panjshir, Afghanistan, in 2021, but later, they assumed that this type of emerald might come from Musakashi, Zambia [40]. Both Hennebois et al. (2022) and Saeseaw et al. (2014, 2015) reported emeralds from this origin [33,41,42]. We compared our samples with those from Musakashi, Zambia, and believed that the Fe-K diagram can be used to distinguish emeralds from Panjshir, Afghanistan, and Musakashi, Zambia, which is consistent with the research conclusion of Saeseaw et al. in 2014. As shown in Figure 22, we can figure out that the emeralds we studied come from Panjshir Valley in Afghanistan rather than from Musakashi in Zambia.

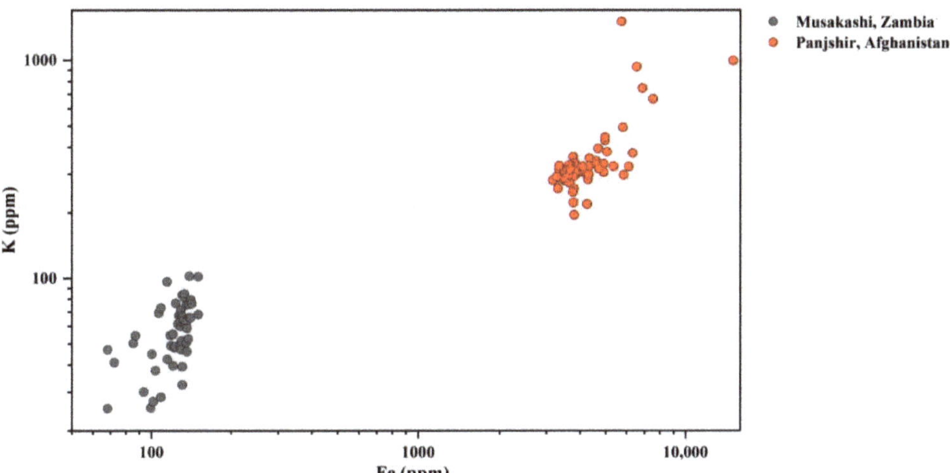

Figure 22. The log binary graph of Fe and K shows that Panjshir emeralds and Kusakashi emeralds are quite different and can be distinguished. Data source: [33,41].

7. Conclusions

Emeralds from the Panjshir Valley in Afghanistan differ slightly in gemological properties from those of other origins and can be distinguished and identified by their serrated, three-phase inclusions containing multiple crystals, type I water-dominated near-infrared spectral absorption, and medium to low Li and Cs, medium to high Sc, and medium to low Na. It is lighter in color, lower in alkali metals, and more complex in gas-liquid three-phase inclusions than the emeralds from the Swat Valley, Pakistan. The main chemical elements responsible for the color of Panjshir Valley emeralds are Cr^{3+}, V^{3+}, and Fe^{3+}, where Cr and V are the most important ones. As a result of the low to moderate alkali metal content, the infrared spectrum indicates that it is a type I water-dominated emerald. In terms of inclusions characteristics, the Afghan Panjshir Valley emerald is similar to the Colombian emerald in that both contain jagged, three-phase inclusions, but the relatively high number of crystals that may be present in the inclusions and the relatively high alkali metal content can determine the origin of the emerald.

Author Contributions: Conceptualization, A.H.S., Q.C. and X.L.; Formal analysis, P.B.; Investigation, Y.L. and X.L.; Data curation, A.H.S. and R.G.; Writing—riginal draft, P.B. and X.L.; Writing—review & editing, Q.C., Y.B., X.G. and X.L.; Supervision, X.L. All authors have read and agreed to the published version of the manuscript.

Funding: The work has been funded by the Hubei Gem & Jewelry Engineering Technology Center (No. CIGTXM-03-202201 and No. CIGTXM-04-S202107), the Fundamental Research Funds for National University, China University of Geosciences (Wuhan) (No. CUGDCJJ202221), the Philosophy and Social Science Foundation of Hubei Province (No. 21G007), NSFC funding (No. 41874105), the Scientific Research Project of Shanghai Jianqiao University (No. 17001SJQ and No. BBYQ2104).

Acknowledgments: The publication of this article in the year 2022 marks the 30th anniversary of the Gemological Institute (Wuhan) and the 70th anniversary of the China University of Geosciences (Wuhan).

Conflicts of Interest: The authors declare no conflict of interest.

References

1. Bariand, P.; Poullen, J.F. The pegmatites of Laghman, Nuristan, Afghanistan. *Mineral. Rec.* **1978**, *9*, 301–308.
2. Neilson, J.B.; Cannon, P.J. *Mineral Evaluation Project Afghanistan: Volume 2, Significant Mineral Occurrences*; United Nations Development Programme: Toronto, ON, Canada, 1977.

3. Bowersox, G.; Snee, L.W.; Foord, E.E.; Seal, R.R. Emeralds of the Panjshir Valley, Afghanistan. *Gems Gemol.* **1991**, *27*, 26–39. [CrossRef]
4. Bowersox, G. The emerald mines of the Panjshir valley, Afghanistan. *InColor* **2015**, 80–87.
5. Bowersox, G.W. A Status Report on Gemstones from Afghanistan. *Gems Gemol.* **1985**, *21*, 192–204. [CrossRef]
6. Giuliani, G.; France-Lanord, C.; Zimmermann, J.L.; Cheilletz, A.; Arboleda, C.; Charoy, B.; Coget, P.; Fontan, F.; Giard, D. Fluid Composition, δD of Channel H_2O, and δ18O of Lattice Oxygen in Beryls: Genetic Implications for Brazilian, Colombian, and Afghanistani Emerald Deposits. *Int. Geol. Rev.* **2010**, *39*, 400–424. [CrossRef]
7. Sinkankas, J. *Emerald and Other Beryls*; Chilton Book, Co.: Radnor, PA, USA, 1981.
8. Forestier, F.H.; Piat, D.H. Emeraudes de Bactriane: Mythe ou réalité, La vallée du Panjshir (Afghanistan). In *L'émeraude Connaissances Actuelles et Prospectives*; Giard, D., Giuliani, G., Cheilletz, A., Fritsch, E., Gonthier, E., Eds.; Association Française de Gemmologie: Paris, French, 1998.
9. Giuliani, G.; Chaussidon, M.; Schubnel, H.J.; Piat, D.H.; Rollion-Bard, C.; France-Lanord, C.; Giard, D.; de Narvaez, D.; Rondeau, B. Oxygen isotopes and emerald trade routes since antiquity. *Science* **2000**, *287*, 631–633. [CrossRef]
10. Krzemnicki, M.S.; Wang, H.A.O.; Buche, S. A New Type of Emerald from Afghanistan's Panjshir Valley. *J. Gemmol.* **2021**, *37*, 474–495. [CrossRef]
11. Hayden, H.H. Notes on the Geology of Chitral, Gilgit, and the Pamirs. *Rec. Geol. Surv. India* **1916**, *45*, 271–335.
12. Bordet, P.; Boutiere, A. Reconnaissance geologique dans I'Hindou Kouch oriental (Badakhchan, Afghanistan). *Bull. De La Soc. De Geol.* **1968**, *10*, 486–496. [CrossRef]
13. Rossovskiy, L.N.; Chmyrev, V.M.; Salakh, A.S. New fields and belts of rare-metal pegmatites in the Hindu Kush (Eastern Afghanistan). *Int. Geol. Rev.* **2009**, *18*, 1339–1342. [CrossRef]
14. Rossovskiy, L.N. Rare-metal pegmatites with precious stones and conditions of their formation (Hindu Kush). *Int. Geol. Rev.* **2010**, *23*, 1312–1320. [CrossRef]
15. Krzemnicki, M.S. New Emerald from Afghanistan. *SSEF Facet.* **2018**, *24*, 12–13.
16. Hsu, T.; Lucas, A. Tucson 2016. *Gems Gemol.* **2016**, *52*, 82–102.
17. Rossovskiy, L.N.; Konovalenko, S.I. South Asian pegmatite belt. *Akad. Nauk SSSR Dokl. Earth Sci. Sect.* **1976**, *229*, 89–91.
18. Giuliani, G.; Groat, L.A.; Marshall, D.; Fallick, A.E.; Branquet, Y. Emerald Deposits: A Review and Enhanced Classification. *Minerals* **2019**, *9*, 105. [CrossRef]
19. Giuliani, G.; Schwarz, D.T. Emerald Deposits- A Review. *Aust. Gemmol.* **2001**, *21*, 17–23.
20. Zong, K.; Klemd, R.; Yuan, Y.; He, Z.; Guo, J.; Shi, X.; Liu, Y.; Hu, Z.; Zhang, Z. The assembly of Rodinia: The correlation of early Neoproterozoic (ca. 900 Ma) high-grade metamorphism and continental arc formation in the southern Beishan Orogen, southern Central Asian Orogenic Belt (CAOB). *Precambrian Res.* **2017**, *290*, 32–48. [CrossRef]
21. Bosshart, G. Emeralds from Colombia (Part 2). *J. Gemmol.* **1991**, *22*, 409–426. [CrossRef]
22. Cui, D.; Liao, Z.; Qi, L.; Zhong, Q.; Zhou, Z. A Study of Emeralds from Davdar, North-Western China. *J. Gemmol.* **2020**, *37*, 374–392. [CrossRef]
23. Hu, Y.; Lu, R. Unique Vanadium-Rich Emerald from Malipo, China. *Gems Gemol.* **2019**, *55*, 338–352. [CrossRef]
24. Wood, D.L.; Nassau, K. Infrared Spectra of Foreign Molecules in Beryl. *J. Chem. Phys.* **1967**, *47*, 2220–2228. [CrossRef]
25. Guo, H.; Yu, X.; Zheng, Y.; Sun, Z.; Ng, M.F.-Y. Inclusion and Trace Element Characteristics of Emeralds from Swat Valley, Pakistan. *Gems Gemol.* **2020**, *56*, 336–355. [CrossRef]
26. Qiao, X.; Zhou, Z.; Schwarz, D.T.; Qi, L.; Gao, J.; Nong, P.; Lai, M.; Guo, K.; Li, Y. Study of the Differences in Infrared Spectra of Emerald from Different Mining Areas and the Controlling Factors. *Can. Mineral.* **2019**, *57*, 65–79. [CrossRef]
27. Mashkovtsev, R.I.; Thomas, V.G.; Fursenko, D.A.; Zhukova, E.S.; Uskov, V.V.; Gorshunov, B.P. FTIR spectroscopy of D2O and HDO molecules in thec-axis channels of synthetic beryl. *Am. Mineral.* **2016**, *101*, 175–180. [CrossRef]
28. Qiao, X.; Zhou, Z.; Nong, P.; Lai, M. Study on the Infrared Spectral Characteristics of H_2O I-type Emerald and the Chontrolling Factors. *Rock Miner. Anal.* **2019**, *38*, 169–178.
29. Huong, L.T.-T.; Häger, T.; Hofmeister, W. Confocal Micro-Raman Spectroscopy: A Powerful Tool to Identify Natural and Synthetic Emeralds. *Gems Gemol.* **2010**, *46*, 36–41. [CrossRef]
30. Bersani, D.; Azzi, G.; Lambruschi, E.; Barone, G.; Mazzoleni, P.; Raneri, S.; Longobardo, U.; Lottici, P.P. Characterization of emeralds by micro-Raman spectroscopy. *J. Raman Spectrosc.* **2014**, *45*, 1293–1300. [CrossRef]
31. Moroz, I.; Roth, M.; Boudeulle, M.; Panczer, G. Raman microspectroscopy and fluorescence of emeralds from various deposits. *J. Raman Spectrosc.* **2000**, *31*, 485–490. [CrossRef]
32. Alonso-Perez, R.; Day, J.M.D. Rare Earth Element and Incompatible Trace Element Abundances in Emeralds Reveal Their Formation Environments. *Minerals* **2021**, *11*, 513. [CrossRef]
33. Saeseaw, S.; Pardieu, V.; Sangsawong, S. Three-Phase Inclusions in Emerald and Their Impact on Origin Determination. *Gems Gemol.* **2014**, *50*, 114–132. [CrossRef]
34. Saeseaw, S.; Renfro, N.D.; Palke, A.C.; Sun, Z.; McClure, S.F. Geographic Origin Determination of Emerald. *Gems Gemol.* **2019**, *55*, 614–646. [CrossRef]
35. Aurisicchio, C.; Conte, A.M.; Medeghini, L.; Ottolini, L.; Vito, C.D. Major and trace element geochemistry of emerald from several deposits: Implications for genetic models and classification schemes. *Ore Geol. Rev.* **2018**, *94*, 351–366. [CrossRef]

36. Groat, L.A.; Giuliani, G.; Marshall, D.D.; Turner, D. Emerald deposits and occurrences: A review. *Ore Geol. Rev.* **2008**, *34*, 87–112. [CrossRef]
37. Groat, L.A.; Giuliani, G.; Stone-Sundberg, J.; Sun, Z.; Renfro, N.D.; Palke, A.C. A Review of Analytical Methods Used in Geographic Origin Determination of Gemstones. *Gems Gemol.* **2019**, *55*, 512–535. [CrossRef]
38. Karampelas, S.; Al-Shaybani, B.; Mohamed, F.; Sangsawong, S.; Al-Alawi, A. Emeralds from the Most Important Occurrences: Chemical and Spectroscopic Data. *Minerals* **2019**, *9*, 561. [CrossRef]
39. Hu, Y.; Lu, R. Color Characteristics of Blue to Yellow Beryl from Multiple Origins. *Gems Gemol.* **2020**, *56*, 54–65. [CrossRef]
40. Krezemnicki, M.S. New Emeralds from Musakashi, Zambia, Appear on the Market. *J. Gemmol.* **2021**, *37*, 769–771. [CrossRef]
41. Pardieu, V.; Detroyat, S.; Sangsawong, S.; Saeseau, S. Tracking emeralds from the Musakashi area of Zambia. *InColor* **2015**, *29*, 18–25.
42. Henneebois, U.; Karampelas, S.; Delaunay, A. Caractéristiques des émraudes de Musakashi, Zambie. *Revue De Gemmol. A. F. G.* **2022**, *216*, 15–17.

Article

Non-Destructive Study of Egyptian Emeralds Preserved in the Collection of the Museum of the Ecole des Mines

Maria Nikopoulou [1,*], Stefanos Karampelas [2,*], Eloïse Gaillou [3,*], Ugo Hennebois [2], Farida Maouche [3], Annabelle Herreweghe [2], Lambrini Papadopoulou [1], Vasilios Melfos [1], Nikolaos Kantiranis [1], Didier Nectoux [3] and Aurélien Delaunay [2]

[1] Department of Mineralogy-Petrology-Economic Geology, School of Geology, Aristotle University of Thessaloniki, 54124 Thessaloniki, Greece
[2] Laboratoire Français de Gemmologie (LFG), 30 rue de la Victoire, 75009 Paris, France
[3] Mineralogy Museum, Mines Paris, PSL University, 75006 Paris, France
* Correspondence: marthonik@geo.auth.gr (M.N.); s.karampelas@lfg.paris (S.K.); eloise.gaillou@mines-paristech.fr (E.G.)

Abstract: In the present study, rough emerald single crystals and rough emeralds in the host rock from the ruins of Alexandria and from the Mount Zabargad in Egypt, preserved in the collection of the museum of the Ecole des Mines (Mines Paris—PSL) since the late 19th or early 20th century, are investigated. All samples were characterized by non-destructive spectroscopic and chemical methods during a week-long loan to the LFG. Raman, FTIR and UV-Vis-NIR spectroscopy revealed that Egyptian emeralds contain H_2O molecules accompanied by relatively high concentrations of alkali ions and are colored by chromium and iron. Additionally, EDXRF showed that the emeralds from Egypt contain up to 84 ppm Rb and low amounts (below 200 ppm) of Cs. Inclusions and parts of the host rock were also observed under optical microscope and analyzed with Raman spectroscopy. Tube-like structures, quartz, calcite, dolomite, albite and phlogopite are associated minerals, and inclusions are identified in these historic emeralds from Egypt. This work will hopefully further contribute to the characterization of emeralds of archaeological significance.

Keywords: emeralds; beryl; Egypt; EDXRF; FTIR; Raman spectroscopy; PL; UV-Vis-NIR

Citation: Nikopoulou, M.; Karampelas, S.; Gaillou, E.; Hennebois, U.; Maouche, F.; Herreweghe, A.; Papadopoulou, L.; Melfos, V.; Kantiranis, N.; Nectoux, D.; et al. Non-Destructive Study of Egyptian Emeralds Preserved in the Collection of the Museum of the Ecole des Mines. *Minerals* **2023**, *13*, 158. https://doi.org/10.3390/min13020158

Academic Editor: Yamirka Rojas-Agramonte

Received: 16 December 2022
Revised: 11 January 2023
Accepted: 17 January 2023
Published: 21 January 2023

1. Introduction

Emerald is the bluish-green to green to yellowish-green variety of beryl (with an ideal formula of $Be_3Al_2SiO_{18}$) colored by chromium and/or vanadium (iron may also contribute to the color); beryl colored solely by iron is classified as green beryl (and not emerald) [1–3]. Emeralds are found in almost all continents but today, the most popular and commercial deposits occur in: Colombia, Zambia and Brazil, with lesser quantities found in Afghanistan, Ethiopia, Madagascar, Russia and Zimbabwe [1–3]. Emerald as a mineral is crystalized in the hexagonal crystal system and is, therefore, birefringent. This means that when light penetrates the crystal (perpendicular to c-axis), it is separated in two rays with different refractive indexes (RIs). The ordinary ray (ω-ray) and the extraordinary ray (e-ray) are n_ω = 1.5705–1.5905 and n_e = 1.5656–1.5975, respectively [4].

The crystal structure of an emerald consists of Be/Si sites surrounded by four atoms of O in tetrahedral coordination and the Al site (known as Y site) surrounded by six O atoms in octahedral coordination [1]. The SiO_4 tetrahedra form six-membered rings parallel to the c-axis (Figure 1a,b). These rings are arranged on top of each other. Viewing a beryl's crystal lattice parallel to c-axis, vacant spaces can be observed, which are called "channels". The studies of beryl crystals with Fourier transform infrared (FTIR) spectroscopy [5] proved that water molecules can be found in the crystal's structure. Water molecules can be classified into two different types (Figure 2): Type I water molecules occur individually in the channels and the symmetry axis is positioned perpendicular to c-axis of the crystal.

Type II water molecules are accompanied by nearby alkali ions (principally Na) in the channels with the water molecule axis parallel to c-axis [6]. Sometimes, the substitution of Al in Y sites [7] with chromophore ions (for example Fe, Cr and V) is responsible for the different variety of colors in beryl. In the case of emerald, the green color is due to the substitution of Al^{3+} with Cr^{3+} and V^{3+} ions, with Fe^{3+} and Fe^{2+} contributing.

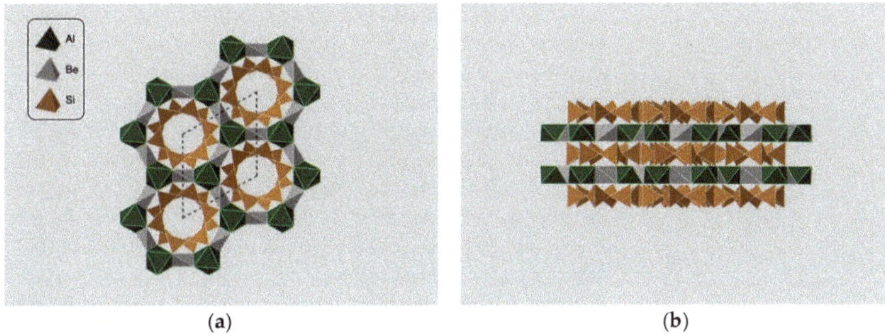

Figure 1. Crystal structure of beryl perpendicular (**a**) and parallel (**b**) to c-axis. Empty channels are mentioned with black dashed line (designed in Crystal Maker).

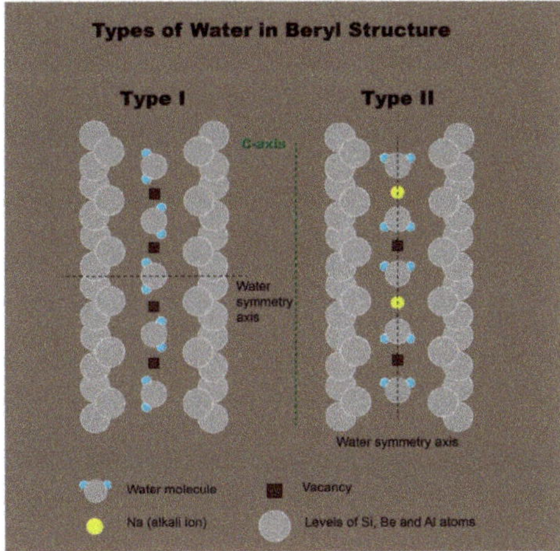

Figure 2. Type I and Type II water molecules in the channels of beryl's structure (Updated from Wood and Nassau, 1968 [5]. Designed in Adobe Illustrator).

The term "Egyptian emeralds" is associated with the legendary and historic emerald source known as "Cleopatra Mines", a term that is used as a marketing tool [8]. The mining of emeralds began in the Wadi Sikait area in Eastern desert of Egypt [9] toward the end of the 1st century BC but most of the mining activity occurred during Roman times [1]. Today, no mining of emeralds occurs in Egypt. The last mining operations were carried out unsuccessfully in the 1920s [10]. The genesis of Egyptian emeralds is linked to the geological history of the southern Eastern Desert. According to references [11], there are at least three generations of beryl in Egyptian emerald deposits. The crystals of the first generation are colorless without zoning and show micro- and macro-cracks produced by brittle deformation. The first-generation crystals are surrounded by a second generation overgrown layer [11]. The beryl crystals in this layer have a light green color with various solid inclusions of mica, phlogopite, albite and quartz. The latest generation of beryl is a Cr-rich emerald with a deep green color and it rarely occurs in quartz veins and aplitic dikes [11]. The edges of this generation have no cracks [11]. Significant deposits of Egyptian emeralds are in Zabara, Wadi Nugrus and Umm Kabu areas with similar geological settings, also known as Eastern Desert deposits in Egypt. Most the beryl crystals in these deposits are of the Proterozoic age and can be categorized into two paragenetic types [12]: (1) emeralds in mica-schists and (2) beryl in granitoids.

The Zabara Mountain has been known for its emerald deposits since the antiquity. In this area, emerald and beryl crystals are found in the contact between gneissic granites or meta-pegmatites (rich in quartz and feldspar) and various types of schists (rich in mica, amphiboles and talc). The granites are S-type [13] and were probably formed by the partial melting of the crust (mainly by metasedimentary rocks). The schists probably originated from the metamorphism of ophiolitic and volcanic rocks. The host rocks of beryl contain abundant fractures and joints filled by quartz veins. Additionally, many fractures in beryl crystals contain quartz. This indicates that the emeralds were crystallized first, followed by quartz. In some places, the deformation caused by later tectonic events resulted in boudinaged metapegmatoid veins in the biotite schists [14]. In the Zabara area, the emerald crystals are plentiful and large (reaching a few cm across), and their colors vary from pale greenish to deep green or bluish-green [15]. Most of the best quality emeralds are restricted to the mica rocks, close to the quartz veins.

The Wadi Nugrus area is manly occupied by mafic and ultramafic rocks (serpentinites, metagabbros and hornblende schists), which contain garnets in places. The intrusion of a Be-enriched magma accompanied by the injection of numerous aplitic and quartz veins in ophiolites and volcanic arc rocks of the Nugrus-Ghadir constitutes the first stage of the geological history of Egyptian emeralds [11]. Most of these rocks have been metasomatized and deformed by later hydrothermal fluids and subsequent metamorphic and tectonic events. Shear zones with a NW–SE direction led to the formation of mylonites and cata-clastites [12]. The emeralds are found between metasediments or various schists in contact with granitic and pegmatite intrusions. Some quartz veins and emerald-bearing schists could be associated with post-magmatic fluids and Na-metasomatism (albitization). All these processes contributed to the final formation of beryl in this region. Most emeralds in Wadi Nugrus reach a few cm across (average 3 cm) the oriented parallel to their prismatic facets [12].

The geology of the Kabu-Um Debaa area is very similar to Wadi Nugrus [10]. Rocks of this region include mica and hornblende schists and serpentinites with garnetiferous magmatic intrusions. The beryliferous zones are found close to quartz veins in contact with mica schists. According to Hassan and El-Shatouri [15], no exposures of the metasedimen-tary rocks were observed in this area. However, some blocks of the psammitic gneiss are observed in the mine dumps, indicating the presence of this rock at shallow depth.

2. Materials and Methods

For this study, the Ecole des Mines (Paris School of Mines-PSL) provided a short-term loan (one week), for scientific purposes, of 8 emerald samples from Egypt to the Laboratoire Français de Gemmologie (LFG) to analyze them using strictly non-destructive methods. The samples and their dimensions are listed in Table 1, as well as Figures 3 and 4. Four of these samples are single crystals (Figure 3), with a characteristic hexagonal prism shape, donated by Émile Bertrand (1844–1909) to the Ecole des Mines. According to the museum records, they come from the ruins of Alexandria [16]. These single crystal samples were also examined with classic gemological tools. The other four samples are incorporated in the host rock (Figure 4). According to the museum references, these samples were given to the museum at the end of the 19th century or at the very beginning of the 20th century. Three of them were provided by the Marquis de Raincourt (1835–1917) and one by Frédéric Cailliaud (1787–1869). Cailliaud was the leader of campaigns in the Zabara area, meaning they searched for the mythic emerald deposits in Egypt, at the beginning of the 19th century [17]. It is more than likely that sample ENSMP 47849a27 comes from one of the expeditions led by Cailliaud.

Table 1. Properties of Egyptian emeralds from Ecole des Mines.

No.	Donor's Name	Wt. (Ct.)	Dimensions (mm)	Density	Color	Shape	Observations
ENSMP 72466_01	Emile Bertrand	1.002	7.25 × 3.73 × 4.33	2.71	Vivid green	Rough	Translucent RI: 1.57-1.58 LWUV-SWUV: Inert
ENSMP 72466_02	Emile Bertrand	2.038	7.29 × 5.43 × 5.55	2.63	Vivid green	Rough	Translucent LWUV-SWUV: Inert
ENSMP 72466_03	Emile Bertrand	3.154	9.94 × 5.69 × 7.38	2.71	Vivid bluish-green	Rough	Translucent LWUV-SWUV: Inert
ENSMP 72466_04	Emile Bertrand	5.392	10 × 7.98 × 8.32	2.77	Vivid bluish-green	Rough	Translucent RI: 1.58 LWUV-SWUV: Inert
ENSMP 47849a26	Marquis de Raincourt	-	34.66 × 29.23 × 30.76	-	Vivid green crystals	Crystals in host rock	Translucent LWUV-SWUV: Inert
ENSMP 47849a27	Frédéric Cailliaud	-	25.93 × 22.74 × 24.52	-	Vivid green crystals	Crystals in host rock	Translucent LWUV-SWUV: Inert
ENSMP 72486	Marquis de Raincourt	-	31.83 × 16.83 × 21.50	-	Vivid bluish-green crystals	Crystals in host rock	Translucent LWUV-SWUV: Inert
ENSMP 13512_5	Marquis de Raincourt	-	85.6 × 57.53 × 50.08	-	Vivid bluish-green crystals	Crystals in host rock	Translucent LWUV-SWUV: Inert

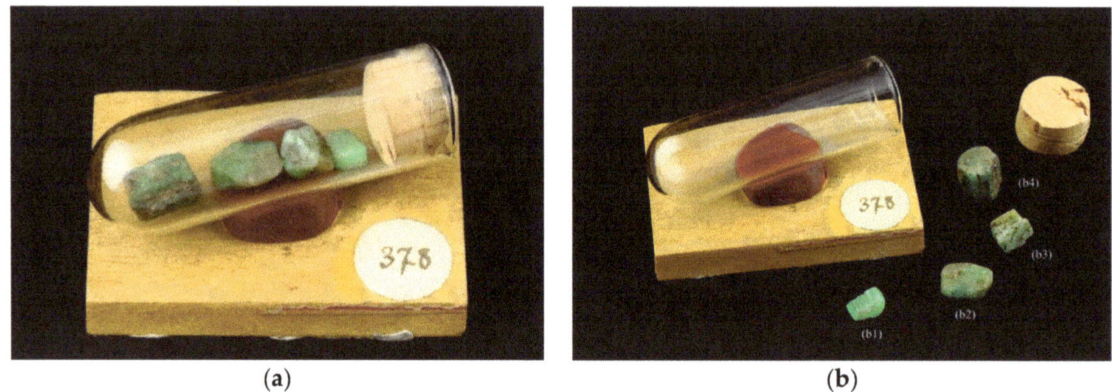

Figure 3. (**a**,**b**) Macroscopic images of Egyptian single-crystal emeralds from Ecole des Mines. (**b1**) ENSMP 72466_01, (**b2**) ENSMP 72466_03, (**b3**) ENSMP 72466_02, (**b4**) ENSMP 72466_04. Photos: Eloïse Gaillou; © Musée de Minéralogie Mines Paris—PSL/E. Gaillou.

Figure 4. Macroscopic images of emerald samples included in the host rock from Ecole des Mines. (**a**) ENSMP 47849a26 (34.66 × 29.23 × 30.76 mm), (**b**) ENSMP 47849a27 (25.93 × 22.74 × 24.52 mm), (**c**) ENSMP 72,486 (31.83 × 16.83 × 21.50 mm), (**d**) ENSMP 13512_5 (85.6 × 57.53 × 50.08 mm). Photos: Eloïse Gaillou; © Musée de Minéralogie Mines Paris—PSL/E. Gaillou.

All samples were examined under a Zeiss Stemi 508 binocular microscope. Refractive index for ENSMP 72466_01 and ENSMP 72466_04 samples is measured by using a refractometer and the distant vision method (also known as spot reading). Hydrostatic balance was used to measure the density of the four samples in Figure 3. Additionally, luminescence reaction was observed for the same four samples under a 6 W ultraviolet (UV) lamp (Vilber Lourmat VL-6.LC) with long-wave ultraviolet (365 nm) and short-wave ultraviolet (254 nm) light, equipped with a CN-6 dark room (10 cm distance between the sample and the lamp).

Raman and photoluminescence (PL) spectra were acquired on the four single crystals, using mobile Raman Spectrometer (GemmoRaman-532SG, Magilabs Oy (Ltd.), Helsinki, Finland)with a 532 nm laser excitation ranging for Raman spectra from 200 to 2000 cm^{-1} (with 1 s exposure time and 4 accumulations) and for PL spectra from 540 to 760 nm (0.3 to 0.4 s exposure time and 30 accumulations). Spectra using micro-Raman Renishaw inVia spectrometer (Renishaw plc, Wotton-under-Edge, Gloucestershire, UK), coupled with an optical microscope and with a 514 nm laser excitation (diode-pumped solid-state laser) from 100 to 1500 cm^{-1} region, were obtained with 10 accumulations and 30s of exposure time. Laser power on the sample was 40 mW for these spectra and the spectral resolution about 2 cm^{-1}. Raman spectra, for the region 3000–3700 cm^{-1}, were obtained with 20 accumulations and 15 s exposure time. Raman spectra on the associated minerals on the four rocks were acquired from 100 to 4000 cm^{-1}, with 1 accumulation and 20 s exposure time. For the PL spectra (500–900 nm), 10 s of exposure time and 1 accumulation with 0.04 mW laser power were used on the sample. A diamond was used for the calibration of both Raman spectrometers by considering its 1331.8 cm^{-1} Raman peak.

FTIR spectra (400–8000 cm^{-1}) were obtained using Nicolet iS5 spectrometer (Thermo Fischer Scientific, Waltham, MA, USA) with 4 cm^{-1} resolution and 500 scans. Visible-near infrared (Vis-NIR) spectra were acquired on the four single crystals from 365 to 1000 nm using a mobile instrument (0.05 to 0.10 s acquisition time and 50 accumulations) with an integrating sphere (Gemmosphere, Magilabs Oy (Ltd.), Helsinki, Finland). Ultraviolet-visible (UV-Vis) spectra were conducted by a Jasco V-630 from 250 to 850 nm, with a data interval (DI) and spectral bandwidth (SBW) of 2 nm and 210 nm/min scan rate. When possible, spectra were acquired parallel and perpendicular to the c-axis (for the single crystals) on the flattest parts of each stone. Due to their size, the rocks and their associated minerals were only studied by using a micro-Raman spectrometer.

For chemical analysis with EDXRF (energy dispersive X-ray fluorescence), sample holders with an aperture of 5 mm diameter were used and specific sets of parameters were optimized for the most accurate analysis of beryl. Various conditions were used for filters and voltage (no filter/4 kV, cellulose/8 kV, aluminum/12 kV, thin palladium/16 kV, medium palladium/20 kV, thick palladium/28 kV, and thick copper/50 kV), with an acquisition time of about 20 min for each sample. All measured iron was calculated as FeO.

3. Results

3.1. Macroscopic, Microscopic and Gemological Observations

The emerald samples are translucent and have a hexagonal prism habit with a vivid bluish-green to vivid green color. Interference colors are due to fissures, as well as various fractures on the surface of the crystals. Emeralds' size ranges from 2 mm to 2 cm and sometimes their growth is interrupted by the intrusion of quartz (identified by micro-Raman spectroscopy; see below) veins (Figure 5a,b). These intrusions may have taken place after the crystallization of the emeralds. Furthermore, some samples (ENSMP 1352_5) reveal boudinage tectonic structures (Figure 4d), which comes in terms with Grundmann and Morteani [14] descriptions. According to their gemological observations, emeralds of Egypt are translucent with RI that ranges from 1.57 to 1.58. All samples are inert under SWUV and LWUV, and their density varies from 2.63 to 2.77.

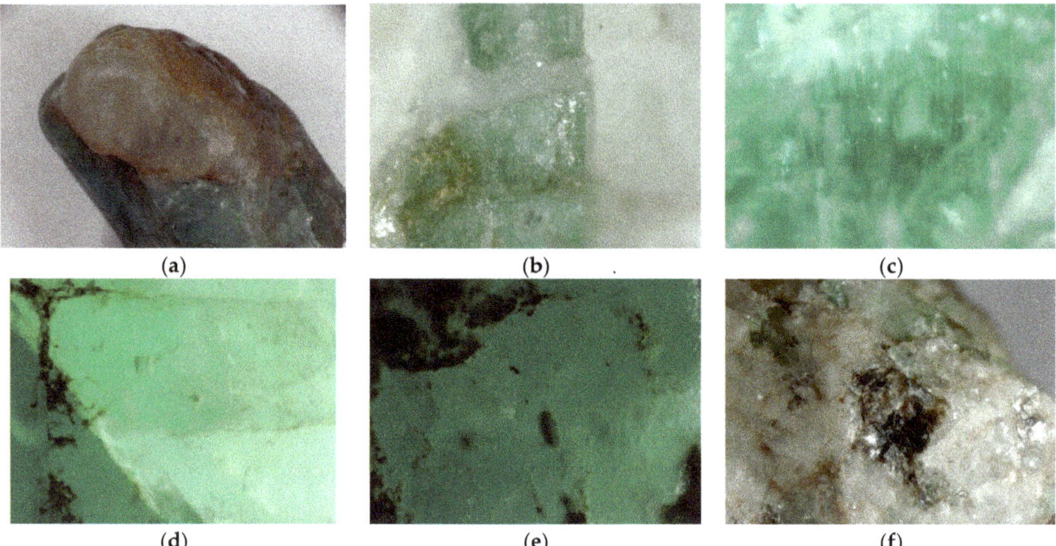

Figure 5. Macroscopic and microscopic characteristics of emeralds from Egypt. (**a**) Quartz crystal which interrupts the growth of an emerald crystal (ENSMP 72466_03); FOV: 8 mm, (**b**) Quartz vein which interrupts the growth of an emerald crystal (ENSMP 47849a26); FOV: 2 mm, (**c**) Growth tubes and rectangular shaped multiphase inclusions parallel to the c-axis (ENSMP 13512_5); FOV: 1 mm, (**d**) Rectangular crystal inclusion in an emerald crystal (ENSMP 72466_02); FOV: 1 mm, (**e**) Macroprismatic crystal inclusion of mica in an emerald crystal (ENSMP 72466_04); FOV: 1 mm, (**f**) Mica, quartz and plagioclase in the host rock of Egyptian emeralds (ENSMP 47849a27); FOV: 4 mm. Microphotos by Ugo Hennebois/LFG; © Musée de Minéralogie Mines Paris—PSL.

Under the microscope, some samples show growth tubes and rectangular shaped multiphase fluid inclusions [17,18] (Figure 5c) along with quartz and mica crystals (identified by micro-Raman spectroscopy; see below). In addition, unidentified single crystal inclusions can be observed with angular or prismatic shapes (Figure 5d) or mica inclusions in macroprismatic shapes (Figure 5e). The minerals of the host rock (Figure 5f), which are observed under the microscope, are: quartz, feldspar (possibly plagioclase) and mica (possibly phlogopite), identified by micro-Raman spectroscopy (see below).

3.2. Raman Spectroscopy

Micro-Raman technique was applied in various ranges: from 200 to 1300 cm^{-1} (Figure 6a,c) and from 3500 to 3700 cm^{-1} (Figure 6b,d). A band at around 686 cm^{-1} is observed due Be-O stretching, at around 1069 cm^{-1} it is linked with Si-O and/or Be-O stretching [3,19–23] and at around 1010 cm^{-1} and above 1100 cm^{-1} it may also be observed on natural emeralds [21] and correspond to stretching vibrations of Si-O and inner vibrations, respectively [3,20–22,24]. Raman bands between 200 and 600 cm^{-1} are related to Si_6O_{18} ring vibrations [3,20,21]. The relative intensity of the Raman bands changes with the change in spectra orientation. Alkali content can be estimated with the full width half maximum (FWHM) of the band at 1069 cm^{-1}; all studied samples demonstrated high alkali content (Table 2) as they have a FWHM > 22 cm^{-1}, ranging from 23 to 27 cm^{-1}, while samples with medium-to-low alkali content have a FWHM < 22 cm^{-1} in the 1069 cm^{-1} band [3,20,25].

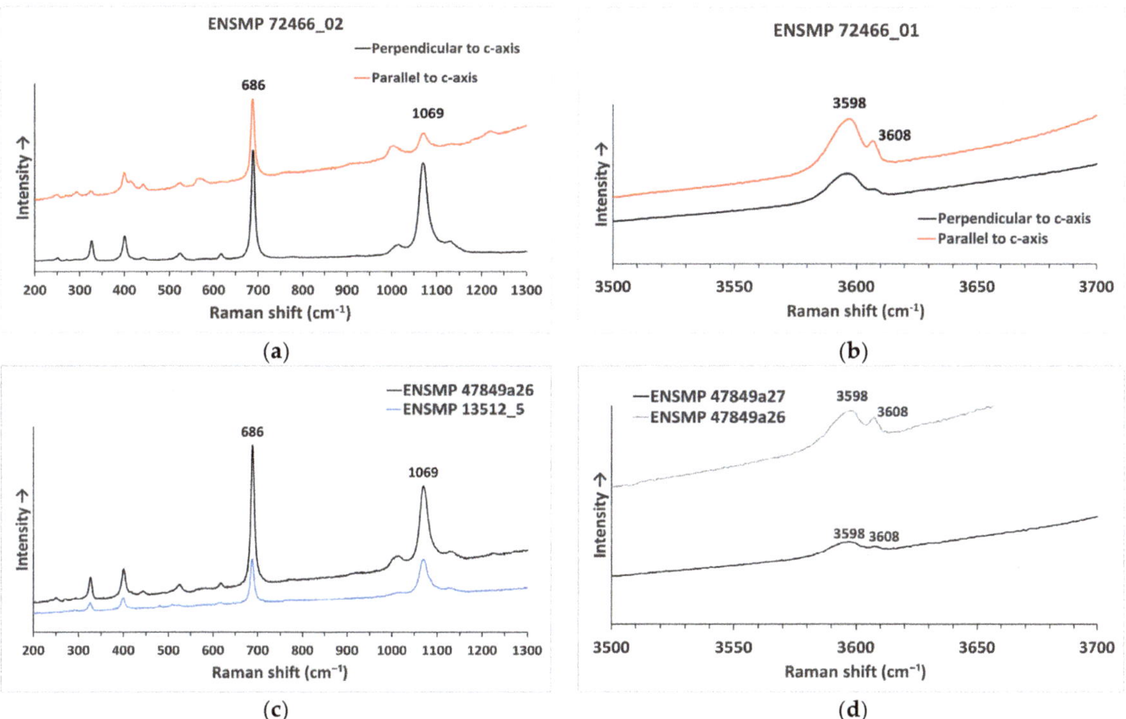

Figure 6. Micro-Raman spectra from Egyptian emeralds (using 514 laser excitation). (a) Raman spectra from 200 to 1300 cm^{-1} of a single-crystal emerald (ENSMP 72466_02) parallel (red line) and perpendicular (black line) to c-axis. Characteristic peaks at 686 and 1069 cm^{-1} are observed, while the 1069 cm^{-1} peak is more intensely parallel to c-axis. (b) Raman spectra from 3500 to 3700 cm^{-1} of a single-crystal emerald (ENSMP 72466_01) parallel (red line) and perpendicular (black line) to c-axis. Peak close to 3608 cm^{-1} is related with Type I water molecule structure. More intense bands at 3598 cm^{-1} are due to Type II water molecule structure with an alkali ion nearby. (c) Raman spectra from 200 to 1300 cm^{-1} of emeralds from samples that are included in the host rock. Spectra were acquired with the laser beam perpendicular to the prism facets. (d) Raman spectra from 3500 to 3700 cm^{-1} of the samples ENSMP 47849a27 (black line) and ENSMP 47840a26 (grey line). The intense band at 3598 cm^{-1} is related with Type II water molecules with an alkali ion nearby.

Table 2. Position and FWHM of the Raman band at around 1069 cm^{-1}, relative intensities of the Raman bands at 3598 cm^{-1} and 3608 cm^{-1} and position of the R1 photoluminescence bands for the samples of different localities. All observations were made using spectra acquired with a laser beam parallel and/or perpendicular to c-axis.

Sample No.	FWHM of Band 1069 cm^{-1}	R1/Rn
ENSMP 72466_01	24.5	0.41
ENSMP 72466_02	27	0.43
ENSMP 72466_03	26	0.35
ENSMP 72466_04	24.5	0.43
ENSMP 13512_5	24.5	0.45
ENSMP 47849a26	23	0.51
ENSMP 47849a27	24.5	0.43
ENSMP 72486	23.5	0.36

Raman spectra area from 3500 to 3700 cm^{-1} also indicates the presence or absence of alkali ions in the crystal lattice based on the height ratio of the two sharp bands situated at around 3600 cm^{-1} [3,6,23,24]. The peak around 3598 cm^{-1} is attributed to water Type II molecules vibrations (with an alkali ion nearby), and the band around 3608 cm^{-1} is attributed to water Type I molecule vibrations (without an alkali ion nearby) [20]. The higher intensity of 3598 cm^{-1} compared to 3608 cm^{-1} in all studied samples further confirms the presence of alkalis in relatively high concentrations [3,6,22,23].

Raman spectra were acquired with a mobile instrument on the following samples: ENSMP 72466_01, ENSMP 72466_02, ENSMP 72466_03 and ENSMP 72466_04. Only the main Raman bands of beryl at 686 cm^{-1} and 1069 cm^{-1} were observed with different relative intensities due to different crystallographic orientations (Figure 7) [19]. Emerald is a mineral that presents strong luminescence phenomena above 2000 cm^{-1} due to the presence of chromium.

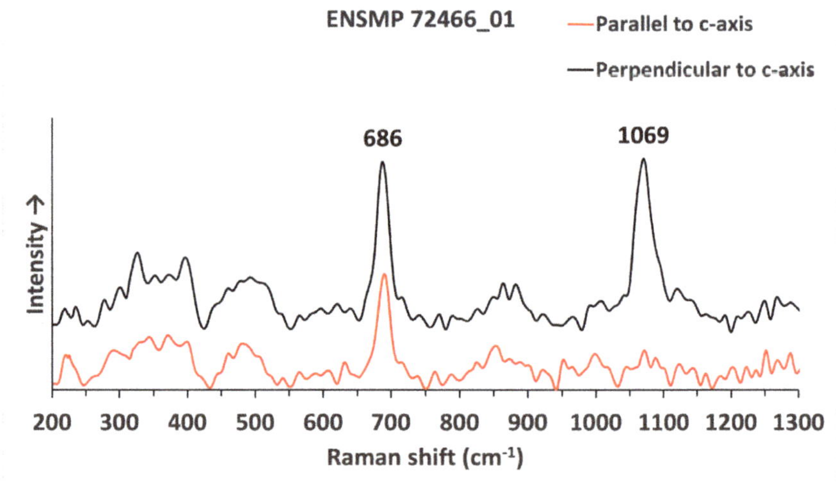

Figure 7. Raman spectra (ENSMP 72466_01) obtained using a mobile instrument with 532 nm laser excitation in the range of 200 to 1300 cm^{-1} parallel (red line) and perpendicular (black line) to the c-axis. The upper spectrum has been vertically offset for better clarity. Characteristic bands at 686 and 1069 cm^{-1} are mentioned. The Raman spectra were baseline corrected.

3.3. PL Spectroscopy

In Figure 8, PL spectra from 650 to 850 nm (Figure 8) are presented. All samples revealed two sharp peaks at around 680 and 684 nm due to Cr^{3+} (known as R2 and R1 lines, respectively), and a broad band centered at around 720–740 nm (known as Rn), which is also linked with traces of chromium [3,26]. The position of R1 band could help to separate natural from synthetic emeralds as there are no synthetic emeralds with R1 > 683.7 nm [3]. Additionally, according to Thompson et al. [27], the relative height ratio of the peaks can help in the determination of high or low concentrations of alkali ions (Table 2). In our study, the height ratio of the R1/Rn ranges between 0.36 and 0.51, revealing that the Egyptian emeralds contain relatively high alkalis while the R1 peak position is above 684 nm in all studied samples. Spectra with the mobile instruments reveal bands at similar positions.

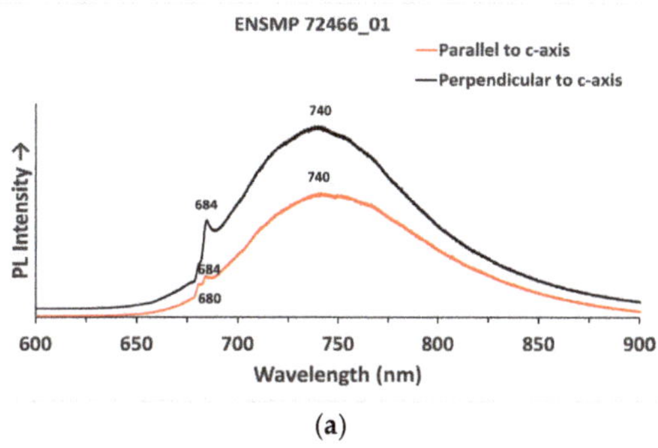
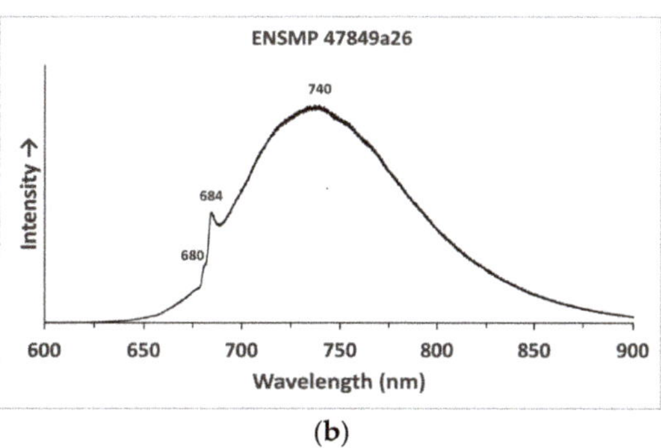

(**a**) (**b**)

Figure 8. PL spectra of Egyptian emeralds (600–900 nm) from micro-Raman instrument. Sharp peaks near 680 nm are linked with Cr^{3+}. Broad band at 740 nm is also associated with the presence of Cr. (**a**) PL spectra of a single-crystal emerald (ENSMP 72466_01) parallel (red line) and perpendicular (black line) to c-axis. The upper spectrum has been vertically offset for better clarity. (**b**) PL spectra of an emerald from a sample, which is included in the host rock (ENSMP 47849a26).

3.4. FTIR Spectroscopy

Due to the irregular surface of the samples, we were not able to acquire FTIR spectra of good quality; in Figure 9, the FTIR spectrum of sample ENSMP 72466_01 is presented. Absorption bands at around 5273 cm^{-1} are linked to Type II water molecules [18]. Around 7095 cm^{-1} (i.e., ca. 1410 nm), a band related mainly to Type II water molecules is also observed with a relatively low signal vs. noise ratio [18,28]. Below 4000 cm^{-1}, the spectra were noisy due to sample surface.

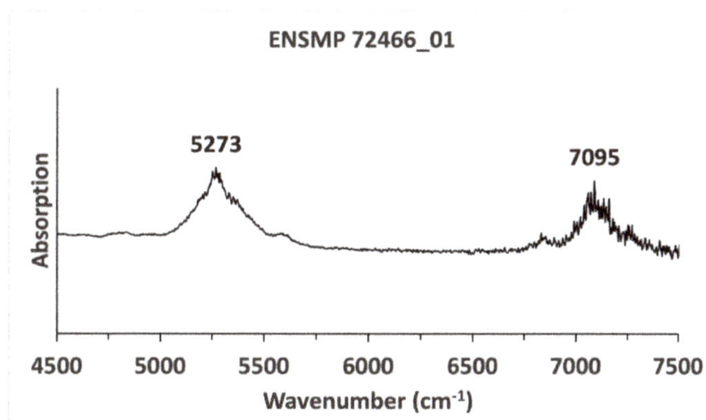

Figure 9. FTIR spectra of a single-crystal (ENSMP 72466_01) Egyptian emerald parallel to c-axis. The intense presence of peaks at 5273 cm^{-1} and 7095 cm^{-1} (about 1410 nm) are related to the presence of Type II water molecules.

3.5. UV-Vis-NIR Spectroscopy

The Egyptian emeralds studied here revealed absorption bands at around 430, 635 and 684 nm, which are mainly linked with Cr^{3+} [3,18,28,29] (Figure 10a,b) and a large band at around 840 nm is related to Fe^{2+} [3,28,29]. All studied samples presented similar absorption bands. In Vis-NIR spectra with the mobile spectrometer (Figure 10b), water-related bands at 960 nm were also observed [30]. An absorption band linked with Fe^{3+} in emeralds might be present at around 375–380 nm [3,18,28], but this cannot be confirmed in the acquired UV-Vis spectra (Figure 10a) due to samples' total absorption in the UV region.

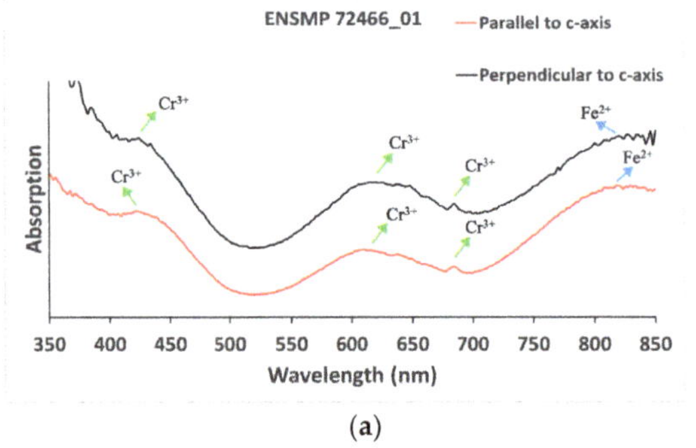
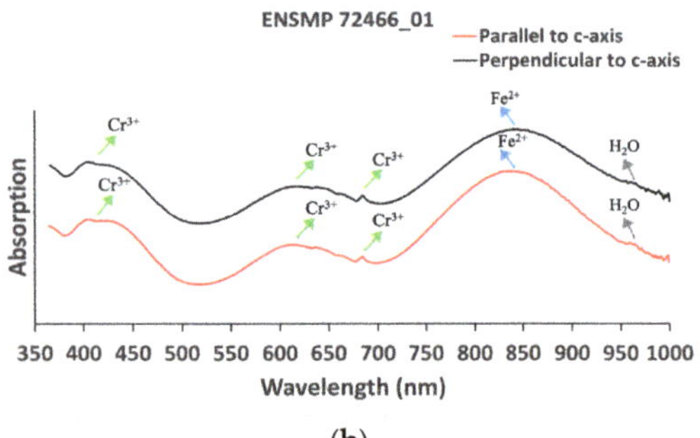

Figure 10. UV-Vis-NIR spectra of an Egyptian emerald single-crystal (ENSMP 72466_01). (**a**) UV-Vis spectra parallel (red line) and perpendicular (black line) to c-axis. The main chromophore elements are chromium and iron. (**b**) Vis-NIR spectra with the mobile instrument parallel (red line) and perpendicular (black line) to c-axis. Additional bands related to water molecules can also be observed at around 960 nm.

3.6. EDXRF

In our study, this method was applied in all samples except ENSMP 1352_5 due to its large size (Tables 3 and 4). All analyses are in ppmw, simply stated as ppm in this work. All samples revealed high concentrations of chromium and iron (up to 2315 and 19,795 ppm, respectively) but relatively low amounts of vanadium (below 665 ppm). Emerald crystals with bluish color (ENSMP 72466_03 and ENSM P72466-04) present higher concentrations of iron (Tables 3 and 4). Detectable amounts of Rb (11 to 84 ppm) were observed in all samples. In most cases, Cs was below the detection limit of the instrument (i.e., ca. 200 ppm) with one sample presenting a concentration slightly above 200 ppm but the chemistry is under the detection limits of the EDXRF instrument.

Table 3. Results of EDXRF analysis of single-crystal Egyptian emerald samples in ppm.

Samples	ENSMP 72466_01 (No. of Analysis: 3)		ENSMP 72466_02 (No. of Analysis: 4)		ENSMP 72466_03 (No. of Analysis: 4)		ENSMP 72466_04 (No. of Analysis: 3)	
Oxides	Min–Max Range	Average	Min–Max Range	Average	Min–Max Range	Average	Min–Max Range	Average
V_2O_3	303–665	551.7	303–563	436	284–414	320.7	172–344	257.3
Cr_2O_3	811–1202	1007.8	941–1735	1293.4	512–1000	715.8	1244–2228	1474.8
FeO	5018–13,348	7676.7	8909–14,057	11,700.3	8168–12,871	9797.3	10,577–19,791	16,233.7
Rb_2O	11–19	15.3	22–31	25	37–84	61	23–42	32.5

Table 4. Results of EDXRF analysis of Egyptian emerald crystals samples in host rock in ppm.

Samples	ENSMP 47849a26 (No. of Analysis: 3)		ENSMP 72486 (No. of Analysis: 3)		ENSMP 47849a27 (No. of Analysis: 1)
Oxides	Min–Max Range	Average	Min–Max Range	Average	Results
V_2O_3	429–506	460	408–475	438	397.3
Cr_2O_3	1568–2315	1935	1788–2120	1996	1873
FeO	2782–10,548	7296.7	11,277–12,052	11,710	7139
Rb_2O	25–53	36.6	48–63	54.3	10

3.7. Associated Minerals

Parts of the host rock were analyzed with micro-Raman technique in order to determine the minerals that take part in the paragenesis of the Egyptian emerald deposits. Micro-Raman spectroscopy revealed the presence of quartz, albite, phlogopite, calcite and dolomite. The peaks in quartz spectrum (Figure 11a) observed at 207 and 466 cm^{-1} are due to symmetric stretching of Si-O while weaker bands located at 265, 355 and 809 cm^{-1} correspond to lattice modes and bending vibrations of Si-O, as well as a weak band at 1082 cm^{-1} that is related to the asymmetric stretching of SiO_4 [31–33].

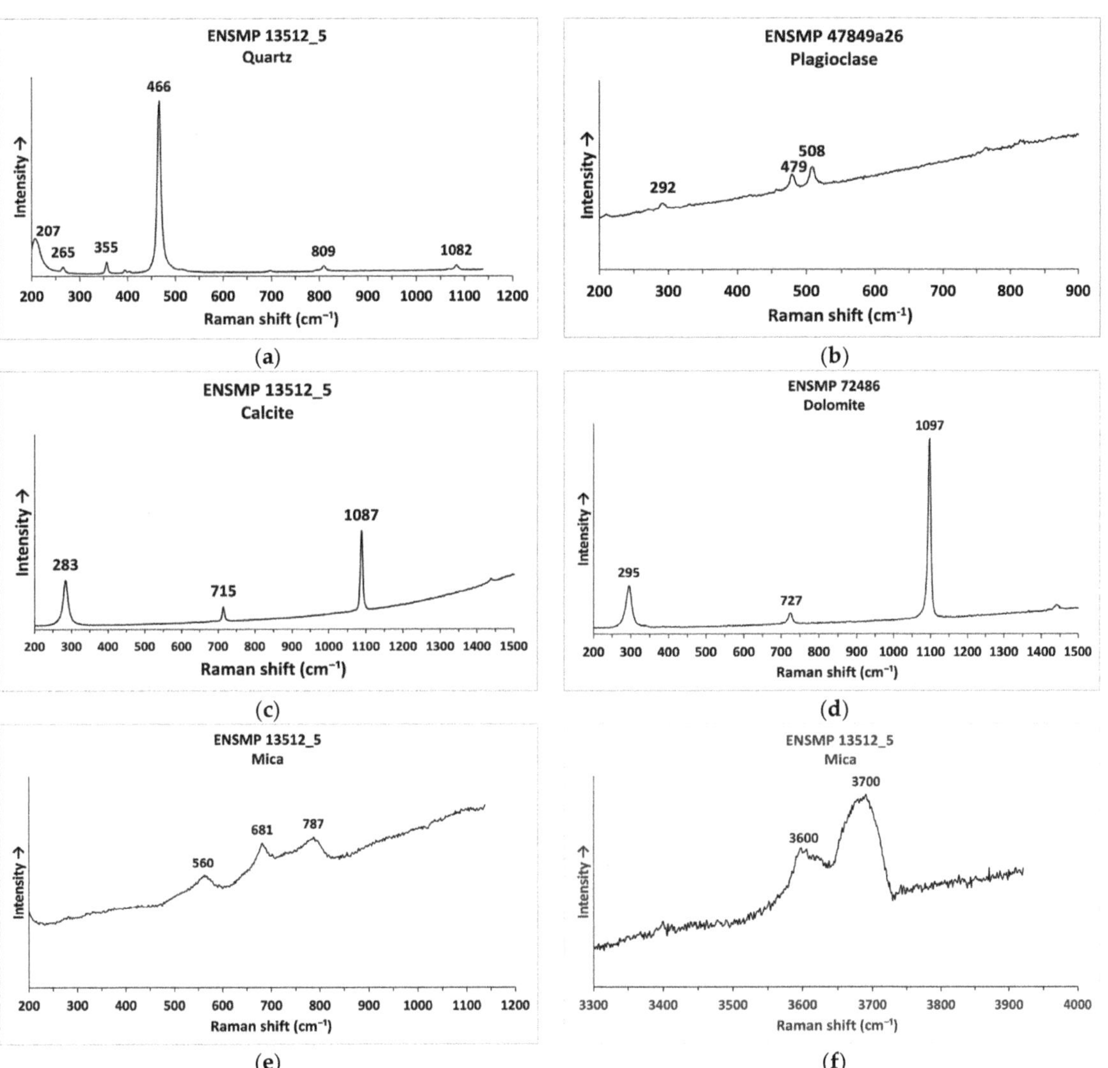

Figure 11. Micro-Raman spectra of minerals from the host rock of the Egyptian emeralds. (**a**) Raman spectra of quartz crystal (ENSMP 13512_5), (**b**) Raman spectra of plagioclase (ENSMP 47849a26), (**c**) Raman spectra of calcite (ENSMP 13512_5), (**d**) Raman spectra of dolomite (ENSMP 72486), (**e**,**f**) Raman spectra of mica (phlogopite) (ENSMP 13512_5).

Feldspar Raman bands (Figure 11b) correspond to plagioclase and possibly to albite with the Raman peaks at 292, 479 and 508 cm^{-1} related to SiO_4 ring vibrations [34,35]. Calcite and dolomite show Raman spectra (Figure 11 c,d) with the following peaks 295, 727 and 1097 cm^{-1} for dolomite and at 283, 715 and 1087 cm^{-1} for calcite due to different vibrations of carbonates [36,37]. The Raman spectrum on a mica inclusion (Figure 11e) shows the vibrations of phlogopite [34] with bands at 560, 681 and 787 cm^{-1} that correspond to O3-T-O1 bend vibrations [35]. The Raman spectrum above 3500 cm^{-1} (Figure 11f) shows large bands at ca. 3600 and 3700 cm^{-1}, which are related to OH-stretching vibrations, possibly in phlogopite [36].

4. Discussion and Conclusions

Spectroscopic, chemical and gemological data on emeralds from Egypt are rare. This study presents such data on some historic rough emeralds from a museum collection, and it is part of a larger ongoing project on the study of historic gemmy material from the Mineralogy Museum of Mines Paris—PSL, in partnership with the LFG. Chemical analysis and spectra of all the studied emerald samples demonstrate similar features. Raman spectroscopy and FTIR spectra revealed that Type II water molecule vibrations, linked with alkali ions, are more intense than Type I water molecule vibrations. High alkali content is probably related with hydrothermal fluids derived from magmatic intrusions. All samples were colored by chromium and iron.

Using Raman spectra and under gemological microscope, the following were found as inclusions and associated minerals: growth tubes with rectangular-shaped multiphase fluid inclusions, quartz, feldspar (plagioclase, possibly albite), mica (likely phlogopite), calcite and dolomite. Quartz and albite are generally associated with granite or pegmatite intrusions and quartz veins, while phlogopite, calcite and dolomite are closely related with metamorphic zones between schists and pegmatite intrusions. All these inclusions and associated minerals are common in Egyptian emeralds [9] and are different from the inclusions in emeralds from other locations around the world, such as some of the very productive mines of gem quality samples in Zambia and Brazil [18,29].

Emerald is a mineral formed in moderate-to-high temperatures and is hosted by diverse rock types of different ages [2]. In our case, the main part of the emerald host rocks are pegmatite dikes with quartz, albite and phlogopite. Moderate Rb and low Cs concentrations were also observed and are probably associated with the S-type granite intrusions. Calcite and dolomite are secondary minerals in the metamorphic zone between the pegmatite quartz veins and the pre-existing schist rocks. Most of the Egyptian emerald deposits in Zabara area are found in the boundaries of these rocks next to the metamorphic zone [12,14].

Based on previous publications, emerald deposits can be classified in two main categories [2,14,37]. The first category is the Type I deposits, which are related with pegmatite or granite intrusions in rock sequences comprising the Archean or Precambrian basement. The second category are the Type II deposits in which the genesis of emeralds is relevant with tectonic events (mainly thrust faults and shear zones). The characterization of Egyptian emerald deposits seems to be doubtful according to the existing bibliography.

Recent studies classify these deposits as Type IID [2], which derive from metamorphic remobilization of Type I deposits. This theory states that initial Type I emerald deposits, which were related to hidden pegmatite or granite intrusions through mafic-ultramafic rocks (mostly amphibolites and schists) underwent later metamorphic and tectonic events that led to at least three generations of beryl crystals in Egypt [11]. Each emerald generation reflects an intense tectonic and metamorphic event [11]. These events are probably related with the presence of Na-plagioclase (low albite) by a metasomatic albitization event [14]. Additionally, the boudinage structure of the emeralds could be a result of extensional tectonic events. Some features, such as the quartz veins, may also be associated with later metamorphic events and the presence of S-type granites from the partial melting of preexisting sedimentary rocks [14]. The above results are consistent with the already existing recent studies; thus, Egyptian emeralds could be characterized as Type IID deposits and interpreted as a hybrid appearance of Type IA deposits that evolved into Type II deposits through tectonic and metamorphic events.

Emeralds have been used in jewelry pieces since the antiquity; however, their origin is still often debatable. It was suggested that emeralds from Hellenistic and Roman periods might be from Egypt, Pakistan and Afghanistan (also known as Bactrian emeralds), as well as Russia (also known as Scythia) and Austria [38,39]. It is noteworthy that there is no written document or clear archaeological evidence of a link between Austrian emeralds and the Romans or Celts [40]. The present work will hopefully further contribute to the characterization of emeralds in the jewelry of archaeological significance.

Author Contributions: M.N. formulated the paper, prepared the experiments, performed data reduction, participated in the data interpretation, drew the figures and wrote the manuscript. S.K., E.G. and U.H. selected the samples, designed the experiments, performed some of the experiments, participated in the data interpretation and edited the manuscript. F.M., A.H., L.P., V.M., N.K., D.N. and A.D. participated in the data interpretation and edited the manuscript. All authors have read and agreed to the published version of the manuscript.

Funding: This research received no external funding.

Data Availability Statement: Not applicable.

Acknowledgments: The authors wish to thank all the colleagues at LFG and Museum of the Ecole des Mines, Mines Paris, PSL University, for their help during this study.

Conflicts of Interest: The authors declare no conflict of interest.

References

1. Groat, L.A.; Giuliani, G.; Marshall, D.D.; Turner, D. Emerald Deposits and Occurences; A Review. *Ore Geol. Rev.* **2008**, *34*, 87–112. [CrossRef]
2. Giuliani, G.; Groat, L.A.; Marshall, D.; Fallick, A.E.; Branquet, Y. Emerald Deposits: A Review and Enhanced Classification. *Minerals* **2019**, *9*, 105. [CrossRef]
3. Karampelas, S.; Al-Shaybani, B.; Mohamed, F.; Sangsawong, S.; Al-Alawi, A. Emeralds from the Most Important Occurrences: Chemical and Spectroscopic Data. *Minerals* **2019**, *9*, 561. [CrossRef]
4. Sinkankas, J. *Emerald and Other Beryls*; Geoscience Press: Prescott, AZ, USA, 1981.
5. Wood, D.L.; Nassau, K. The Characterization of Beryl and Emerald by Visible and Infrared Absorbtion Spectroscopy. *Am. Mineral.* **1968**, *53*, 777–800.
6. Huong, L.T.; Häger, W.; Hofmeister, W. Confocal Micro-Raman Spectroscopy: A Powerful Tool to Identify Natural and Synthetic Emeralds. *Gems Gemol.* **2010**, *64*, 36–41. [CrossRef]
7. Wang, H.; Shu, T.; Chen, J.; Guo, Y. Characteristics of Channel-Water in Blue-Green Beryl and Its Influence on Colour. *Crystals* **2022**, *12*, 435. [CrossRef]
8. Ogden, J. Cleopatra's Emerald Mines: The Marketing of a Myth. *J. Gemmol.* **2022**, *38*, 156–170. [CrossRef]
9. Jennings, R.H.; Kammerling, R.C.; Kovaltchouk, A.; Calederon, G.P.; El Baz, M.K.; Koivula, J.I. Emeralds and Green Beryls of Upper Egypt. *Gems Gemol.* **1993**, *29*, 100–115. [CrossRef]
10. Hume, W.F. *Geology of Egypt*; Goverment Press: Cairo, Egypt, 1934; Volume 2, p. 300.
11. Giuliani, G. *Émeraudes, Tout Un Monde*; Les Editions du Piat, Galvenas-F-43200: Saint-Julien-de-Pinet, France, 2022.
12. Gawad, A.; Ene, A.; Skublov, S.G.; Gavrilchik, A.K.; Ali, M.A.; Ghoneim, M.M.; Nastavkin, A.V. Trace Element Geochemistry and Genesis of Beryl from Wadi Nugrus, South Eastern Desert, Egypt. *Minerals* **2022**, *12*, 206. [CrossRef]
13. Soliman, M.M. Ancient Emerald Mines and Beryllium Mineralization Associated with Precambrian Stanniferous Granites in the Nugrus-Zabara Area, Southeastern Desert, Egypt. *Arab Gulf Res.* **1986**, *4*, 529–548.
14. Grundmann, G.; Morteani, G. Multi-Stage Emerald Formation during Pan-African Regional Metamorphism: The Zabara, Sikait, Umm Kabo Deposits, South Eastern Desert of Egypt. *J. Afr. Earth Sci.* **2008**, *50*, 168–187. [CrossRef]
15. Hassan, M.A.; El-Shatouri, H.M. Beryl Occurrences in Egypt. *Min. Geol.* **1976**, *26*, 253–262. [CrossRef]
16. Gaillou, E.; Maouche, F.; Barthe, A.; Nectoux, D.; Lechartier, M. Émeraudes Historiques de La Collection de l'École Nationale Supérieure Des Mines de Paris. *Le Règne Minéral* **2022**, *in press*.
17. Cailliaud, F. *Voyage à l'Oasis de Thèbes et Dans Les Déserts Situés à l'Orient et à l'Occident de La Thébaïde (Histoire)*; Hachette Livre-BNF: Paris, France, 2022; ISBN 978-2-01-351722-5. (In French)
18. Saeseaw, S.; Pardieu, V.; Sangsawong, S. Three-Phase Inclusions in Emerald and Their Impact on Origin Determination. *Gems Gemol.* **2014**, *50*, 114–132. [CrossRef]
19. Jehlička, J.; Culka, A.; Bersani, D.; Vandenabeele, P. Comparison of Seven Portable Raman Spectrometers: Beryl as a Case Study. *J. Raman Spectrosc.* **2017**, *48*, 1289–1299. [CrossRef]
20. Bersani, D.; Azzi, G.; Lambruschi, E.; Barone, G.; Mazzoleni, P.; Raneri, S.; Longobardo, U.; Lottici, P.P. Characterization of Emeralds by Micro-Raman Spectroscopy. *J. Raman Spectrosc.* **2014**, *45*, 1293–1300. [CrossRef]
21. Moroz, I.; Roth, M.; Boudeulle, M.; Panczer, G. Raman Microspectroscopy and Fluorescence of Emeralds from Various Deposits. *J. Raman Spectrosc.* **2000**, *31*, 485–490. [CrossRef]

22. Huong, L.T. Microscopic, Chemical and Spectroscopic Investigations on Emeralds of Various Origins. Ph.D. Thesis, Fachbereich Chemie, Pharmazie und Geowissenschaften der Johannes Gutenberg-Universität Mainz, Mainz, Germany, 2008.
23. Hagemann, H.; Lucken, A.; Bill, H.; Gysler-Sanz, J.; Stalder, H.A. Polarized Raman Spectra of Beryl and Bazzite. *Pgysiscs Chem. Miner.* **1990**, *17*, 395–401. [CrossRef]
24. Moroz, I.; Panczer, G.; Roth, M. Laser-Induced Luminescence of Emeralds from Different Sources. *J. Gemmol.* **1998**, *26*, 357–363. [CrossRef]
25. Huong, L.T.T.; Hofmeister, W.; Häger, T.; Karampelas, S.; Kien, N.D.T. A Preliminary Study on the Seperation of Natural and Synthetic Emeralds Using Vibrational Spectroscopy. *Gems Gemol.* **2014**, *50*, 287–292. [CrossRef]
26. Wood, D.L. Absorption, Fluorescence, and Zeeman Effect in Emerald. *J. Chem. Phys.* **1965**, *42*, 3404–3410. [CrossRef]
27. Thompson, D.B.; Kidd, J.D.; Åström, M.; Scarani, A.; Smith, C.P. A Comparison of R-Line Photoluminescence of Emeralds from Different Origins. *J. Gemmol.* **2014**, *34*, 334–343. [CrossRef]
28. Mathieu, V.M. IR and UV-Vis Spectroscopy of Gem Emeralds, a Tool to Differentiate Natural, Synthetic and/or Treated Stones? Master's Thesis, Universiteit Gent, Ghent, Belgium, 2009.
29. Saeseaw, S.; Renfro, N.D.; Palke, A.C.; Sun, Z.; McClure, S.F. Geographic Origin Determination of Emerald. *Gems Gemol.* **2019**, *55*, 614–646. [CrossRef]
30. Araújo Neto, J.F.; Brito Barreto, S.; Carrino, T.A.; Müller, A.; Lira Santos, L.C.M. Mineralogical and Gemological Characterization of Emerald Crystals from Paraná Deposit, NE Brazil: A Study of Mineral Chemistry, Absorption and Reflectance Spectroscopy and Thermal Analysis. *Braz. J. Geol.* **2019**, *49*, 684–698. [CrossRef]
31. Buzgar, N.; Apopei, A.I.; Diaconu, V.; Buzatu, A. The Composition and Source of the Raw Material of Two Stone Axes of Late Bronze Age from Neamț County (Romania)—A Raman Study. *An. Stiintifice Ale Univ. "Al. I. Cuza" Din Iasi AUI Ser. Geol.* **2013**, *59*, 5–22.
32. Bersani, D.; Aliatis, I.; Tribaudino, M.; Mantovani, L.; Benisek, A.; Carpenter, M.A.; Gatta, G.D.; Lottici, P.P. Plagioclase Composition by Raman Spectroscopy. *J. Raman Spectrosc.* **2018**, *49*, 684–698. [CrossRef]
33. Sharma, S.K.; Misra, A.K.; Ismail, S.; Singh, U.N. Remote Raman Spectroscopy of Various Mixed and Composite Mineral Phases at 7.2 m Distance. In Proceedings of the 37th Lunar and Planetary Science Conference, League City, TX, USA, 13–17 March 2006.
34. Wang, A.; Freeman, J.; Jolliff, B.L. Understanding the Raman Spectral Features of Phyllosilicates. *J. Raman Spectrosc.* **2015**, *46*, 829–845. [CrossRef]
35. McKeown, D.A.; Bell, M.I.; Etz, E.S. Raman Spectra and Vibrational Analysis of the Trioctahedral Mica Phlogopite. *Am. Minerol.* **1999**, *84*, 970–976. [CrossRef]
36. Tlili, A.; Smith, D.C.; Beny, J.M.; Boyer, H. A Raman Microprobe Study of Natural Micas. *Mineral. Mag.* **1989**, *53*, 165–179. [CrossRef]
37. Schwarz, D.; Giuliani, G. Emerald Deposits-A Review. *Aust. Gemmol.* **2001**, *21*, 17–23.
38. Giuliani, G.; Chaussidon, M.; Schubnel, H.-J.; Piat, D.H.; Rollion-Bard, C.; Rrance-Lanord, C.; Giard, D.; de Narvaez, D.; Rondeau, B. Oxygen Isotopes and Emerald Trade Routes Since Antiquity. *Science* **2000**, *287*, 631–633. [CrossRef] [PubMed]
39. Calligaro, T.; Dran, J.-C.; Poirot, J.-P.; Querre, G.; Salomon, J.; Zwaan, J.C. PIXE/PIGE Characterization of Emeralds Using an External Micro-Beam. *Nucl. Instrum. Methods Phys. Res. Sect. B Beam Interact. Mater. At.* **2000**, *161–163*, 769–774. [CrossRef]
40. Schmetzer, K. History of Emerald Mining in the Habachtal Deposit of Austria, Part II. *Gems Gemol.* **2022**, *58*, 18–46. [CrossRef]

Article

Update on Emeralds from Kagem Mine, Kafubu Area, Zambia

Ran Gao [1,2], **Quanli Chen** [1,2,3,*], **Yan Li** [1,2] **and Huizhen Huang** [1,2]

[1] Gemological Institute, China University of Geosciences, Wuhan 430074, China; gaoran@cug.edu.cn (R.G.); yanli@cug.edu.cn (Y.L.); huizhenhuang_0920@163.com (H.H.)
[2] Hubei Province Gem & Jewelry Engineering Technology Research Center, Wuhan 430074, China
[3] School of Jewelry, West Yunnan University of Applied Sciences, Baoshan 679100, China
* Correspondence: chenquanli_0302@163.com; Tel.: +86-183-2722-5368

Abstract: Kagem emerald mine in Zambia is deemed to the largest open-pit emerald mine with extremely high economic value and market share in the world. To meet the market demand for tracing the origin of emeralds, 30 emeralds from the region were tested, and some discoveries were made compared to previous studies. This study provides a full set of data through standard gemological properties, inclusions, color characteristics, advanced spectroscopic and chemical analyses, including Raman, micro micro-UV-Vis-NIR, FTIR, and LA-ICP-MS. The most common inclusions in Kagem emeralds are two-phase inclusions, which exhibit elongated, hexagonal, oval, irregular shapes or appear as negative crystals with incomplete hexagonal prism. These inclusions consist of H_2O or $H_2O + CO_2$ (liquid) and $CO_2 + N_2$ or $CO_2 + N_2 + CH_4$ (gas). Mineral inclusions typically include actinolite, graphite, magnetite, and dolomite. Black graphite encased in actinolite in Kagem emeralds is first reported. The FTIR spectrum of Kagem emeralds reveals that the absorption of type II H_2O is stronger than that of type I H_2O, indicating the presence of abundant alkali metals, which was confirmed through chemical analysis. Kagem emeralds contain high levels of Na (avg. 16,440 ppm), moderate-to-high Cs (avg. 567 ppm), as well as low-to-moderate levels of K (avg. 185 ppm) and Rb (avg. 14 ppm) concentrations.

Keywords: emerald; Kagem Mine; gemological characteristics; origin traceability

Citation: Gao, R.; Chen, Q.; Li, Y.; Huang, H. Update on Emeralds from Kagem Mine, Kafubu Area, Zambia. *Minerals* **2023**, *13*, 1260. https://doi.org/10.3390/min13101260

Academic Editors: Stefanos Karampelas and Emmanuel Fritsch

Received: 27 July 2023
Revised: 25 September 2023
Accepted: 26 September 2023
Published: 27 September 2023

1. Introduction

Emerald is the green variety of beryl colored by chromium and vanadium, and its ideal chemical formula is $Be_3Al_2SiO_{18}$. The traditional classification system for emerald deposits has been expanded upon by Giuliani et al. (2019) [1]. Emerald occurrences and deposits were reclassified into two main types: tectonic magmatic-related (Type I) and tectonic metamorphic-related (Type II), and further subdivided into seven sub-types based on the host rock types. The Kafubu deposit in Zambia was classified as Type IA, associated with mafic-ultramafic rocks.

Zambia is one of the most significant emerald sources worldwide, second only to Colombia. Kagem emerald mine is located in the Ndola Rural Emerald Restricted Area and lies south of Kitwe and west of Ndola, in Zambia's Copperbelt Province (Figure 1). Kagem Mine is considered the world's largest single open-pit emerald mine, accounting for approximately 25%–30% of global emerald production, with a potential mine life of 22 to 2044 years [2]. Besides the Kagem Mine, there are other larger mines in the Kafubu area, including Kamakanga, Grizzly, and Chantete [3,4].

Bank (1974) [5] proposed that the necessary chromic element in emeralds, chromium, comes from the magnetite in talc-magnetite schist. Koivula (1982) [6] was the first to report on the presence of tourmaline as inclusion in Zambian emeralds. Sliwa & Nguluwe (1984) [7] described the geological setting of Zambian emeralds. Seifert et al. (2004a) [8] first provided quantitative geochemical, petrological, and mineralogical data on the major rock types of the Kafubu emerald deposits. Seifert et al. (2004b) [9] conducted a comprehensive study of the environmental impact of the emerald mines in the Kafubu area. Zwaan et al. (2005) [3] first reported the Musakashi area, a new emerald deposit initially worked by local miners in 2002 [10]. In 2014, GIA Lab conducted a field exploration of the history, region geology, mining methodology, processing, operation, and auction of the Kagem Mine and published a detailed report [11]. However, detailed studies on the inclusion, spectroscopy, and chemical composition of Kagem emerald still need to be completed [12].

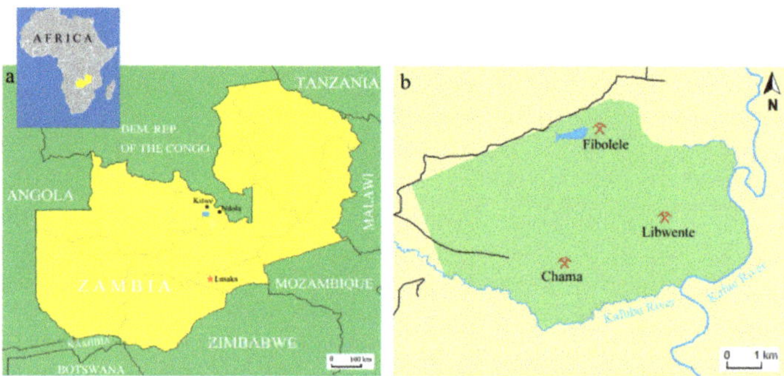

Figure 1. (**a**) Zambia is a landlocked country located in the southern part of Africa, the capital is Lusaka. Kagem Mine (blue rectangle) is located in north-central Zambia, near the Kafubu River in the Ndola Rural Emerald Restricted Area (NRERA); (**b**) There are three operating pits in the Kagem Mine: Chama, Libwente, and Fibolele. Picture: Ran Gao.

We received a batch of emerald samples from the Kagem Mine, Zambia in September 2022. This article provides a brief overview of the history and region geology of the Kagem Mine and a detailed research of gemological properties, inclusions, UV-Vis-NIR, FTIR, and chemical composition analysis. These analyses can effectively differentiate our materials from those sourced from other significant emerald localities.

2. Geological Setting

The Kagem emerald concession covers an area of approximately 46 square kilometers within a 200-square-kilometer productive zone and comprises the current operating Chama open pit mine and the bulk sampling pits at Libwente and Fibolelem [3].

The Kafubu area is located geologically at the center of the transcontinental Pan-African belts in central-southern Africa. The Crustal evolution is thought to be related to three successive orogenic events: the Ubendian, Irumide and Lufilian (Pan-African) [13–15]. The Kafubu emerald deposits occur within metamorphic rocks of the Muva Supergroup that date back to around 1700 Ma [million years ago] [16,17]. Muva Supergroup overlays the basement granite gneisses, consisting of footwall mica schist, talc-magnetite schist, amphibolite, and quartz-mica schist from bottom to top. The entire crustal domain subsequently underwent folding, thrusting, and metamorphism during the Pan-African orogeny, peaking at 530 Ma [18].

Emerald mineralization of the Kagem Mine is hosted by the talc-magnetite schist, which contains talc-chlorite-actinolite-magnetite schists (Figure 2) [7,14]. These schists were identified as metamorphosed komatiites—Mg-rich ultramafic rocks [9]. During the late stages of the Pan-African orogeny, the talc-magnetite schist was intruded by beryllium-rich pegmatite dykes (typically 2–10 m thick) [19,20]. These pegmatites appear as feldspar-quartz-muscovite bodies or as minor quartz-tourmaline veins and are believed to be related to neighbouring granitic rocks [7]. The steeply inclined pegmatite dykes and quartz tourmaline veins typically trend north to south or northwest to southeast.

The emerald mineralization in the Kagem Mine results from the interaction between metabasites and pegmatites and their accompanying hydrothermal fluids [9,19]. Emeralds are found in the phlogopite reaction zone (typically 0.5–3 m wide) between the talc-magnetite schist and the pegmatites and in the quartz-tourmaline veins, which produces gem-quantity emeralds, but only a tiny percentage [3]. In the reaction zone, talc-magnetite schist provided the important chromophore of chromium, while the pegmatites provided the beryllium. Emerald was formed under the proper pressure, temperature conditions (400–600 MPa and 590–630 °C) and chemical environment around 500 Ma [9,21]. The area subsequently underwent intense shearing and folding during the Lufilian orogeny, which may account for the fracturing and opacity of many emerald crystals [7].

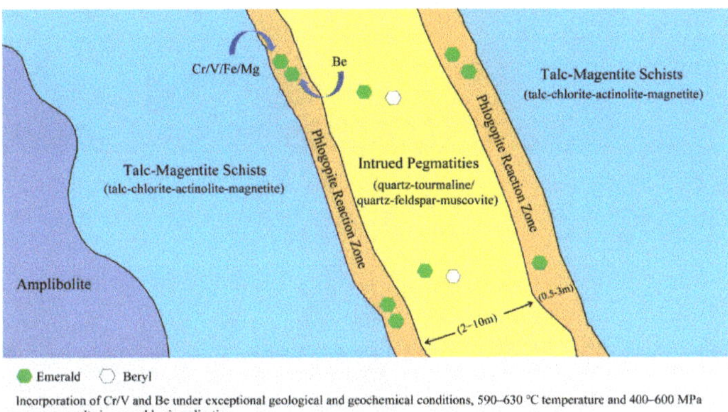

Figure 2. Schematic representation of the metasomatic reaction zone between the talc-magnetite schists and the pegmatite formed common beryl and gem-quality emeralds. Modified from [22].

3. History

There are two major emerald areas in Zambia, including Musakashi and Kafubu areas. The Musakashi area was first reported in 2005. The emeralds from this area were characterized by their intense bluish-green color and multiphase inclusions, which were similar to the emeralds found in Muzo, Colombia [3,10,23]. Beryl was first discovered in the Kafubu area in 1928 by two geologists named Dicks and Baker. Some small exploration works were carried out during the 1940s and 50s' [7]. Zambia's Geological Survey Department mapped the Miku area and verified the deposit in 1971 [13,14]. After new occurrences were discovered in the 1970s, the Kafubu area became a significant producer of fine-quality emeralds. Because of the dramatic expansion of production and extensive illegal mining, the Government relocated the local population and established a restricted zone called the Ndola Rural Emerald Restricted Area [9]. In 1980, Kagem Mining Ltd. (55% owned by the Zambian Government) was authorized to explore and mine the Kafubu area in the same year [9].

In 2004, the British public-listed Gemfields Resources PLC began systematic exploration near the Pirala mine, south of the Ndola River, and discovered significant emerald deposits. Gemfields was awarded a management contract there in 2007. In the following year, Gemfields acquired 75% ownership of the mine, the remainder being held by the Government [11].

Kagem is primarily an open-pit mine, which presents the advantage of providing accessibility to every carat of emerald (Figure 3). After the surrounding rock is removed, miners utilize hammers and chisels to recover the emeralds. Any extracted production goes into a red production box (Figure 4). Since 2010, the Kagem Mine has been responsible for approximately 50% of emerald production in Zambia. Despite the high production volume, only a small portion of gem-quality emeralds are available for exportation In 2022, the Kagem Mine produced a total of 37.2 million carats of emeralds and beryl, including 259,500 carats of premium emeralds [24].

Figure 3. The Kagem Mine is the world's largest open-pit mine for emeralds and is located in the Ndola Rural District, Copperbelt Province, Zambia. Photo by Xiangjie Xiao.

Figure 4. (**a**) Miners use hammers and chisels to extract emeralds from the rock in the reaction zone; (**b**) The collected emeralds were in a secure box. Photo by Xiangjie Xiao.

Colored gemstone auctions have become a major source of revenue for the Gemfields Group. The company established its proprietary grading system to assess each gem according to its individual characteristics (size, color, shape and clarity) and implemented a pioneering auction and trading platform. The auctions were divided into two quality ranges: one for higher quality (HQ) emeralds and one for commercial quality (CQ, formerly known as lower-quality before 2016) emeralds [2]. The auctions are held in multiple cities, including Jaipur, Johannesburg, Lusaka, London, Dubai and Singapore [24]. Gemfields had held 43 auctions of emerald and beryl mined at Kagem up to November 2022 and generated $899 million in aggregate revenues [24]. The per-carat price for HQ and CQ emeralds in 43 auctions has been recorded (Figure 5a,b). The specific auction mix and the quality of the lots offered at each auction vary in characteristics such as size, color and clarity due to changes in mined production and market demand. Overall, the price of emeralds showed a clear upward trend, with HQ emeralds increasing from an initial $4.40 per-carat (July 2009) to a peak of $155.90 (May 2022) per-carat and CQ emeralds rising from an initial $0.31 per-carat (January 2010) to a highest $9.37 per-carat (April 2022). The value and demand for emeralds are rising steadily.

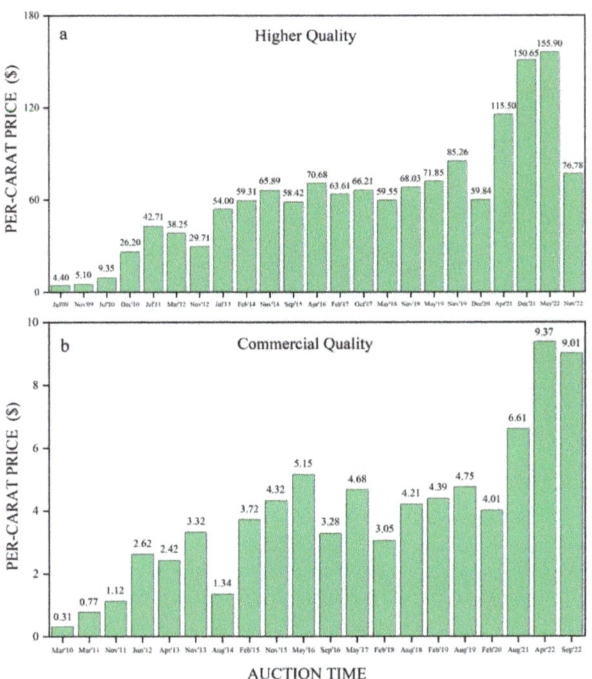

Figure 5. (**a**) The per-carat price (calculated as: received/total weight sold) for higher quality emeralds from the Kagem Mine at auctions has been tracked since July 2009; (**b**) The per-carat price for commercial quality emeralds from the Kagem Mine at auctions has been recorded since March 2010. Auction data from [24].

4. Materials and Methods

A total of 30 transparent faceted emeralds (K1-30) were acquired from a gem trader who had visited the Kagem Mine and collected them (Figure 6). According to the GIA's classification scheme, the emeralds we obtained belong to E-type samples [25]. There was no visible color zoning in the speciments although few contained some dark inclusions, and a few dark inclusions were visible, and five crystals were treated with oil and wax fillings.

Standard gemological properties were obtained on all the samples. Refractive indices and birefringence were obtained with a gem refractometer (FGR-003A, FABLE, Shenzhen, China). UV fluorescence was examined under a UV lamp with long-wave (365 nm) and short-wave (254 nm) light in a darkened box. We also tested their reaction under the Chelsea filter (FCF-25, FABLE, Shenzhen, China). Dichroism was observed and photographed under a polarizing film (FID-1, FABLE, Shenzhen, China). Specific gravity was determined by the hydrostatic method. Internal features were observed and photographed by a Leica M205A microscopic system. In some cases, a polarizing microscope was used as well.

Inclusions were identified using a JASCO NRS-7500 Series Confocal Raman Microscope (JASCO, Tokyo, Japan) with 532 nm and 457 nm lasers at the Gemmological Institute, China University of Geosciences, Wuhan. Solid inclusions were identified in the 2000–100 cm^{-1} range with the 532 nm laser using a grating of 600 grooves/mm. Two-phase inclusions were identified in the 4000–100 cm^{-1} range with the 457 nm laser using a grating of 600 grooves/mm. The laser power was around 10 mW. Three scans with 15 s integration time for each scan were taken for a single spectrum. The Raman shift was calibrated with monocrystalline silicon (at 521 cm^{-1}).

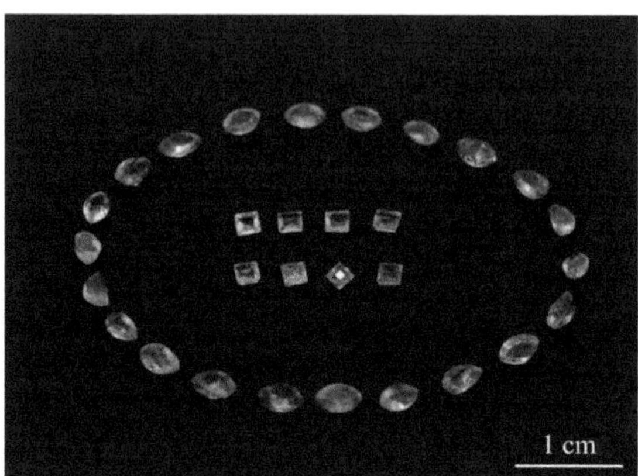

Figure 6. Thirty faceted emeralds from the Kagem Mine, Zambia (the average weight was 0.15 ct). Photo by Ran Gao.

UV-Vis-NIR spectra were recorded by a Jasco MSV–5200 micro-spectrometer (JASCO, Tokyo, Japan) in the range of 300–900 nm, at a scan speed of 200 nm/min at the Gemmological Institute, China University of Geosciences, Wuhan. The optical aperture was set at 100 nm and 0.5 nm data interval. The black cross center of the faceted sample, when viewed through a polarizing mirror, indicates the C-axis. Polarized spectra of each oriented sample were collected to obtain ordinary ray (o-ray) and extraordinary ray (e-ray) absorption spectra.

Fourier-transform infrared spectroscopy (FTIR) was performed using a Bruker VERTEX 80 FTIR spectrometer (BRUKER OPTICS, Billerica, MA, USA), using 32 scans and 4 cm^{-1} spectral resolution at the Gemmological Institute, China University of Geosciences, Wuhan. The scanning ranges were 9000–4000 cm^{-1} in transmission mode.

Trace element contents were analyzed by laser ablation-inductively coupled plasma-mass spectrometry (LA-LCP-MS) using an Agilent 7700 ICP-MS combined with a GeoLas 193 nm laser (Agilent, Singapore) at Wuhan SampleSolution Technology Co., Ltd., Wuhan, China. We set the laser fluence at 9 J/cm^2 with a 6 Hz repetition rate and the laser spot size at 44 μm diameter. Each analysis incorporated a background acquisition time of approximately 20–30 s, followed by 50 s of ablation. Element concentrations were calibrated against multiple reference materials (BCR-2G, BHVO-2G, and BIR-1G) without using an internal standard, and Al was chosen as the normalizing element [26]. Standard reference material NIST 610 glass was also applied to time-drift correction. The standard for LA-ICP-MS measurements is that the calibration values of the monitored reference materials agree within the error range within the recommended values. Quality control deviations: Major elements within 5% uncertainty and trace elements within 10% uncertainty. Two to three spots per sample were analyzed.

5. Results

5.1. Gemological Properties

The gemological properties of emeralds from the Kagem Mine are summarized in Table 1. The emeralds ranged from green to blueish green, and some displayed an attractive saturated bluish-green color. The refractive indices varied from 1.578–1.591 for n_e and 1.589–1.597 for n_o with birefringence between 0.004 and 0.008. These values are higher than most of the significant emerald deposits. Specific gravity values ranged from 2.67 to 2.86. The emeralds from the Kagem Mine were typically inert to long- and short-wave UV radiation. The emeralds showed no reaction under the Chelsea filter. Dichroism was medium to strong yellowish green (o-ray) and bluish green (e-ray).

Table 1. Gemological properties of emeralds from Kagem Mine, Zambia.

Color	Light to medium bluish green; typically, a saturated green with a medium tone
Clarity	Very slightly to heavily included
Refractive indices	$n_o = 1.589$–1.597; $n_e = 1.578$–1.591
Birefringence	0.004–0.008
Specific gravity	2.67–2.86
Pleochroism	Medium to strong yellowish green (o-ray) and bluish green (e-ray)
Fluorescence	Inert to long- and short-wave UV radiation
Chelsea filter	No reaction
Visible Spectrum	Distinct lines at~680 nm; and complete absorption <430 nm
Internal features	Two-phase inclusions with a gas bubble, display elongated, hexagonal, oval, or irregular shapes, were common Densely distributed black mineral crystals, usually with a hexagonal or rectangular outline Colorless transparent crystals Strong iridescent colors in fissures Mineral inclusions: needle-like actinolite; clusters of black magnesite; colorless dolomite; jagged or oval graphite encased in actinolite

5.2. Microscopic Characteristics

Two-phase inclusions with a gas bubble were commonly found in the Kagem emeralds, displaying various shapes: rectangular, elongated, hexagonal, oval, or irregular shapes (Figure 7a). Some two-phase inclusions occurred as isolated negative crystals along a healed fissure plane, with an incomplete hexagonal prism shape (Figure 7b). At room temperature, the gas bubbles typically occupy approximately one-third of the volume of the cavity hosting the two-phase inclusions. Raman analysis confirmed that the liquid phase of the two-phase inclusions primarily consisted of water or a liquid mixture of H_2O and CO_2. The gas components included $CO_2 + N_2$ or $CO_2 + N_2 + CH_4$ (Figure 8). There was a two-phase inclusion with liquid and solid phases, and the latter was identified as dolomite (peaks at 174, 296, and 1095 cm^{-1}) by Raman analysis (arrow 1 in Figure 7a). It is worth noting that a hexagonal multi-phase inclusion, containing a gas bubble and an obvious solid phase in an aqueous solution, is rare in Kagem emeralds (arrow 2 in Figure 7a).

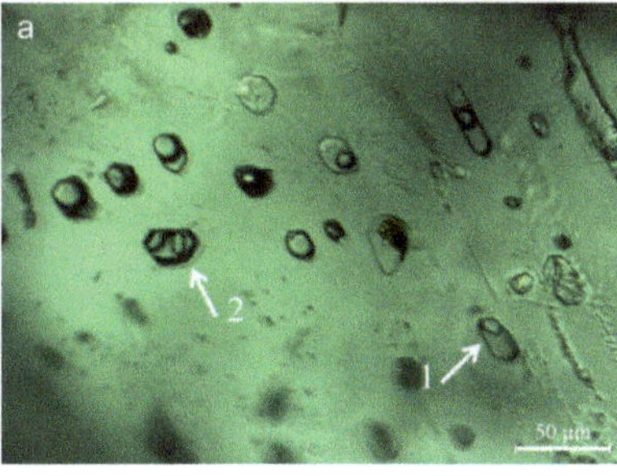
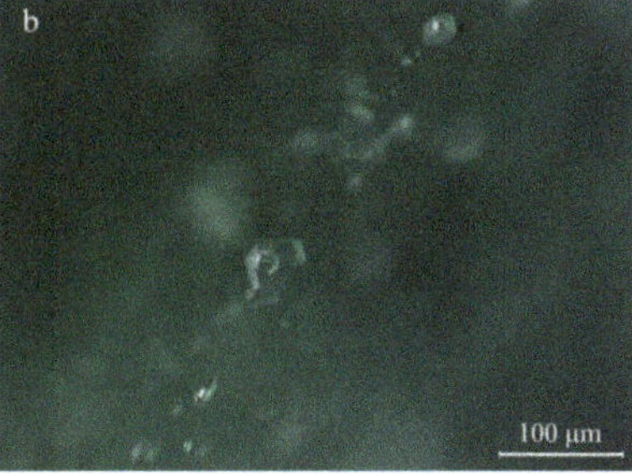

Figure 7. (**a**) Two-phase inclusions with various shapes, a two-phase inclusion with solid phase and a hexagonal three-phase inclusion (indicated by white arrows 1, 2); (**b**) Two-phase inclusions appear as negative crystals with an incomplete hexagonal prism. Photomicrographs by Ran Gao.

Various mineral inclusions were observed in Kagem emeralds under the standard gemological microscope. Colorless needle-like minerals of varying lengths were seen in many samples and identified as actinolite through Raman analysis (Figure 9a). Black jagged minerals, identified by Raman analysis as graphite, were visible in the actinolite needles (Figure 9a). Occasionally, graphite appeared in oval or irregular shapes. Densely distributed black mineral crystals (Figure 9b), which appeared hexagonal or rectangular under high magnification, were identified using Raman spectroscopy as magnetite. Similar magnetite inclusions were also found in emeralds from Davdar, China [27]. A colorless transparent crystal was identified as dolomite via Raman spectroscopy (Figure 9c). Partly healed fissures displayed intense iridescent colors (Figure 9d).

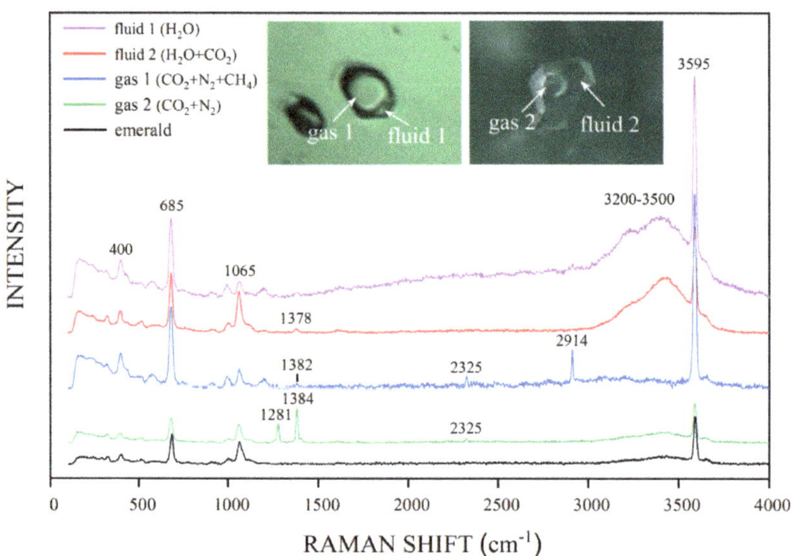

Figure 8. Two-phase inclusions with a gas bubble in emeralds from Kagem Mine. Emerald: peaks at 400, 685 and 1065 cm^{-1}; CO$_2$: peaks at 1281 and 1378–1384 cm^{-1}; N$_2$: peak at 2325 cm^{-1}; CH$_4$: peak at 2914 cm^{-1}; H$_2$O in the inclusion: bands from 3200–3500 cm^{-1}; H$_2$O in structure channel: peak at 3595 cm^{-1} [28,29]. Spectra are offset vertically for clarity. Photomicrographs by Ran Gao.

5.3. UV-Vis-NIR Spectroscopy

Representative UV-Vis-NIR absorption spectra of the emeralds from the Kagem Mine are illustrated in Figure 10. Although the intensities varied among the samples, all exhibited the same bands for the ordinary and extraordinary rays. Specifically, the ordinary ray (o-ray) showed bands at 372, 430, 610, 637, and 840 nm, as well as a doublet at 681 and 684 nm. The bands at 430, 610, and 840 nm were broad and the positions were estimated. The extraordinary ray (e-ray) displayed bands at 372, 427, 625, 640, 662, 684, and 840 nm, the absorbance of a peak at 372 nm was weaker than its o-ray counterpart.

The bands at 430 and 610 nm (o-ray) and 427, 625 and 640 nm (e-ray), as well as the peak at 684 nm (e-ray), indicated the presence of Cr^{3+}, which causes the green color in emerald [30,31]. Additional, weaker peaks at 637, 662 and the doublet at 681 and 684 nm were also associated with the presence of Cr^{3+}. The broad absorption bands from 600 to 750 nm were possibly linked to Fe^{3+}-Fe^{2+} charge transfer [32]. Moreover, the peak at 372 nm indicated the presence of Fe^{3+}, and the band at 840 nm was caused by Fe^{2+} [3,10,30,33], showing weaker intensity than the bands caused by Cr^{3+}. The UV-Vis-NIR spectra feature differs from some previous [10,34,35], but is similar to other scholars' findings [3,36].

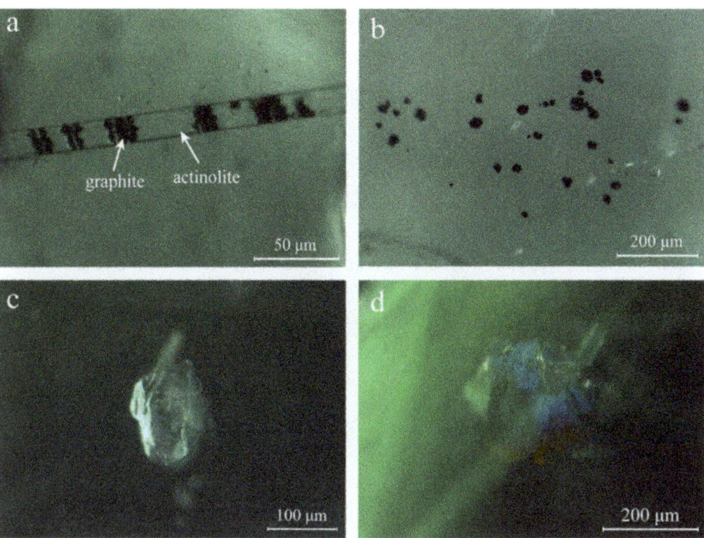

Figure 9. Various mineral inclusions were observed in Kagem emeralds. (**a**) Black jagged graphites were wrapped inside a colorless needle-like growth tube, identified using Raman spectroscopy as actinolite; (**b**) Cluster of black magnetite appeared in the form of hexagons or rectangles under high magnification; (**c**) An isolated colorless crystal was consistent with the Raman spectrum of dolomite; (**d**) Obvious iridescent colors in partly healed fissures. Photomicrographs by Ran Gao.

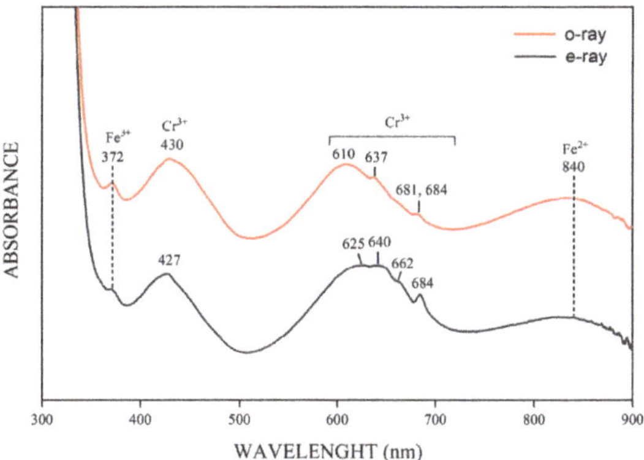

Figure 10. Representative UV-Vis-NIR spectra were collected on sample K-3, which has 2652 ppm Cr, 267 ppm V, and 8759 ppm Fe. The spectra indicate the presence of Cr^{3+}, Fe^{2+}, and Fe^{3+}. The intensities or positions of the Chromophore elements varied in different spectral orientations. The spectra are offset vertically for clarity.

5.4. Infrared Spectrometry

The near-infrared (NIR) spectrum of the emeralds was mainly related to the existing mode of H_2O molecules in the channel (Figure 11). H_2O molecules in the c-channels have been mainly classified as type I or type II water with their twofold axis perpendicular or parallel to the crystal c-axis, respectively [37].

In the NIR range of 8000–4000 cm^{-1}, the absorption spectrum bands of emeralds are predominantly attributed to the combined frequency and double-frequency vibration of structural water [38]. The most obvious band at 5274 cm^{-1} was caused by the combined frequency of bending (v_2) and antisymmetric stretching (v_3) modes of type I or II H_2O molecules [38,39]. The sharp bands at 7097–7075 cm^{-1} doublet, slightly weak bands at 6840 and 7268 cm^{-1}, and two small shoulder bands at 5340 and 5205 cm^{-1} near the 5274 cm^{-1} band were assigned to type II H_2O molecules, which were associated with alkali ions in the channels of the emerald structure. The relatively strong band at 7140 cm^{-1} was related to the overtone frequency absorption of type I water, whereas other spectral bands of type I water, such as 7275 and 6820 cm^{-1} [38,40], were too weak to be discernible.

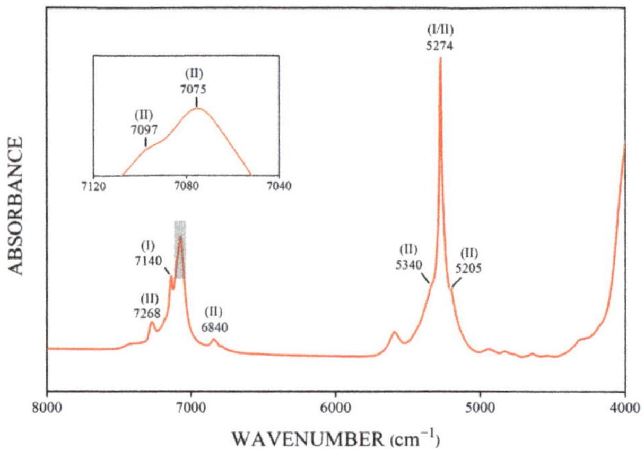

Figure 11. The representative FTIR spectrum of Kagem emeralds shows some significant bands caused by the vibrations of type I or type II H_2O molecules in the channels of the emerald structure. This graph displays that the spectral band intensities of type II H_2O molecules are higher than those of type I H_2O molecules in Kagem emeralds.

5.5. Trace Element Analysis

Twenty faceted emeralds from the Kagem Mine were analyzed by LA-ICP-MS, three spots for each sample and calculated their average values. The results are shown in Table 2.

Table 2. Chemical composition (in ppm) of Kagem emeralds, obtained by LA-ICP-MS.

Samples	Element	Min–Max	Average (SD)	Median	LOD
	Li	105.2–250.1	164.4 (31.46)	165.6	1.05
	Na	13,810–17,410	16,440 (730.0)	16,640	23.89
	Mg	13,350–17,230	15,650 (725.3)	15,730	4.17
	K	121.0–283.9	184.9 (35.83)	179.2	25.99
	Sc	96.78–1534	435.8 (298.6)	340.0	0.49
	Ti	3.967–51.28	16.17 (8.190)	14.75	1.07
Kagem emeralds	V	196.8–528.8	307.2 (77.95)	293.1	0.25
30 samples	Cr	539.6–4844	2034 (1010)	1758	6.26
	Fe	4822–14,160	9713 (2181)	9875	59.31
	Ni	6.845–37.43	19.49 (6.771)	20.22	5.61
	Zn	BDL–15.86	3.766 (3.608)	3.225	2.02
	Ga	2.916–50.49	31.15 (10.02)	34.10	0.40
	Rb	9.381–20.78	14.29 (2.246)	14.29	0.75
	Cs	128.4–1052	567.3 (268.6)	503.6	0.37

Data was rounded to 4 Significant Figures; SD = standard deviations; LOD = limit of detection; BDL = below detection limit.

The Kagem emeralds for this study tended to have abundant alkali metals Li, Na, K, Rb, and Cs. Total alkali element concentrations ranged from 14,227 to 18,634 ppm (avg. 17,405 ppm). The dominant alkali metal was Na, which ranged from 13,812 to 17,408 ppm, averaging 16,437 ppm. The emeralds also contained a significant amount of Mg and Fe, ranging from 13,348 to 17,225 ppm (avg. 15,645 ppm) and from 4822 to 14,159 ppm (avg. 9713 ppm), respectively. Compared to the concentrations of Sc we measured (97–1534 ppm, avg. 436 ppm), analyses of emeralds by Saeseaw et al. (2014) [10] display relatively low Sc (12–75 ppm, avg. 31 ppm) in emeralds from Kafubu area. Cr (540–4844 ppm, avg. 2034 ppm), the most crucial chromophore in emeralds, was higher than V (197–529 ppm, avg. 307 ppm).

6. Discussion

6.1. Inclusion Characteristics of Kagem Emeralds

The rectangular and hexagonal shapes of the two-phase inclusions found in Kagem emerald were consistent with the multi-phase inclusion characteristics observed in schist-hosted emerald deposits, such as those found in Brazil, Russia, and most African deposits. We updated the component of the gas phase in two-phase inclusions. It is not monolithic and can contain not only CO_2 but also occasionally N_2 and CH_4. Multi-phase inclusions with a solid phase were rare. H_2O-NaCl-CO_2 type inclusions were not found in our samples. Black graphite encased in actinolite was first observed in Kagem emeralds. The occurrence of abundant black magnetite and transparent dolomite further demonstrates a direct association between the internal characteristics of gemstones and their geological background. The talc-magnetite schist contained high concentrations of Fe and Mg.

6.2. Fe^{2+} Absorption Band in UV-Nis-NIR Spectroscopy

The Fe content in Table 2 represents the total Fe acquired by LA-ICP-MS, including Fe^{3+} and Fe^{2+}. Kagem emeralds showed the weaker Fe^{2+}-related absorption band in the NIR region compared to the bands caused by Cr^{3+}. However, a lack of correlation was found between the intensity of the Fe^{2+}-related band and the total Fe content [41]. Divalent metal ions (Me^{2+}), such as Mg^{2+} and Fe^{2+}, can substitute for Al^{3+} in the emerald structure. In order to maintain charge balance, the channels must be occupied by alkali ions such as Na^+, K^+, Rb^+, and Cs^+ (mainly Na^+) [1,42], as follows:

$$Al^{3+} = Me^{2+} + Na^+, \tag{1}$$

Moreover, the linear positive correlation between Na^+ content and the sum of divalent cations is approximately 1:1 [41], thereby Fe^{2+} concentration can be calculated by the following equations:

$$Na^+{}_{ppma} = Fe^{2+}{}_{ppma} + Mg^{2+}{}_{ppma}, \tag{2}$$

$$X_{ppma} = X_{ppm} \frac{E_{am}}{X_{am}} \tag{3}$$

To clarify, ppma represents parts per million by atom, E_{am} represents the average atomic mass of emerald chemical formula, and X_{am} represents the atomic mass of elements (Na, Mg and Fe).

The proportion of Fe^{2+} in the total Fe we calculated varied from 5.37% to 59.33%. A part of Fe^{2+} generated intervalence charge transfer with Fe^{3+}, and this portion of Fe^{2+} did not affect the band in the NIR region. Additionally, the content of Cr^{3+} was a crucial factor to consider. We found a positive relationship between the Fe^{2+} to Cr^{3+} content ratio and the intensity ratio of the Fe^{2+}-related band at 840 nm to the Cr^{3+}-related band at 610 nm (o-ray). We speculated that the weaker band of Fe^{2+} in our samples was due to the low Fe^{2+} to Cr^{3+} value ratio. The concentration of Fe^{2+} can be roughly determined based on the content of bivalent cations and alkali metals. However, when discussing the relative strength of the Fe^{2+}-related absorption band, the concentration of Cr^{3+} and V^{3+} should also be considered.

6.3. Channel Water Types of Kagem Emerald

Alkali metal ions, water and other macromolecules that exist within the structural channel of emeralds can maintain the charge balance and structural stability of minerals. The concentration of alkali metal ions affects the type of emerald channel water. In the presence of alkali ions, the axis of symmetry of adjacent H_2O molecules is changed from perpendicular to parallel to the c-axis due to the electrostatic attraction between the charged cation and the oxygen of the H_2O molecule [39].

Kagem emeralds, dominated by Na, contain abundant alkali metal ions (avg. 17,405 ppm). The NIR spectra of Kagem emeralds (Figure 11) revealed that the bands for type II H_2O at 7097–7075 cm^{-1} doublet were more intense than those for type I H_2O at 7140 cm^{-1}. Moreover, the shoulder bands at 5340 and 5205 cm^{-1} were closer to 5274 cm^{-1} and were more obvious with higher proportions of type II H_2O. As mentioned above, Kagem emeralds dominate type II H_2O and the Kagem Mine belongs to one of three mineralization environment types—alkali-rich type (other types include alkali-poor and transition) [38]. Similar emeralds were found in Swat Valley, Pakistan, Goias, Brazil, and most African deposits.

6.4. Origin Traceability Analysis

A log-log plot of Rb versus Cs content can separate the Kagem mine from emeralds of all other significant deposits (Figure 12). Kagem emeralds exhibit low-to-moderate Rb values and moderate-to-high Cs, with a slight overlap with emeralds from Ural, Russia. Emeralds from Malipo, China, and Chitral, Pakistan display high Cs content, while Zimbabwean and Madagascan samples exhibit high Rb value. The emeralds from Colombia show very low Rb and Cs values. The plotted points of the two China deposits can be easily separated by their Cs content.

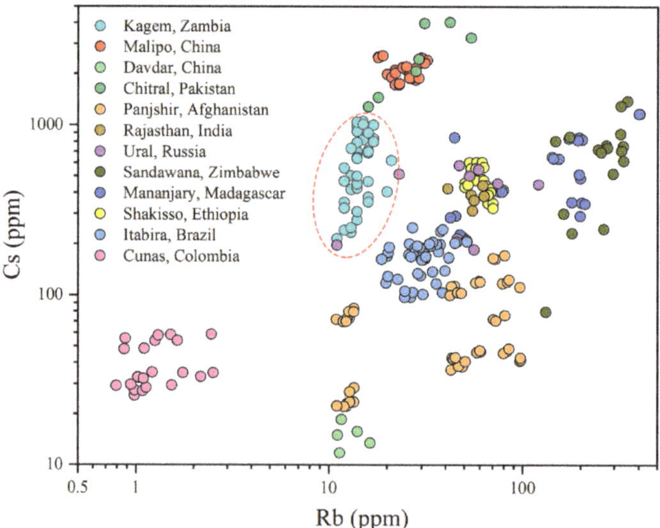

Figure 12. A log-log plot displays the Rb versus Cs contents in emeralds from twelve different localities. Kagem emeralds (within the red dashed) show moderate-to-high Cs values and low-to-moderate Rb. Other sources are from [42–51].

A better separation between the Kagem Mine and other significant emerald localities can be made by plotting Na versus K (Figure 13). Kagem emeralds display high Na values combined with low-to-moderate K, despite a slight overlap with emeralds from Sandawana, Zimbabwe and Shakisso, Ethiopia, which have relatively high K. Emeralds from Madagascar, Ethiopia, Zimbabwe, and India, all of which belong to Type IA occurrences [1], contain high Na value over 10,000 ppm. The emeralds from Panjshir, Afghanistan exhibit a broad range of Na and K values in this plot. A log-log plot of Fe versus Zn was created to better distinguish our samples from those in Zimbabwe and Ethiopia (Figure 14). The samples from Kagem show relatively low Zn value.

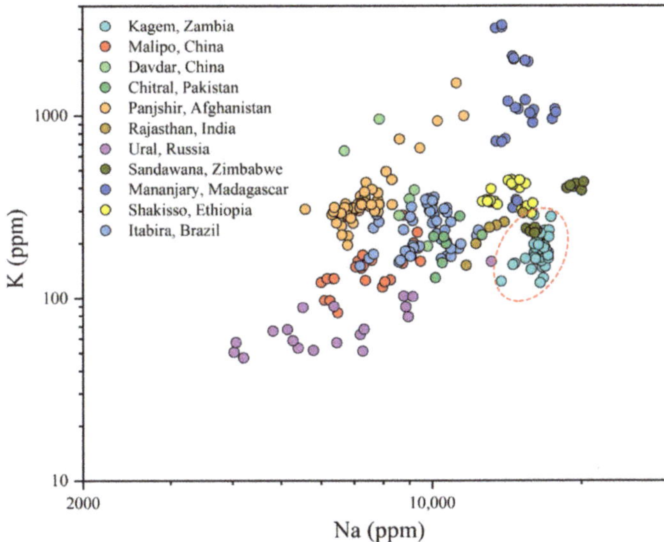

Figure 13. A Na vs K plot shows the relatively high Na and low-to-moderate K concentrations in Kagem emeralds (within the red dashed) compared to the other significant deposits. Sources for some of the data are the same as in Figure 12.

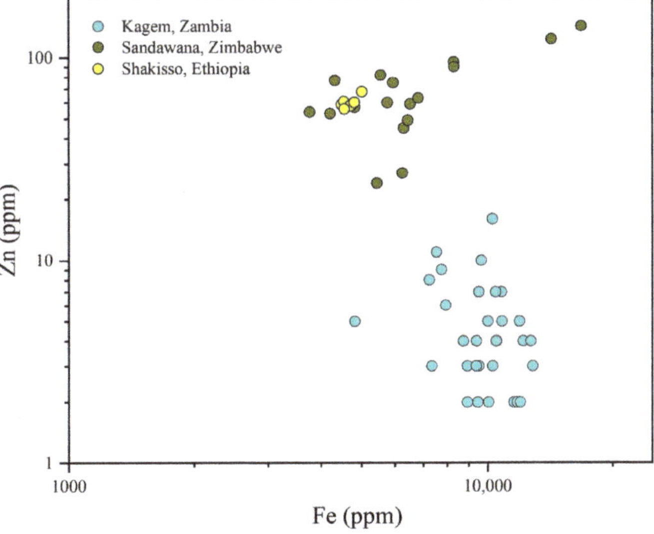

Figure 14. The log binary graph of Fe and Zn shows that Kagem emeralds are not similar to those from Zimbabwe and Ethiopia in this data and can be distinguished. Data source: [43,51].

7. Conclusions

The origin tracing of emeralds requires sufficient samples as support, and new features have been discovered with the advancement of research methods. On the basis of previous research, this study provides new discoveries on Kagem emeralds, including the inclusion, spectral characteristics and chemical analyses.

Two-phase inclusions are common in Kagem emeralds, with the fluid phase containing H_2O or liquid mixtures of $H_2O + CO_2$, and the gas phase consisting of combinations of $CO_2 + N_2$ or $CO_2 + N_2 + CH_4$. A special two-phase inclusion with a solid phase has been identified as dolomite. Raman spectroscopy analysis revealed that Kagem emeralds contain a variety of solid inclusions, including actinolite, graphite, magnetite, and dolomite. Kagem emeralds typically show a conspicuous band caused by Fe^{2+} in the NIR region, and its relative intensity is affected by various elements. Kagem emeralds contain moderate-to-high Cs content (avg. 567 ppm) and low-to-moderate Rb (avg. 14 ppm) content. Additionally, they tend to exhibit high Na values (avg. 16,440 ppm) combined with low-to-moderate K (avg. 185 ppm). The projection diagram of Rb vs. Cs, Na vs. K, and Fe vs. Zn can distinguish the Kagem Mine from other significant emeralds deposits.

Author Contributions: Conceptualization, R.G. and Q.C.; formal analysis, R.G.; investigation, R.G. and Q.C.; writing—original draft preparation, R.G.; writing—review and editing, Q.C., Y.L. and H.H.; supervision, Q.C. All authors have read and agreed to the published version of the manuscript.

Funding: This research was funded by the Hubei Gem & Jewelry Engineering Technology Center (No. CIGTXM-03-202201 and No. CIGTXM-04-S202107), the Fundamental Research Funds for National University, China University of Geosciences (Wuhan) (No. CUGDCJJ202221), the Philosophy and Social Science Foundation of Hubei Province (No. 21G007), NSFC funding (No. 41874105).

Data Availability Statement: All supporting data and computational details are available on written request. These data are stored by the main author of this article.

Acknowledgments: We are grateful to Xiangjie Xiao for providing emerald samples and photographs of the deposit.

Conflicts of Interest: The authors declare no conflict of interest.

References

1. Giuliani, G.; Groat, L.A.; Marshall, D.; Fallick, A.E.; Branquet, Y. Emerald Deposits: A Review and Enhanced Classification. *Minerals* **2019**, *9*, 105. [CrossRef]
2. Gemfields Group Limited 2022. Interim Report 2022. Available online: https://senspdf.jse.co.za/documents/2022/jse/isse/GMLE/Interim22.pdf (accessed on 13 September 2022).
3. Zwaan, J.C.; Seifert, A.; Vrána, S.; Laurs, B.; Anckar, B.; Simmons, W.; Falster, A.; Lustenhouwer, W.; Koivula, J.; Garcia-Guillerminet, H. Emeralds from the Kafubu Area, Zambia. *Gems Gemol.* **2005**, *41*, 116–148. [CrossRef]
4. Zhang, Y.; Yu, X.-Y. Inclusions and Gemological Characteristics of Emeralds from Kamakanga, Zambia. *Minerals* **2023**, *13*, 341. [CrossRef]
5. Bank, H. The emerald occurrence of Miku, Zambia. *J. Gemmol.* **1974**, *14*, 8–15.
6. Koivula, J. Tourmaline as an Inclusion in Zambian Emeralds. *Gems Gemol.* **1982**, *18*, 225–227. [CrossRef]
7. Sliwa, A.S.; Nguluwe, C.A. Geological setting of Zambian emerald deposits. *Precambrian Res.* **1984**, *25*, 213–228. [CrossRef]
8. Seifert, A.; Vrána, S.; Lombe, D.K.; Mwale, M. Impact Assessment of Mining of Beryl Mineralization on the Environment in the Kafubu Area; Problems and Solutions. *Geol. Surv. Czech Rep.* **2004**, *1*, 63.
9. Seifert, A.; Žáček, V.; Vrána, S.; Pecina, V.; Zacharias, J.; Zwaan, J.C. Emerald mineralization in the Kafubu area, Zambia. *Bull. Geosci.* **2004**, *79*, 1–40.
10. Saeseaw, S.; Pardieu, V.; Sangsawong, S. Three-Phase Inclusions in Emerald and Their Impact on Origin Determination. *Gems Gemol.* **2014**, *50*, 114–132. [CrossRef]
11. Hsu, T.; Lucas, A.; Pardieu, V.; Gessner, R. A Visit to the Kagem Open-pit Emerald Mine in Zambia. *Gems Gemol.* Available online: http://www.gia.edu/gia-news-research-kagem-emerald-mine-zambia (accessed on 22 July 2022).
12. Behling, S.; Wilson, W.E. The kagem emerald mine: Kafubu area, Zambia. *Mineral. Rec.* **2010**, *41*, 59–67.
13. Hickman, A.C.J. The Miku emerald deposit. *Geol. Surv. Zambia Econ. Rep.* **1972**, *27*, 35.
14. Hickman, A.C.J. The Geology of the Luanshya Area. *Geol. Surv. Zambia Rep.* **1973**, *46*, 32.
15. Tembo, F.; Kambani, S.; Katongo, C.; Simasiku, S. *Draft Report on the Geological and Structural Control of Emerald Mineralization in the Ndola Rural Area, Zambia*; University of Zambia, School of Mines: Lusaka, Zambia, 2000; p. 72.

16. Daly, M.; Unrug, R. The Muva Supergroup of Northern Zambia: A craton to mobile belt sedimentary sequence. *Trans. Geol. Soc. S. Afr.* **1982**, *85*, 155–165.
17. Waele, B.D.; Wingate, M.T.D.; Mapani, B.; Fitzsimons, I. Untying the Kibaran knot: A reassessment of Mesoproterozoic correlations in southern Africa based on SHRIMP U-Pb data from the Irumide belt. *Geology* **2003**, *31*, 509–512. [CrossRef]
18. John, T.; Schenk, V.; Mezger, K.; Tembo, F. Timing and PT Evolution of Whiteschist Metamorphism in the Lufilian Arc-Zambezi Belt Orogen (Zambia): Implications for the Assembly of Gondwana. *J. Geol.* **2004**, *112*, 71–90. [CrossRef]
19. Cosi, M.; Debonis, A.; Gosso, G.; Hunziker, J.; Martinotti, G.; Moratto, S.; Robert, J.P.; Ruhlman, F. Late proterozoic thrust tectonics, high-pressure metamorphism and uranium mineralization in the Domes Area, Lufilian Arc, Northwestern Zambia. *Precambrian Res.* **1992**, *58*, 215–240. [CrossRef]
20. Parkin, J. Geology of the St. Francis Mission and Musakashi River Areas. Explanation of Degree Sheet 1226 NE and SE Quarters and Part of 1126 SE Quarter. *Geol. Surv. Zambia Rep.* **2000**, *80*, 49.
21. Holland, T.; Blundy, J. Non-ideal interactions in calcic amphiboles and their bearing on amphibole-plagioclase thermometry. *Contrib. Mineral. Petrol.* **1994**, *116*, 433–447. [CrossRef]
22. Gemfields Group Limited 2015. A Competent Persons Report on the Kagem Emerald Mine, Zambia. Available online: https://www.gemfieldsgroup.com/wp-content/uploads/2018/08/U6512_Kagem_Report_v14R.pdf (accessed on 22 September 2022).
23. Pardieu, V.; Detroyat, S.; Sangsawong, S.; Saeseaw, S. In Search of Emeralds from the Musakashi Area of Zambia. *InColor* **2015**, 6–12.
24. Gemfields Group Limited 2023. Operational Market Update For the year ended 31 December 2022. Available online: https://www.gemfieldsgroup.com/operational-market-update-31-december-2022/?upf=dl&id=13426 (accessed on 12 February 2023).
25. Vertriest, W.; Palke, A.C.; Renfro, N.D. Field gemology: Building a research collection and understanding the development of gem deposits. *Gems Gemol.* **2019**, *55*, 490–511. [CrossRef]
26. Liu, Y.; Hu, Z.; Gao, S.; Günther, D.; Xu, J.; Gao, C.; Chen, H. In situ analysis of major and trace elements of anhydrous minerals by LA–ICP–MS without applying an internal standard. *Chem. Geol.* **2008**, *257*, 34–43. [CrossRef]
27. Cui, D.; Liao, Z.; Qi, L.; Zhou, Z. Update on inclusions in emeralds from Davdar, China. *Gems Gemol.* **2020**, *56*, 543–545.
28. Frezzotti, M.L.; Tecce, F.; Casagli, A. Raman spectroscopy for fluid inclusion analysis. *J. Geochem. Explor.* **2012**, *112*, 1–20. [CrossRef]
29. Hu, Y.; Lu, R. Color Characteristics of Blue to Yellow Beryl from Multiple Origins. *Gems Gemol.* **2020**, *56*, 54–65. [CrossRef]
30. Wood, D.L.; Nassau, K. The characterization of beryl and emerald by visible and infrared absorption spectroscopy. *Am. Mineral.* **1968**, *53*, 777–800.
31. Schmetzer, K.; Berdesinski, W.; Bank, H. Über die Mineralart Beryll, ihre Farben und Absorptionsspektren. *Z. Der Dtsch. Gemmol. Ges.* **1974**, *23*, 5–39.
32. Schmetzer, K. Letter: Comment on analysis of three-phase inclusions in emerald. *Gems Gemol.* **2014**, *50*, 316–319.
33. Platonov, A.N.; Taran, M.N.; Minko, O.E.; Polshyn, E.V. Optical absorption spectra and nature of color of iron-containing beryls. *Phys. Chem. Miner.* **1978**, *3*, 87–88.
34. Saeseaw, S.; Renfro, N.; Palke, A.; Sun, Z.; McClure, S. Geographic Origin Determination of Emerald. *Gems Gemol.* **2019**, *55*, 614–646. [CrossRef]
35. Weng, X. The study on Gemological and Mineralogical Characteristics of Zambia Emerald. Master's Thesis, China University of Geosciences, Beijing, China, 2014. (In Chinese with English Abstract).
36. Pardieu, V.; Sharma, A.; Sangsawong, S.; Raynaud, V.; Luetrakulprawat, S. Emeralds from Zambia. *Gamma Gaz.* **2021**, *1*, 38–61.
37. Wood, D.; Nassau, K. Infrared Spectra of Foreign Molecules in Beryl. *J. Chem. Phys.* **1967**, *47*, 2220–2228. [CrossRef]
38. Qiao, X.; Zhou, Z.; Schwarz, D.; Qi, L.; Gao, J.; Nong, P.; Lai, M.; Guo, K.; Li, Y. Study of the Differences in Infrared Spectra of Emerald from Different Mining Areas and the Controlling Factors. *Can. Mineral.* **2019**, *57*, 65–79. [CrossRef]
39. Zheng, Y.; Yu, X.; Xu, B.; Zhang, T.; Wu, M.; Wan, J.; Guo, H.; Long, Z.; Chen, L.; Qin, L. Water Molecules in Channels of Natural Emeralds from Dayakou (China) and Colombia: Spectroscopic, Chemical and Crystal Structural Investigations. *Crystals* **2022**, *12*, 331. [CrossRef]
40. Mashkovtsev, R.; Thomas, V.; Fursenko, D.; Zhukova, E.; Uskov, V.; Gorshunov, B. FTIR spectroscopy of D2O and HDO molecules in the c-axis channels of synthetic beryl. *Am. Mineral.* **2016**, *101*, 175–180. [CrossRef]
41. Hu, Y. Characteristics of Crystal Chemistry and Its Application to Nomenclature and Classification of Beryl. Ph.D. Thesis, China University of Geosciences, Wuhan, China, 2022. (In Chinese with English Abstract).
42. Cui, D.; Liao, Z.; Qi, L.; Qian, Z.; Zhou, Z. A Study of Emeralds from Davdar, North-Western China. *J. Gemmol.* **2020**, *37*, 374–392. [CrossRef]
43. Zwaan, J.C. Gemology, geology and origin of the Sandawana emerald deposits, Zimbabwe. *Scr. Geol.* **2006**, *131*, 1–211.
44. Bidny, A.; Baksheev, I.; Popov, M.; Anosova, M. Beryl from deposit of the Ural Emerald Belt, Russia: ICP-MS-LA and infrared spectroscopy study. *Moscow Univ. Geol. Bull.* **2011**, *66*, 108–115. [CrossRef]
45. Hu, Y.; Lu, R. Unique Vanadium-Rich Emerald from Malipo, China. *Gems Gemol.* **2019**, *55*, 338–352. [CrossRef]
46. Karampelas, S.; Alshaybani, B.; Mohamed, F.; Sangsawong, S.; Al-Alawi, A. Emeralds from the Most Important Occurrences: Chemical and Spectroscopic Data. *Minerals* **2019**, *9*, 561. [CrossRef]

47. Romero-Ordóñez, F.H.; González-Duran, A.F.; Garcia-Toloza, J.; Cohen, J.; Cedeño-Ochoa, C.; González, H.; Sarmiento, L. Mineralogy and Fluid Inclusions of the Cunas Emerald Mine, Maripí, Boyacá, Colombia. *Earth Sci. Res. J.* **2021**, *25*, 139–156. [CrossRef]
48. Chen, Q.; Bao, P.; Li, Y.; Shen, A.; Gao, R.; Bai, Y.; Gong, X.; Liu, X. A Research of Emeralds from Panjshir Valley, Afghanistan. *Minerals* **2022**, *13*, 63. [CrossRef]
49. Hanser, C.S.; Gul, B.; Häger, T.; Botcharnikov, R. Emerald from the Chitral Region, Pakistan: A New Deposit. *J. Gemmol.* **2022**, *38*, 234–252. [CrossRef]
50. Qin, L.-J.; Yu, X.-Y.; Guo, H.-S. Fluid Inclusion and Chemical Composition Characteristics of Emeralds from Rajasthan Area, India. *Minerals* **2022**, *12*, 641. [CrossRef]
51. Huang, Z.; Li, G.; Weng, L.; Zhang, M. Gemological and Mineralogical Characteristics of Emerald from Ethiopia. *Crystals* **2023**, *13*, 233. [CrossRef]

Article

Purple-Violet Gem Spinel from Tanzania and Myanmar: Inclusion, Spectroscopy, Chemistry, and Color

Jinlin Wu [1], Xueying Sun [2], Hong Ma [1], Peiying Ning [1,*], Na Tang [1], Ting Ding [1], Huihuang Li [1], Tianyang Zhang [1] and Ying Ma [1]

[1] National Gemstone Testing Center Shenzhen Lab. Company Ltd., Shenzhen 518000, China
[2] Guild Gem Laboratories, Shenzhen 518000, China
*   Correspondence: ningpy@ngtc.com.cn

Abstract: Purple-violet gem spinels from Tanzania and Myanmar have been investigated for their gemological, spectroscopic, chemical, and colorimetric characteristics. Samples TS and MS both had a purple hue with a pinkish or brownish secondary tone and medium–strong saturation. We identified a number of inclusions, including dolomite, phlogopite, and forsterite in Tanzanian spinel and magnesite, apatite, baddeleyite, anhydrite, pyroxene, and graphite in Myanmar spinel. Tanzanian spinels have slightly lower FWHM (full width at half maximum) values of the 406 cm^{-1} line in the Raman spectrum and the Cr^{3+} zero phonon line in the PL spectrum compared to samples from Myanmar. Fe, Mn, Cr, V, and Zn are proved as useful discriminators to distinguish these two geographic locations. UV-Vis-NIR spectra and CIE L*a*b* parameters are compared with trace element chemistry. Both samples are colored by Fe^{2+}, with minor Fe^{3+}, Cr^{3+}, and V^{3+}. Cr, V, and Fe are combined to influence the hue angle and lightless of purple spinels from Tanzania. However, due to the relatively stable content in Myanmar samples, Fe shows a minor effect on these two parameters. It is worth noting that all inclusion scene, spectral, and chemical characteristics, as well as the comparison presented in this study are of a limited number of samples from Tanzania and Myanmar.

Keywords: purple-violet spinel; inclusion; chemistry; Tanzania; Myanmar

---

1. Introduction

Spinel minerals belong to a large group of compounds with cubic symmetry (space group Fd3̄m). "Normal" spinel is usually represented by the formula AB_2O_4, where A (Mg, Fe^{2+}, Zn, Mn, Ni, Co, Cu, Ge) and B (Al, Fe^{3+}, Cr, V, Ti) cations generally occupy the tetrahedral coordination (T) and octahedral coordination (O) (Figure 1). "Invert" spinel is described as $B(AB)O_4$ with an octahedral–tetrahedral disorder of A and B cations. The intermediate combination of both is considered "disordered". The inversion parameter describes the degree of disorder and varies from 0 (completely normal) to 1 (completely inverted). Most gem spinels are primarily the "normal" spinel $MgAl_2O_4$ [1–5].

The composition as well as transition metal cation distribution of spinel minerals have a strong influence on their physical properties and offers a wide range of colors, mainly pink to red and purple, orange, violet to blue, green, and even black [3,6–9]. Most gem-quality spinels have compositions close to $MgAl_2O_4$ sensu stricto, which can be used as a gemstone when it has good quality and beautiful colorations [2].

Geologically, gem-quality spinels from Asia and eastern Africa are recovered from various geological settings, mainly marble-hosted deposits. Famous sources of gem-quality purple spinels locate worldwide, including in East Africa (i.e., Madagascar and Tanzania) and Asia (i.e., Myanmar, Tajikistan, Sri Lanka, and Vietnam) [10]. Red spinel from Myanmar and cobalt blue spinel from Vietnam are particularly sought after in the gem marketplace and have been studied thoroughly in previous studies [2,11–16]. However, there is little

Citation: Wu, J.; Sun, X.; Ma, H.; Ning, P.; Tang, N.; Ding, T.; Li, H.; Zhang, T.; Ma, Y. Purple-Violet Gem Spinel from Tanzania and Myanmar: Inclusion, Spectroscopy, Chemistry, and Color. *Minerals* **2023**, *13*, 226. https://doi.org/10.3390/min13020226

Academic Editors: Stefanos Karampelas and Emmanuel Fritsch

Received: 13 December 2022
Revised: 31 January 2023
Accepted: 31 January 2023
Published: 4 February 2023

---

research on purple-violet spinel specifically. Until relatively recently with the new finding of attractive purple spinels in Afghanistan and new deposits in Vietnam [17–20], purple spinel was often marginalized commercially and has been gradually pursued by consumers.

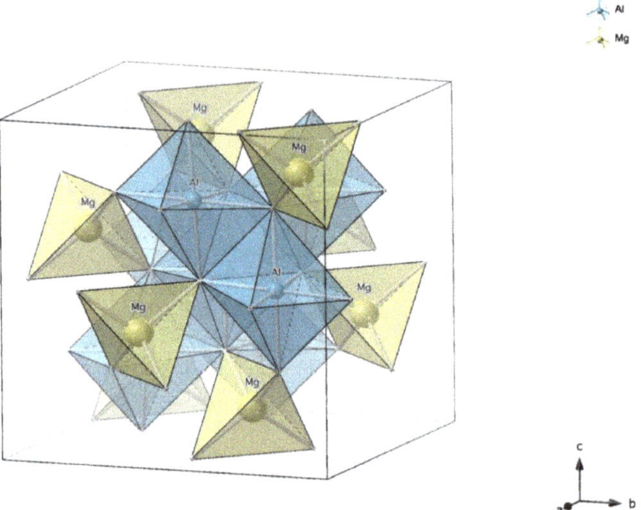

Figure 1. Schematic diagram of "normal" spinel $MgAl_2O_4$.

In this study, gem-quality purple-violet spinels collected from Tanzania and Myanmar (and local gem markets) were analyzed thoroughly and compared using gemological characteristics, internal features, spectroscopic data, and multivariate statistical analysis of chemical data. The origin of their coloration is also discussed.

2. Geological Location

2.1. Tanzania

In eastern Africa, the primary deposits are located in marbles that belong to the Neoproterozoic metamorphic Mozambique Belt [2]. Gem spinels were discovered in marbles near Matombo and Mahenge in the Morogoro region in the late 1980s [14,21]. Ipanko mine and the nearby secondary deposits produced fine stones in 2000 and became famed in 2007 [21]. Spinel is associated with calcite, dolomite, pargasite, blue apatite, phlogopite, graphite, clinohumite, chlorite, and pyrite [22].

2.2. Myanmar

The gem-quality spinel deposits of Myanmar are located within the Himalayan orogenic belt, which was formed by the collision between the Indian plate and the Eurasian plate [12,23]. The Mogok area, a major source of gem-quality spinels, is situated in the central part of the Mogok Metamorphic Belt (Figure 2) and mainly consists of upper amphibolite to granulite facies marbles, calc–silicate rocks, gneisses, and quartzite [24–28]. Spinels are mined from primary deposits (marbles) and secondary deposits such as alluvial and eluvial–deluvial placers, as well as karstic sinkholes and caverns [29].

Figure 2. Map of spinel locations in Tanzania and Myanmar (modified from Google Earth).

3. Materials and Methods

A total of 23 faceted spinel samples from Tanzania (TS-1 to TS-15) and Myanmar (MS-1 to MS-8), ranging from 0.38 to 0.95 ct, were collected from a trusted dealer and analyzed for this study.

All the samples were tested using standard gemological instruments for their refractive index (RI) and long- and short-wave UV fluorescence (365 nm and 254 nm wavelength, respectively). Specific gravity (SG) was determined using the hydrostatic method with an electronic balance. Microscopic observations and photomicrography of internal features were recorded with a VHX-2000 super depth-of-field microscope (max. magnification 500×).

Raman and photoluminescence (PL) spectra were collected with a Renishaw in Via Raman microspectrometer under the following instrumental conditions: Raman spectra of the host spinels and the inclusions were acquired from 100 to 2000 cm^{-1} using a 785 nm laser and 532 nm (500 mW laser output power), respectively, with an acquisition time of 10 seconds, a grating of 1200 grooves/mm and 1800 grooves/mm, and about 1 cm^{-1} resolution. The instrument was calibrated using the 520.00 ($\pm$0.2) cm^{-1} line of silicon. The spectrum of each sample was collected at several locations on the sample. All the spectra were processed, and the peaks were fitted by the software Origin. Photoluminescence (PL) spectra were recorded from 535 to 800 nm using an excitation wavelength of 532 nm (0.005 mW laser output power), with an acquisition time of 20 seconds and a grating of 1800 grooves/mm.

Laser ablation–inductively coupled plasma–mass spectrometry (LA-ICP-MS) analyses of samples were performed using a PerkinElmer 350 D ICP-MS spectrometer (NexION 350, Perkin Elemer, Waltham, MA, USA) with an NWR-213 laser ablation system (213 nm ablation wavelength, 20 Hz frequency with energy 25 $\pm$ 1 J/cm^2, and a spot size of 44 μm). Reference materials included two NIST glasses (SRM 610 and SRM 612) and three USGS glasses (BHVO-2G, BIR-1G, and BCR-2G). Each analysis incorporated a background acquisition of approximately 20–30 s followed by 50 s of data acquisition from the sample.

Ultraviolet–visible (UV-Vis) absorption spectra were collected from 300 nm to 800 nm using a PerkinElmer Lambda 950 spectrometer (Lambda 950, Perkin Elemer, Waltham, MA, USA) at room temperature, with a slit width of 2.0 nm, a data interval of 1.0 nm, an integration time of 10 s, a scan speed of 266.75 nm/min, a light source conversion wavelength of 319.20 nm, and a detector switching wavelength of 860.80 nm. Then, the integrating sphere reflectivity method was applied to quantitatively calculate the color of the purple spinels in the even color space CIELAB 1976.

Fluorescence spectra were analyzed by a QSpec Gem-3000 spectrophotometer (GEM 3000, BiaoQi Optoelectronics, Guangzhou, China) with the following test conditions: integration time of 150 ms, an average number of 10, smoothing width of 2 nm, wavelength collected from 450 nm to 900 nm, long-wave UV excitation (365 nm), room temperature.

4. Results

4.1. Gemmological Characteristics

All the samples in this study (Figure 3) are transparent with few inclusions. The spinels from Tanzania were purple, pinkish-purple, brownish-purple, and violet in medium tone and medium–strong saturation. They exhibited various reactions when exposed to long-wave (365 nm) ultraviolet light, including inert-to-strong red/orange/green fluorescence. Samples from Myanmar had a medium tone, medium–moderate saturation, a purple hue with a pinkish or brownish secondary tone, and violet, as well as weak-to-strong red/orange fluorescence (Table 1).

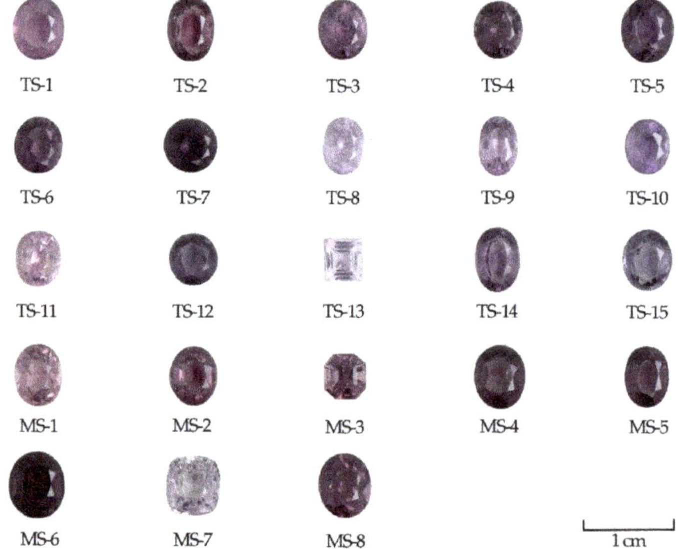

Figure 3. Purple-violet spinel samples in this study (TS: Tanzanian spinel; MS: Myanmar spinel).

4.2. Inclusions

Microscopic observations revealed distinct internal characteristics in Tanzanian purple spinels. The fine dust-like exsolved particles are the most classical features, similar to the pink-red spinel from Tanzania. They scattered throughout the stone (Figure 4a) or were accompanied by negative crystals of octahedral or strongly distorted octahedral shapes of varying sizes (Figure 4b). Occasionally, the oriented long or short needles presented iridescence under fiber optic illumination (Figure 4c,d). As previously mentioned by Chankhantha (2021) and Schmetzer (1992) [30,31], these needles or particles were possibly caused by the exsolution of högbomite, giving samples a somewhat cloudy appearance. Since Tanzanian spinels formed in marbles, it was not surprising to find abundant carbonates as the most common mineral inclusions. Dolomite is often included in negative crystals aligned in a plane (Figure 4e) or presents a whitish, turbid, or 'frosted' aspect with black materials attached (Figure 4f,h). Closely related is the impurity in the marble-host rock, phlogopite [$KMg_3Si_3AlO_{10}(F; OH)_2$], which is present in spinels from nearly all the sources.

Phlogopite is typically present as colorless, subhedral crystals (Figure 4g). Figure 4f reveals a colorless, subhedral olivine (forsterite) $[Mg_2SiO_4]$ (Figure 4f) in one sample, which is extremely rare in spinel and was reported only once in Tanzanian spinel [32].

Table 1. Summary of gemological properties of Tanzanian and Myanmar spinels.

Sample Number	Color	Cut-Shape	Weight/ct	Dimensions/mm	RI	UV Fluorescence Long Wave (365 nm)	Mineral Inclusions *
TS-1	pinkish-purple	modified brilliant-oval	0.54	5.95 × 4.90 × 2.27	1.715	orange, medium	
TS-2	brownish-purple	modified brilliant-oval	0.62	6.25 × 4.46 × 2.45	1.717	red, medium	
TS-3	purple	modified brilliant-oval	0.64	5.70 × 4.62 × 3.18	1.717	red, strong	
TS-4	brownish-purple	modified brilliant-oval	0.53	5.48 × 4.61 ×2.85	1.715	red, medium	
TS-5	purple	modified brilliant-oval	0.73	6.26 × 5.05 × 2.98	1.716	green, weak	
TS-6	purple	modified brilliant-oval	0.56	5.34 × 4.58 × 2.83	1.715	inert	
TS-7	purple	modified brilliant-round	0.55	5.18 × 5.10 × 2.91	1.718	inert	Dolomite (10/15) Phlogopite (3/15) Forsterite (1/15)
TS-8	violet	modified brilliant-oval	0.45	5.47 × 3.83 × 2.70	1.712	green, strong	
TS-9	violet	modified brilliant-oval	0.44	5.77 × 3.72 × 2.57	1.714	orange, medium	
TS-10	violet	modified brilliant-oval	0.52	5.23 × 4.20 × 3.12	1.714	green, weak	
TS-11	pinkish-purple	modified brilliant-oval	0.52	5.35 × 4.30 × 3.06	1.715	orange, medium	
TS-12	violet	modified brilliant-oval	0.46	5.05 × 4.62 × 2.39	1.716	green, weak	
TS-13	violet	step- rectangular	0.38	4.15 × 3.58 × 2.63	1.713	green, strong	
TS-14	purple	modified brilliant-oval	0.57	6.24 × 4.28 × 2.50	1.714	inert	
TS-15	violet	modified brilliant-oval	0.60	5.95 × 4.73 × 2.58	1.714	orange, weak	
MS-1	pinkish-purple	modified brilliant-oval	0.64	5.95 × 4.54 × 3.12	1.717	red, strong	
MS-2	brownish-purple	modified brilliant-oval	0.53	5.69 × 4.55 × 2.66	1.715	red, strong	
MS-3	brownish-purple	step- octagonal	0.41	4.35 × 4.00 × 3.16	1.714	red, medium	Apatite (1/8) Magnesite (6/8) Anhydrite (1/8) Baddeleyite (1/8) Graphite (2/8) Pyroxene (1/8)
MS-4	brownish-purple	modified brilliant-oval	0.64	6.14 × 4.89 × 2.71	1.715	red, strong	
MS-5	purple	modified brilliant-oval	0.60	5.87 × 4.16 × 2.88	1.716	red, medium	
MS-6	purple	modified brilliant-oval	0.95	6.39 × 5.35 × 3.66	1.717	red, medium	
MS-7	violet	modified brilliant-cushion	0.86	5.78 × 5.14 × 3.57	1.714	orange, medium	
MS-8	purple	modified brilliant-oval	0.74	6.14 × 4.98 × 3.30	1.716	red, strong	

* Mineral inclusions identified by Raman spectroscopy.

Myanmar purple spinels in this study are relatively internally clean. Among the inclusions seen in the Myanmar spinel were colorless mineral inclusions identified as apatite, magnesite, anhydrite, and pyroxene, irregularly shaped black graphite, and reddish-brownish baddeleyite. Apatite $[Ca_5(PO_4)_3(F, Cl, OH)]$ inclusions are present in a variety of other habits ranging from substantially prismatic, subhedral to rounded (Figure 5a–c). Sometimes, they were accompanied by attached black irregular graphite. Magnesite appeared with a "frosted halo" around (Figure 5d), and a subhedral mineral belonging to the pyroxene supergroup shows a distinct set of cleavage planes (Figure 5e). A tiny colorless anhydrite $[CaSO_4]$ was discovered (Figure 5f), which easily hydrates and then combines into gypsum. A reddish-brownish, prismatic baddeleyite $[ZrO_2]$ with a tension crack is a surprise in this study (Figure 5g). To date, baddeleyite and anhydrite have only been discovered in Myanmar spinel [13]. Therefore, the combination of subhedral baddeleyite and anhydrite may indicate the locality of Myanmar.

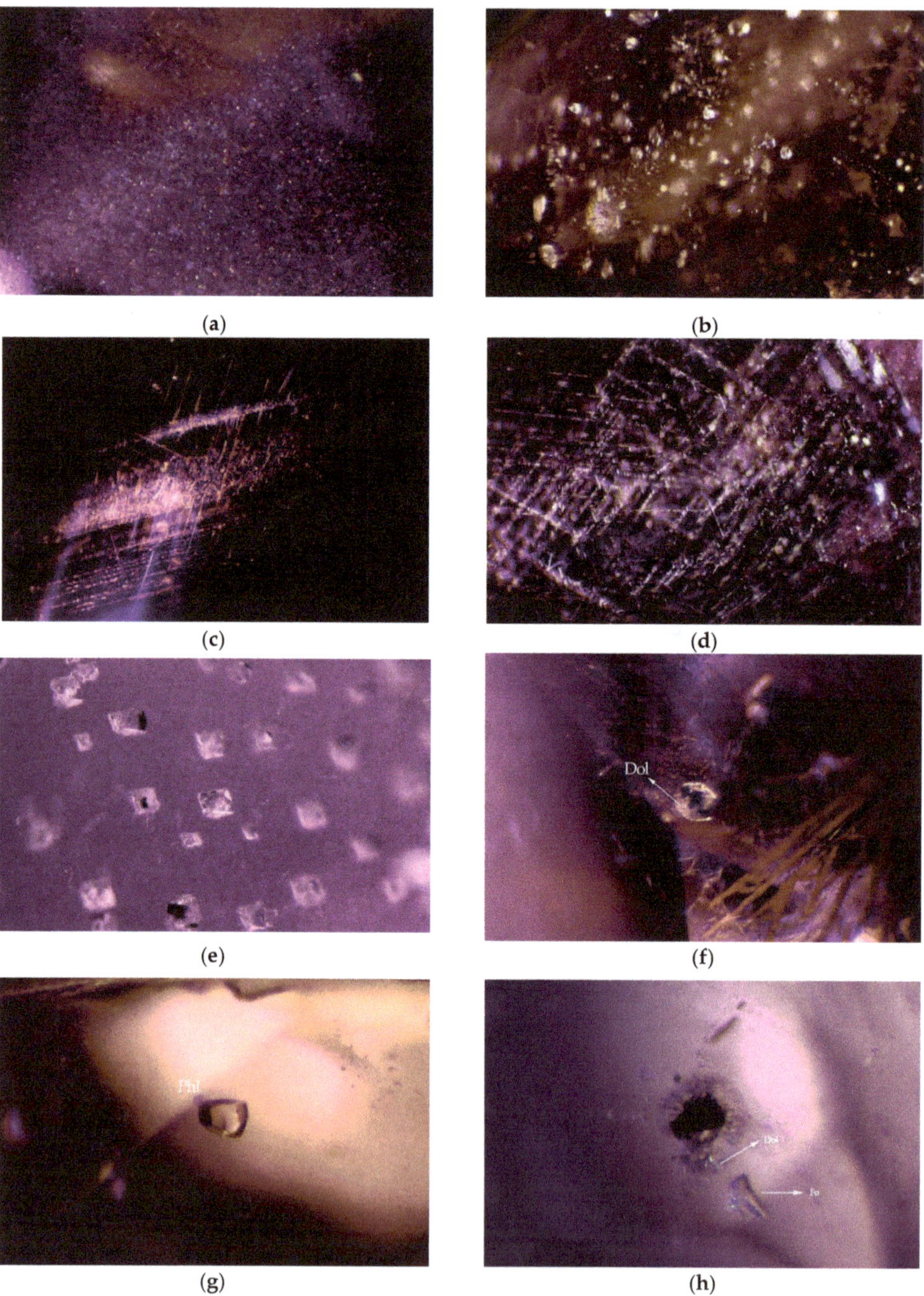

Figure 4. Inclusion scenes in purple spinels from Tanzania: (**a**) scattered tiny particles (40×); (**b**) negative crystals with fine particles (40×); (**c**) oriented, long iridescent needles (40×); (**d**) short needles (40×); (**e**) dolomite aligned in a plane (600×); (**f**) a subhedral, whitish dolomite (Dol) in contact with black mineral (80×); (**g**) an isolated, colorless phlogopite (Phl) (80×); (**h**) a colorless, irregular dolomite (Dol), and a subhedral forsterite (Fo) (80×).

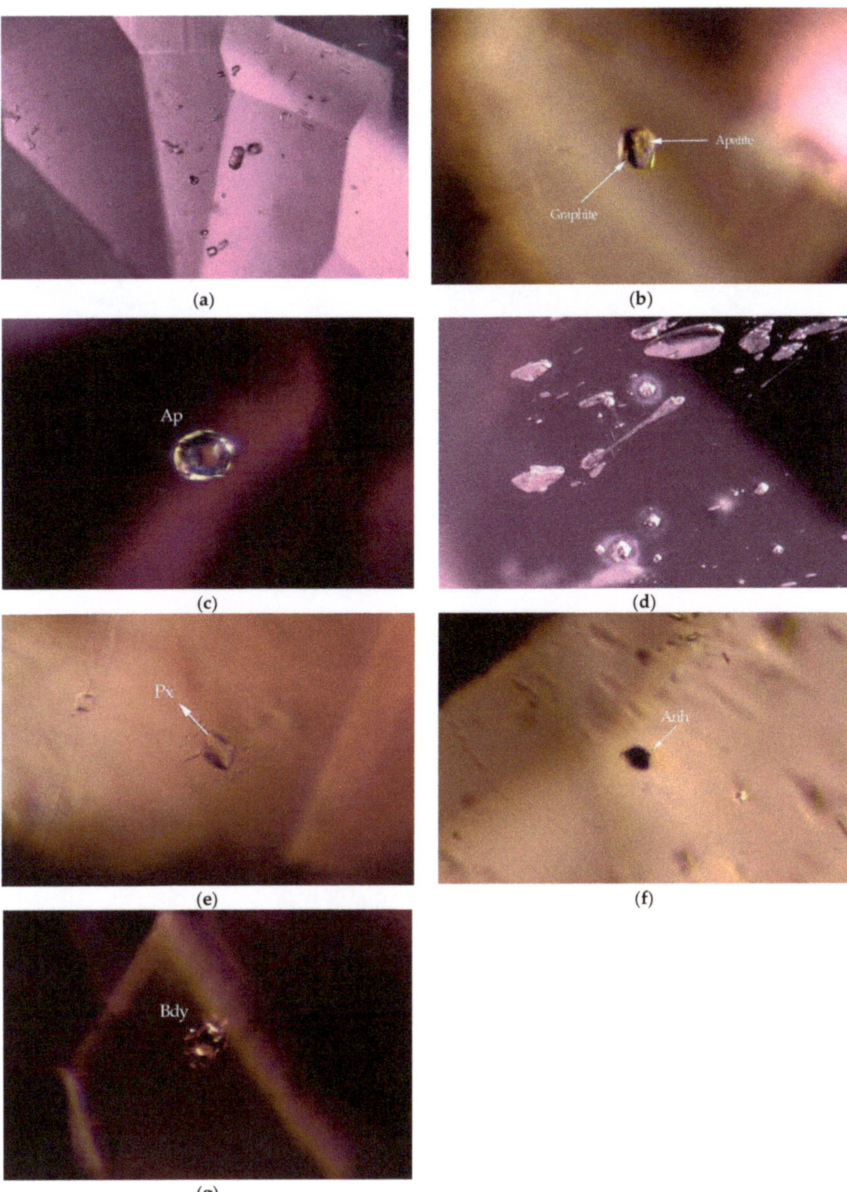

Figure 5. Inclusion scenes in purple spinels from Myanmar: (**a**) groups of euhedral apatite (Ap) (200×); (**b**) a single subhedral apatite (Ap) is associated with graphite (C) (80×); (**c**) isolated rounded apatite (80×); (**d**) several magnesite minerals included in negative crystals (200×); (**e**) a subhedral pyroxene (Px) was surrounded by some tiny colorless crystals (80×); (**f**) tiny colorless anhydrite (Anh) (80×); (**g**) reddish-brownish baddeleyite (Bdy) crystal associated with a tension crack (80×).

Raman spectroscopy allowed the identification of several mineral inclusions, and the representative Raman spectra are shown for these mineral inclusions analyzed in our samples, together with a spectrum of the host spinel in Figure 6.

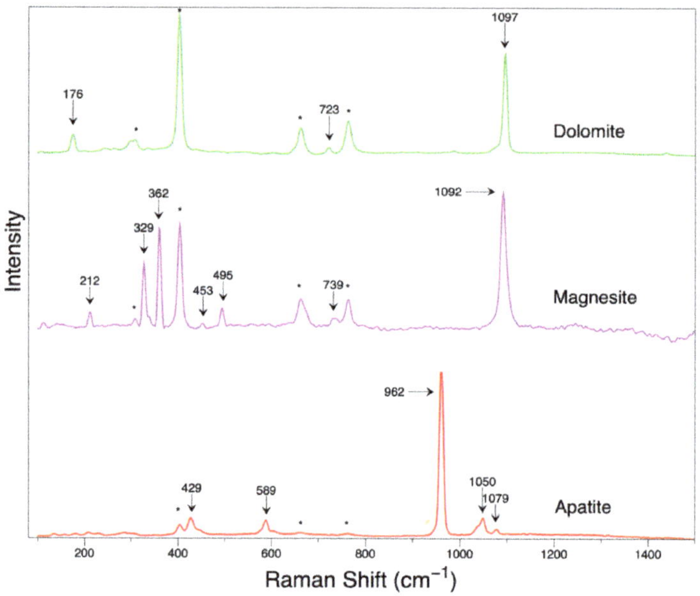

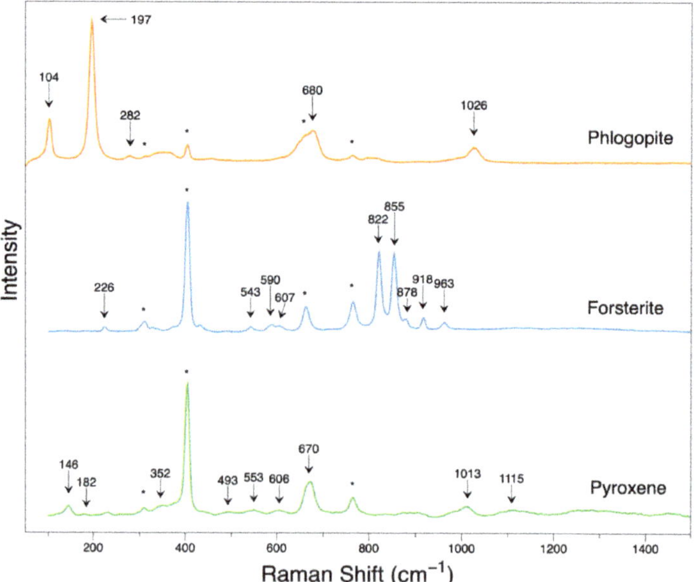

Figure 6. *Cont.*

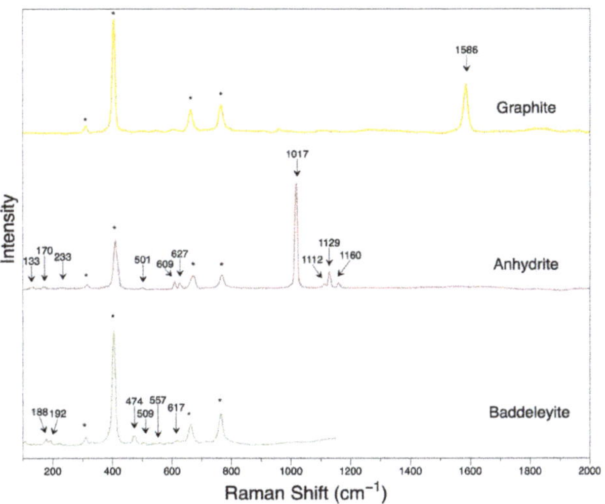

Figure 6. Representative Raman spectra are shown for mineral inclusions analyzed in our samples. Peaks in the inclusion spectra that are marked with an asterisk (*) are from the host spinel. All the spectra are stacked for clarity.

4.3. Raman and PL Spectroscopy

Raman spectra of spinels from Myanmar and Tanzania were collected in the spectral range of 100–2000 cm^{-1} (Figure 7). All the samples exhibit four intense and well-defined bands at around 312 cm^{-1}, 406 cm^{-1}, 665 cm^{-1}, and 767 cm^{-1}. These peaks are assigned to the $T_{2g}(1)$, E_g, $T_{2g}(2)$, and A_{1g} mode [33]. The most prominent feature is the narrow line at about 406 cm^{-1}.

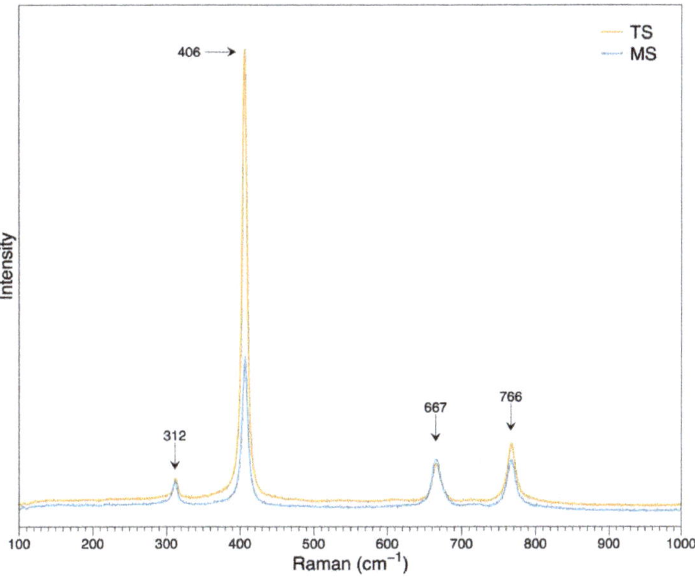

Figure 7. Raman spectra of spinels from Tanzania and Myanmar.

In addition, photoluminescent (PL) emission spectra were also recorded (Figure 8).

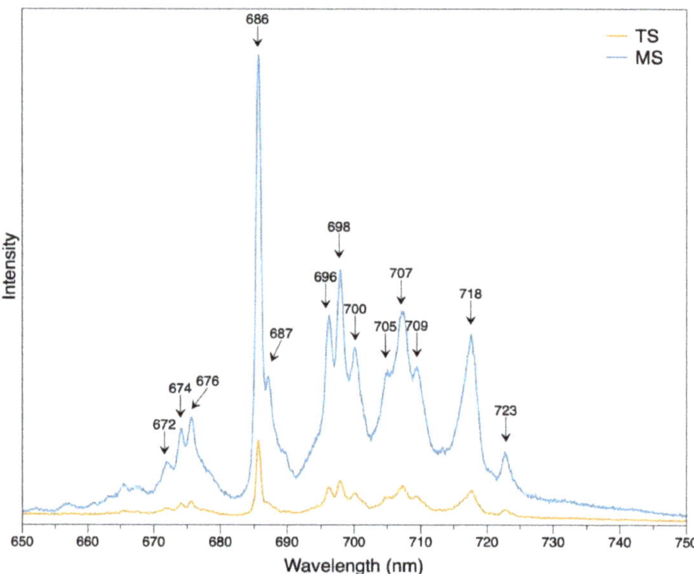

Figure 8. PL spectra of spinels from Tanzania and Myanmar.

The PL spectrum of spinel is comprised of a strong zero phonon line at approximately 686 nm, vibronic sidebands of that line, and other lines associated with Cr^{3+} pairs. The sharp and defined chromium emission features verified that the stone was natural and unheated. Heat treatment typically broadens and shifts the position of PL peaks [4,34,35].

4.4. Chemical Fingerprinting

Trace element analysis is a powerful tool for the origin determination of spinels. Spinels appear to be sensitive to slight changes in their geological environment, which induce unique trace element signatures for stones from different geographic localities [12]. Table 2 summarizes the results for selected elements as determined by LA-ICP-MS. By carefully analyzing the chemical profiles, plotting elements and their ratios in a 2D diagram is applied to distinguish these samples from Tanzania and Myanmar.

Red-pink spinels from Tanzania are enriched in Zn [2], while the Zn values in purple-violet samples in this study are variable, ranging from 188 to 16424 ppmw. The most characteristic feature of Tanzanian spinels is their enrichment in Fe and Mn relative to Myanmar spinels (Figure 9). The Fe/V-Mn-Cr ternary diagram separates the two sources with only a small overlap (Figure 10). The Cr-Mn vs. Fe/Cr diagram shows that the majority of Tanzanian spinel sits in the Cr<Mn box with a higher Fe/Cr ratio from 10 to 1110; on the contrary, all the spinels from Myanmar sit in the Cr>Mn box with Fe/Cr ratio lower than 50 (Figure 11).

Table 2. Chemical composition by LA-ICP-MS of purple spinels from Tanzania and Myanmar *.

Elements (ppmw)	Tanzania (*n* = 15)	Myanmar (*n* = 8)	Detection Limit	Previous Study **
Li	2.9–322 (40.2)	2.9–17.6 (8.1)	0.01–0.07	–
Be	3.9–24.2 (8.5)	1.1–12.8 (5.9)	0.07–0.29	–
K	0.1–25.9 (3.9)	bdl.–6.1 (2.1)	0.06–0.92	–
Ti	21.7–562 (139)	44.5–528 (223)	0.29–1.45	0
V	29.3–380 (132)	41.7–768 (496)	0.06–0.41	408
Cr	10.7–1094 (236)	92.4–1138 (665)	0.93–1.64	342
Fe	5565–18,060 (10,660)	4480–6359 (5466)	0.98–2.39	7140
Mn	79.6–635 (202)	9.54–178.62 (42.73)	0.12–0.46	0
Zn	188–16,424 (2277)	458–6553 (2023)	0.21–0.92	2030
Ga	115–301 (209)	62.9–157 (105)	0.03–0.10	–
Ni	0.3–13.2 (4.5)	1.2–42.1 (12.7)	0.04–0.16	–

* Numbers in parentheses are median values; bdl. = below detection limit. ** Representative data of purple spinel from Tanzania from Giuliani et al. (2017) [2].

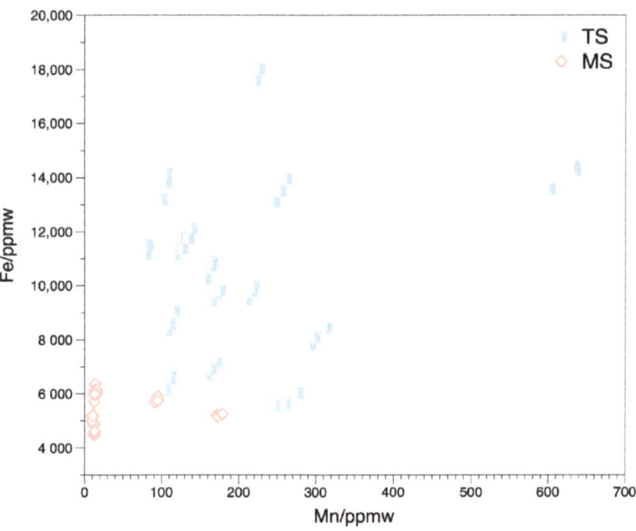

Figure 9. The scatter diagram of Fe vs. Mn reveals that Tanzanian purple spinels can be separated from Myanmar samples.

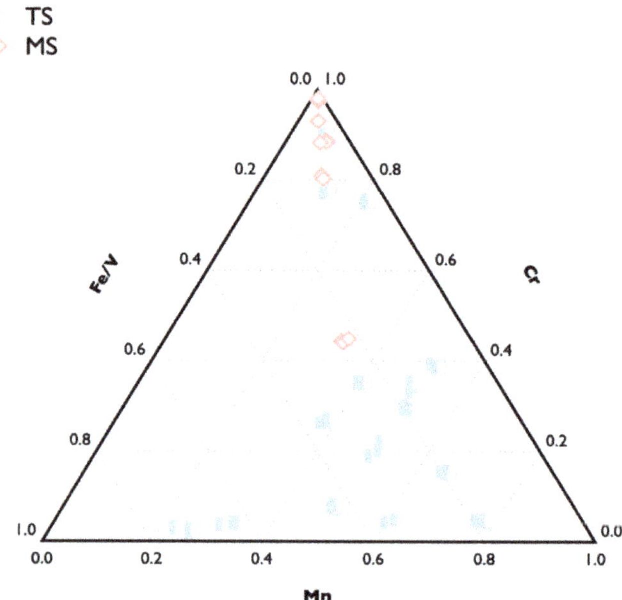

Figure 10. The Fe/V+Mn+Cr ternary diagram shows that the data clusters for the two spinel occurrences can be distinguished to a great extent.

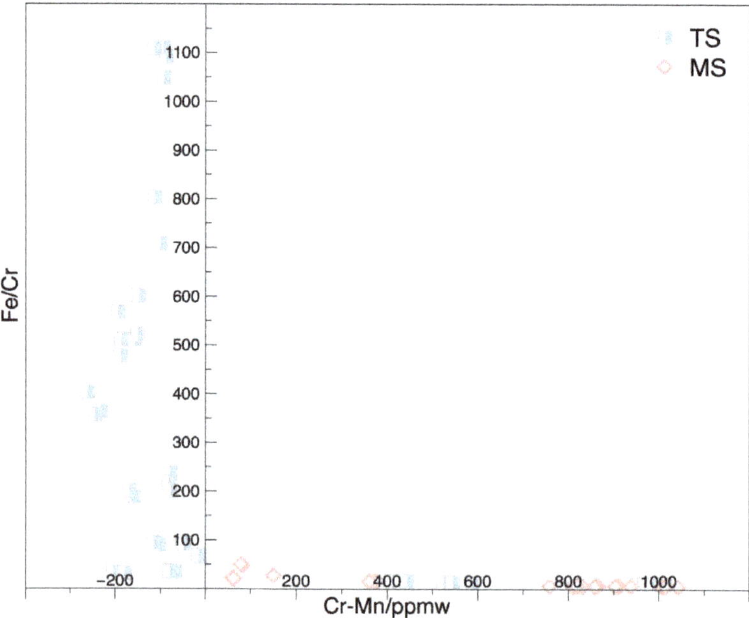

Figure 11. The scatter diagram plots Cr-Mn vs. Fe/Cr for purple spinels from Tanzania and Myanmar.

4.5. UV-Vis-NIR Spectroscopy

The UV-Vis-NIR spectra of all the samples show strong similarities (Figure 12), characterized by featured bands attributed to the spin-forbidden transition of ^5E (D)$\rightarrow^3T_2$ (H) of $^TFe^{2+}$ at ~556 nm [6]. The $^TFe^{2+}$ spin-forbidden transitions ^5E (D)$\rightarrow^3E$ (D) and ^5E (D)$\rightarrow^3T_2$

(G) are responsible for the absorption bands around ~372 nm and ~386 nm, respectively. Moreover, these absorption bands may be amplified by $^{T}Fe^{2+}$–$^{M}Fe^{3+}$ ECP transitions [6,20]. Two broad bands at ~458 nm and ~476 nm may be assigned principally to spin-forbidden $^{6}A_{1g}{\rightarrow}^{4}A_{1g}$, $^{4}E_{g}$ transitions of isolated $^{M}Fe^{3+}$ ions, possibly intensified by ECP interactions and by spin-forbidden transitions of $^{T}Fe^{2+}$ [6]. The absorption around ~540 nm may be caused by spin-allowed d–d transitions $^{3}T_{1}$ (F)$\rightarrow^{3}T_{2}$ (F) in V^{3+} at the M sites or spin-allowed electronic d–d transitions $^{4}A_{2g}{\rightarrow}^{4}T_{2g}$ (F) in Cr^{3+} at the M sites [6,9,20,36–41].

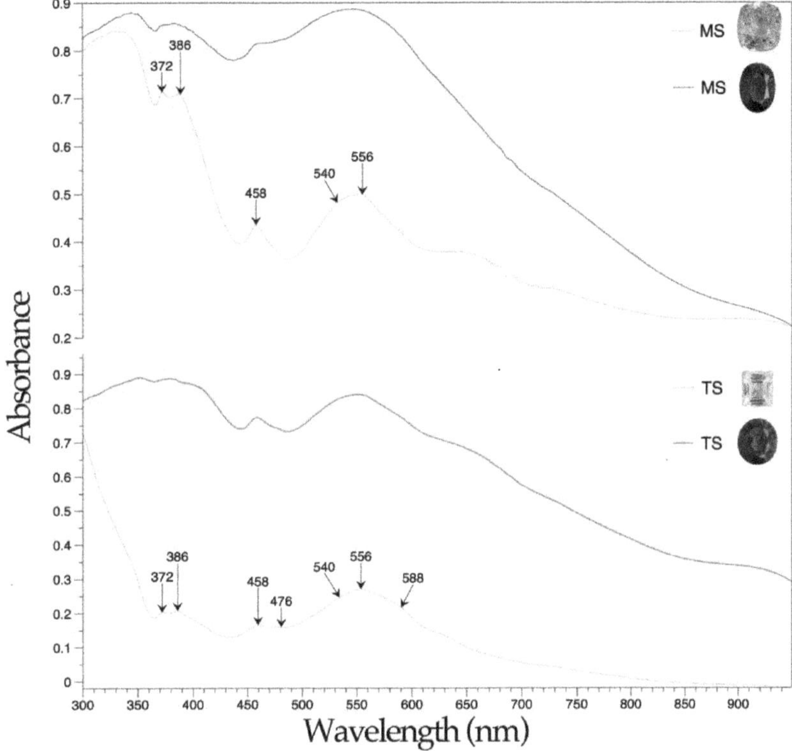

Figure 12. The representative UV-Vis-NIR absorption spectra of spinel samples TS and MS show absorption peaks related to Fe^{2+}, Fe^{3+}, Cr^{3+}, and V^{3+}.

In addition, a strong Cr-related signature with a closely spaced group of fluorescent lines at around 700 nm was observed in MS samples. Fe, Cr, and V are the main chromophores of these stones, resulting in the purple color. Further, comparing the samples with different saturation, higher saturation stones showed stronger absorption and a rapid increase in absorption at wavelengths below 400 nm.

4.6. Quantitative Characterization of Color

Modern colorimetry is mainly developed from the CIE 1976 L*a*b* color system. The lightness, L*, indicates a transition from the darkest black at L* = 0 to the brightest white at L* = 100. The coordinates, a* and b*, stand for neutral gray when the values at a* = 0 and b* = 0. a* represents the red/green opponent colors, with red at the positive axis value and green at the negative axis value. b* represents the yellow/blue opponent colors, with yellow at the positive axis value and blue at the negative axis value. The saturation of each color is proportional to the absolute value of the axis value. Parameters a* and b* jointly determine the chroma C* and hue angle h° is derived from a* and b*, which reflects color characteristics more in line with the habit of color description in daily life. The formulas are as follows:

$$C^* = \sqrt{a^{*2} + b^{*2}}, \tag{1}$$

$$h^* = \arctan \frac{b^*}{a^*} \tag{2}$$

The color parameters of each sample were quantitatively characterized by applying CIE D65 light source and N9 Munsell neutral background as testing conditions. The results are shown in Table 3. The absence of the negative half axis of a* and positive half axis b* indicates that the color of purple spinel from both origins is controlled by red and blue. The combination of these red and blue features is responsible for the purple color. The experimental results show that the color parameters L*, a*, and C* are in the similar range for samples from two origins, while b* and h° show differences.

Table 3. Colorimetric coordinates L*, a*, and b* of purple spinels.

Parameters	Tanzania	Myanmar
L*	40.11–69.92	40.70–60.45
a*	0.92–12.89	3.61–11.28
b*	−11.20−−0.75	−3.68−−0.71
h°	−65.43−−15.38	−31.28−−3.68
C*	1.73–16.62	2.62–11.75

The color coordinates a* and −b* of TS samples show a significant negative linear correlation with $R^2 = 0.8623$. In contrast, no distinct relationship is observed between a* and b* in MS samples (Figure 13). Moreover, a* and −b* of TS spinels are both positively correlated with its C*(Figure 14a,b), indicating that the chroma is controlled by both red and blue tones, and the influence degree of red is much stronger. MS samples show a similar relationship (Figure 14a), except that −b* has less relativity with C* due to the smaller variance of b*(Figure 14b).

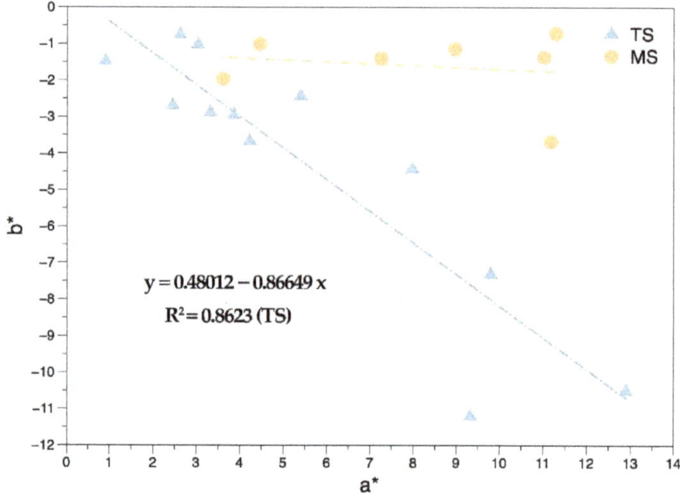

Figure 13. The color coordinate a* is inversely proportional to b* for TS samples, while a* barely depends on b* for MS samples.

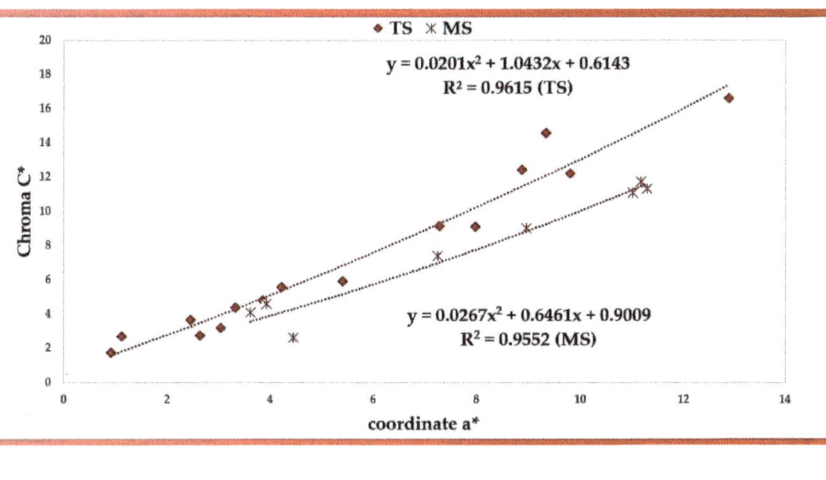

(a)

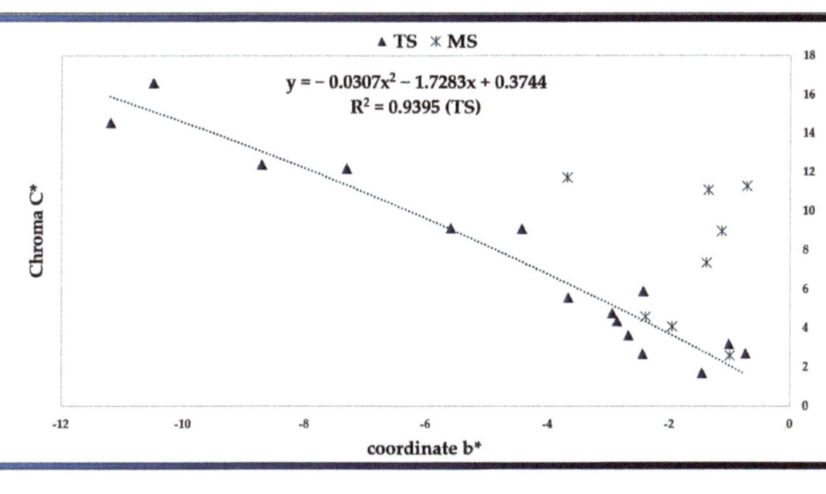

(b)

Figure 14. The color analysis of purple-violet spinel. (**a**) A highly positive correlation between the color coordinates a* and its chroma C*. (**b**) A positive correlation between the color coordinate −b* and its chroma C* for TS and a weak correlation between these two parameters for MS.

5. Discussion

5.1. Color and Fluorescence

Gem-quality spinel occurs in a variety of colors based on the trace elements present within the stone. Most of the (orangy) red-pink spinels attribute their color to Cr and V concentrations [6,28]. Equally high contents of Cr and V will cause red, while lower Cr and higher V may induce a more orange hue [41]. Blue colors in spinels are mainly caused by various electronic processes in Fe and Co cations. Other colors, magenta, purple, and green, are mainly caused by a high concentration of Fe, especially the Fe^{2+}/Fe^{3+} ratio [41]. Moreover, Mn is known to act as a yellow chromophore [42].

Thus, the relationship between Cr/V/Fe contents and hue angle/lightness is analyzed. We focused on the Tanzanian samples first. There is a significant positive correlation between Cr/V contents and hue angle, while the relationship between Fe content and hue is the reverse (Figure 15). Moreover, the lightness lowers with the increase of these elements.

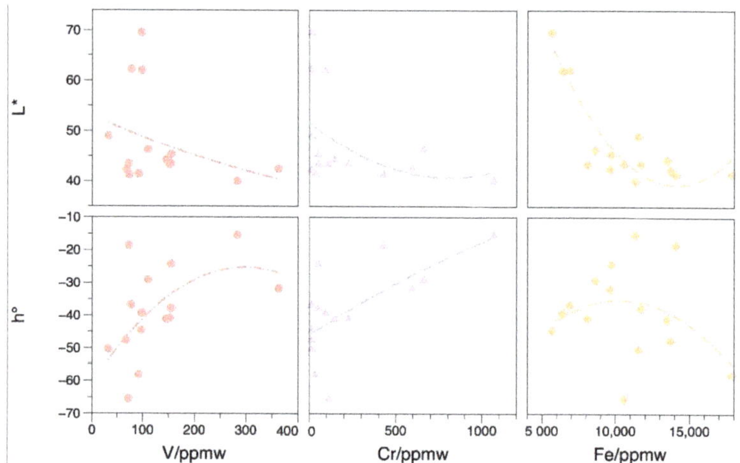

Figure 15. The relationship between vanadium/chromium/iron and the lightness/hue angle in Tanzanian spinels.

For the Myanmar spinels, the relationship between Cr and hue angle is similar (Figure 16). Different from TS samples, V and Fe contents have little impact on these two parameters. We speculated that its Fe content (4480–6359 ppmw) is too stable to influence the lightness and hue angle. This Fe concentration can cause a certain and stable blue color, then the variation of the hue is controlled by the various red color caused by the difference in Cr values.

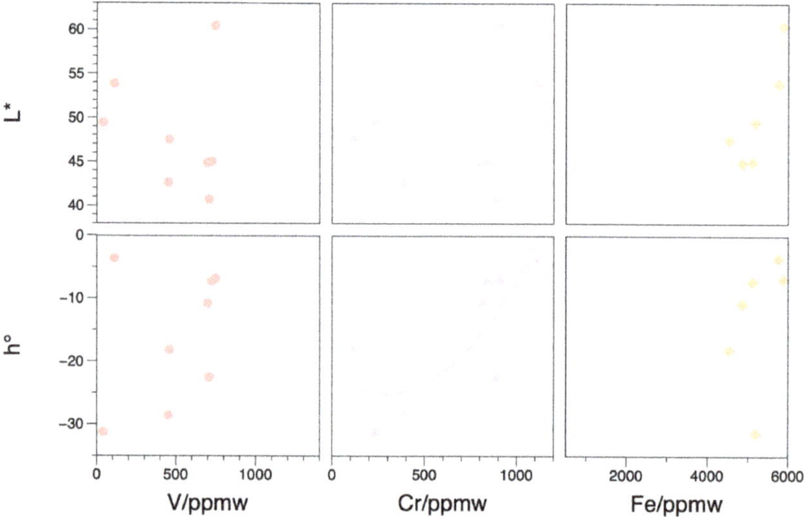

Figure 16. The relationship between V/Cr/Fe concentrations (ppmw) and the lightness/hue angle in Myanmar spinels.

Most spinel samples in this study showed a red under long-wave UV excitation (365 nm). Of note were striking strong green and medium orange fluorescence in several Tanzanian samples. The green luminescence in spinel is attributed to tetrahedral Mn^{2+} [43–45], while red fluorescence is related to the presence of Cr^{3+} [46,47]. Therefore, the spinel presents red fluorescence while Cr dominates and green fluorescence when Mn plays a leading role. When the impact of both elements is very similar, orange color, a mixture of red and green, appears. The representative fluorescence spectra with green, orange, and red luminescence were shown (Figure 17). An emission peak centered at about 512 nm is related to the presence of Mn. The green luminescence sample shows the strongest Mn-related peak, whereas the red fluorescence sample only displays a series of peaks around 700 nm. The orange luminescence sample shows a weak peak at ~512 nm and several strong peaks near 700 nm.

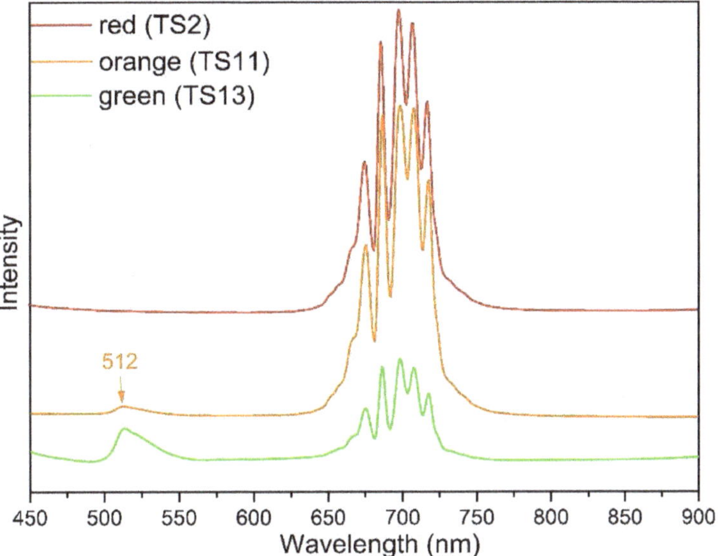

Figure 17. Fluorescence spectra of spinel samples with green, orange, and red luminescence. The peak at ~512 nm corresponds to the green luminescence.

Furthermore, we compared the Cr/Mn ratio and found that the value of spinels with green fluorescence is apparently lower than that of red fluorescence ones (Figure 18). The orange fluorescence is somewhere in between, which was consistent with our hypothesis. When comparing the two provenances, the Cr/Mn ratio is higher in Myanmar spinels with orange and red fluorescence. The existence of Fe may restrain the intensity of the fluorescence. Due to the difference in the elements affecting the fluorescence color, green and red fluorescence with different intensities were analyzed separately. The Fe/Mn ratio and the intensity of the green reaction were anticorrelated (Figure 19a). Similarly, the intensity of the red reaction increases with the decreasing Fe/Cr ratio (Figure 19b).

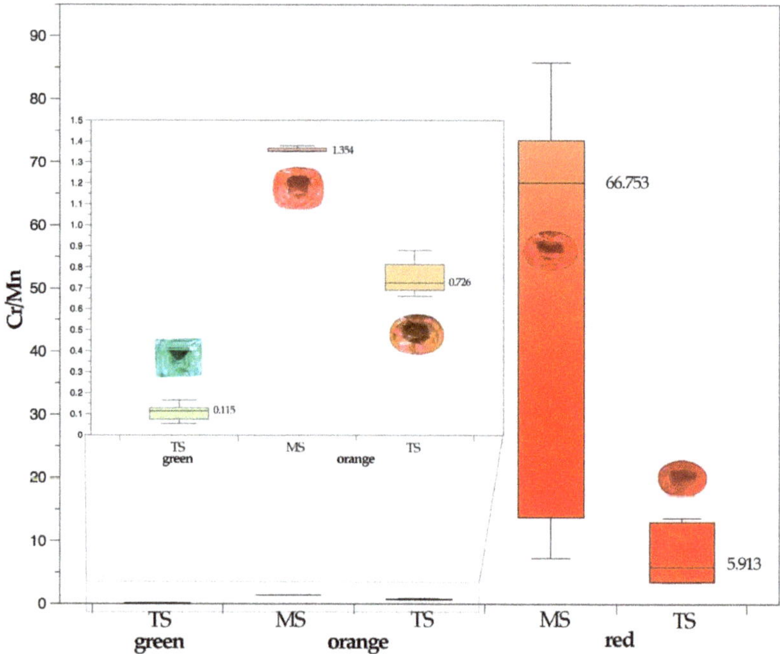

Figure 18. The relationship between Cr/Mn ratio and fluorescence with various colors. The values in the figure are the average value of each group. The values of the Cr/Mn ratio from high to low in order are red, orange, and green fluorescence. In addition, the Cr/Mn ratios of orange and red fluorescence are apparently higher in Myanmar samples.

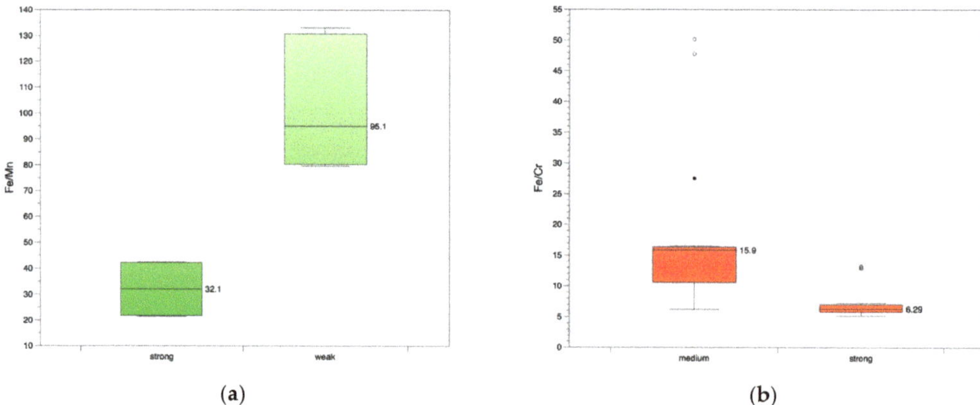

(a)

(b)

Figure 19. (**a**) The Fe/Mn ratio affects the intensity of green fluorescence. The higher the Fe/Mn ratio, the lower the intensity. (**b**) The negative relationship of Fe/Cr ratio and the intensity of red fluorescence. The values in the figure are the average value of each group.

5.2. FWHM (Full Width at Half Maximum)

The cation disordering information, e.g., the rearrangement of some of the cations in the unit cell, can be reflected in the shape and width of the 406 cm^{-1} peak in the Raman spectrum and the 686 nm peak in the PL spectrum, including broadening and shoulder development [48,49]. The above chemistry part has shown the impurity difference between the spinel from Tanzania and Myanmar. The cation distribution between T and M sites in the spinel structure is highly sensitive to temperature, pressure, oxygen fugacity, and bulk rock and fluid compositions [3,50,51]. Although both belong to "normal" spinel, the different relative degrees of cation disorder can be expected in these spinels due to the different geological environment, especially temperature. Higher temperature will affect the cation substitution between T and M positions to some extent and promote the order-disorder phase transition. The FWHM of 406 cm^{-1} peak in the Raman spectrum increases obviously with the degree of the order–disorder phase transition [49–52].

According to the previous literature, the FWHM (full width at half maximum) of the 406 cm^{-1} line in the Raman spectrum is in the range of 6.8–10.6 cm^{-1} and varies in different origins in natural unheated gem spinels [4]. Based on this finding, the obtained 406 cm^{-1} lines underwent Lorentz fitting, and the FWHM values of this line were calculated (Figure 20a).

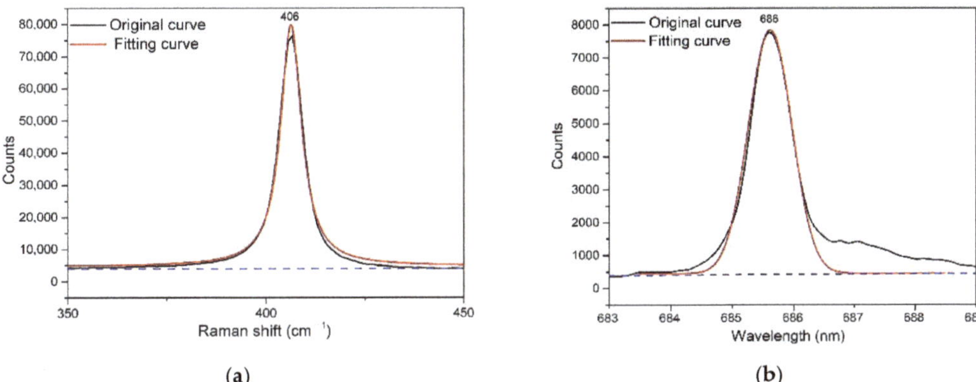

Figure 20. (**a**) The Lorentz fitting of the 406 cm^{-1} line of TS10. (**b**) The Lorentz fitting of the 686 nm line of TS10.

The values of Myanmar are distinctly higher (Figure 21a). The FWHM values of Tanzanian spinels are obviously lower than Myanmar samples and are in the range of 6.56–7.76 and 8.67–9.86, respectively. Thus, the FWHM serves as a good indicator to differentiate purple spinels from Tanzania and Myanmar.

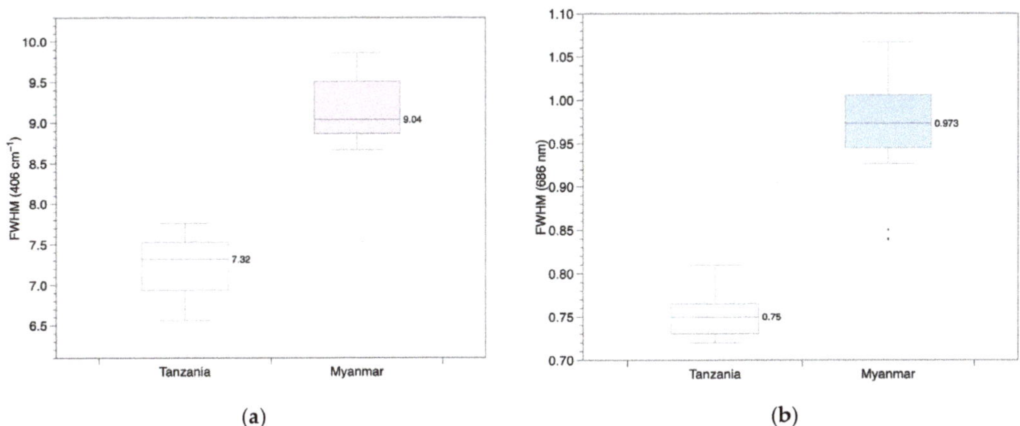

Figure 21. (**a**) The FWHM of the 406 cm^{-1} peak in the Raman spectrum in spinels from Tanzania and Myanmar. The values of the former are distinctly lower than those of the latter, with no overlapping. (**b**) The FWHM of the 686 nm peak in the PL spectrum in spinels from Tanzania and Myanmar.

The FWHM of this sharp Cr^{3+} zero phonon line of the PL spectrum proves to be origin-dependent [4]. Similarly, we performed Lorentz fitting of this peak at ~686 nm and calculated the according FWHM values (Figure 20b). It is not surprising that the FWHM values for Myanmar samples are higher (0.839–1.067) than those for Tanzanian spinels (0.720–0.810) (Figure 21b), which is consistent with the Raman analysis.

Due to the limited quantity of samples in this study, more samples with similar color from other sources need to be collected and analyzed to better confirm this conclusion in the further study.

6. Conclusions

Spinels from Tanzania (TS) and Myanmar (MS) both had a purple hue with a pinkish or brownish secondary tone, medium–strong saturation, and red/orange fluorescence in UV (365 nm). Moreover, green fluorescence was observed only in sample TS due to its relatively higher Mn concentration. Fine dust-like exsolved particles, oriented needles, dolomite, and forsterite are typical of Tanzania spinel. Myanmar spinels contain various mineral inclusions, including magnesite, apatite, baddeleyite, anhydrite, pyroxene, and graphite. Although carbonate is frequent in spinel, it occurs as dolomite in Tanzanian spinel and magnesite in Myanmar spinel. The FWHM values of the 406 cm^{-1} line in the Raman spectrum and the Cr^{3+} zero phonon line in the PL spectrum are relatively higher in sample MS than TS. Tanzanian spinels are characterized by extremely richer Fe and Mn concentrations, as well as a higher Fe/Cr ratio, compared to Myanmar samples. The purple color of all the samples is caused by Fe^{2+}, with minor Fe^{3+} and Cr^{3+}. Fe and Cr/V have a prominent, opposite correlation with the hue angle or chroma of purple spinel from Tanzania. In contrast, V and Fe hardly affect these two parameters in Myanmar samples. All inclusion scene, spectral, and chemical characteristics, as well as the comparison in this study are limited to the small number of samples from Tanzania and Myanmar. More samples from these mines, as well as other sources such as Vietnam, Afghanistan, and Pakistan, are needed to extend this study and to perform proper origin determination.

Author Contributions: Conceptualization, J.W., X.S. and P.N.; methodology, J.W. and X.S.; validation, J.W., T.Z., and Y.M.; formal analysis, J.W., X.S. and H.M.; investigation, J.W., Y.M. and N.T; data curation, J.W. and N.T.; writing—original draft preparation, J.W. and X.S.; writing—review and editing, J.W., P.N. and H.M.; supervision, T.D., H.L. and T.Z.; project administration, J.W. and H.L.; funding acquisition, T.D. and P.N. All authors have read and agreed to the published version of the manuscript.

Funding: This research was funded by NGTC Research Foundation, grant number NGTCSZ2022003.

Data Availability Statement: Not applicable.

Acknowledgments: Luoyuan Liu is thanked for his assistance in LA-ICP-MS testing. We also thank four anonymous reviewers and the academic editor for their constructive comments, which helped us to improve the manuscript.

Conflicts of Interest: The authors declare no conflict of interest.

References

1. D'Ippolito, V.; Andreozzi, G.B.; Bersani, D.; Lottici, P.P. Raman fingerprint of chromate, aluminate and ferrite spinels. *J. Raman Spectrosc.* **2015**, *46*, 1255–1264. [CrossRef]
2. Giuliani, G.; Fallick, A.E.; Boyce, A.J.; Pardieu, V.; Pham, V.L. Pink and red spinels in marble: Trace elements, oxygen isotopes, and sources. *Can. Mineral.* **2017**, *55*, 743–761. [CrossRef]
3. D'Ippolito, V.; Andreozzi, G.B.; Hålenius, U.; Skogby, H.; Hametner, K.; Günther, D. Color mechanisms in spinel: Cobalt and iron interplay for the blue color. *Phys. Chem. Miner.* **2015**, *42*, 431–439. [CrossRef]
4. Saeseaw, S.; Wang, W.Y.; Scarratt, K.; Emmett, J.L.; Douthit, T.R. Distinguishing Heated Spinels from Unheated Natural Spinels and from Synthetic Spinels: A Short Review of On-Going Research. GIA. 2009. Available online: https://www.gia.edu/doc/Heated-spinel-Identification-at-April-02-2009.pdf (accessed on 15 November 2022).
5. Bosi, F.; Biagioni, C.; Pasero, M. Nomenclature and classification of the spinel supergroup. *Eur. J. Mineral.* **2019**, *31*, 183–192. [CrossRef]

6. Andreozzi, G.B.; D'Ippolito, V.; Skogby, H.; Hålenius, U.; Bosi, F. Color mechanisms in spinel: A multi-analytical investigation of natural crystals with a wide range of coloration. *Phys. Chem. Miner* **2018**, *46*, 343–360. [CrossRef]

7. Malsy, A.K.; Karampelas, S.; Schwarz, D.; Klemm, L.; Armbruster, T.; Tuan, D.A. Orangey-red to orangey-pink gem spinels from a new deposit at Lang Chap (Tan Huong-Truc Lau), Vietnam. *J. Gemmol.* **2012**, *33*, 19–27. [CrossRef]

8. Shigley, J.E.; Stockton, C.M. Cobalt-blue gem spinels. *Gems Gemol.* **1984**, *20*, 34–41. [CrossRef]

9. Schmetzer, K.; Haxel, C.; Amthauer, G. Colour of natural spinels, gahnospinels and gahnites. *J. Mineral. Geochem.* **1989**, *2*, 159–180.

10. Peretti, A.; Kanpraphai-Peretti, A.; Güther, D. World of magnificent spinel: Provenance and identification. *Contrib. Gemol.* **2015**, *11*, 285–292.

11. Pardieu, V. Hunting for "Jedi" spinels in Mogok. *Gems Gemol.* **2014**, *50*, 46–57. [CrossRef]

12. Malsy, A.; Klemm, L. Distinction of gem spinels from the Himalayan Mountain Belt. *Chimia* **2010**, *64*, 741–746. [CrossRef] [PubMed]

13. Phyo, M.M.; Bieler, E.; Franz, L.; Balmer, W.; Krzemnicki, M.S. Spinel from Mogok, Myanmar—A Detailed Inclusion Study by Raman Microspectroscopy and Scanning Electron Microscopy. *J. Gemmol.* **2019**, *36*, 418–435. [CrossRef]

14. Giuliani, G.; Fallick, A.E.; Boyce, A.J.; Pardieu, V.; Pham, V.L. Pink and Red Gem Spinels in Marble and Placers. *Incolor* **2019**, *43*, 14–28.

15. Balocher, F. The spinels of Mogok: A brief overview. *Incolor* **2019**, *43*, 34–39.

16. Chauviré, B.; Rondeau, B.; Fritsch, E.; Ressigeac, P.; Devidal, J.-L. Blue spinel from the Luc Yen District of Vietnam. *Gems Gemol.* **2015**, *51*, 2–17. [CrossRef]

17. Boehm, E.W. Gem Notes: Purple spinel from Badakhshan, Afghanistan. *J. Gemmol.* **2017**, *35*, 694–697.

18. Blauwet, D. Gem News: Spinel from northern Vietnam, including a new mine at Lang Chap. *Gems Gemol.* **2011**, *47*, 60–61.

19. Hänsel, S.; Lhuaamporn, T.; Suphan, C. Gem Notes: Update on Purple Spinel from Afghanistan. *J. Gemmol.* **2021**, *37*, 678–680. [CrossRef]

20. Belley, P.M.; Palke, A.C. Purple Gem Spinel from Vietnam and Afghanistan: Comparison of Trace Element Chemistry, Cause of Color, and Inclusions. *Gems Gemol.* **2021**, *57*, 228–238. [CrossRef]

21. Pardieu, V.; Hughes, R. Spinel: Resurrection of a classic. *InColor* **2008**, *2*, 10–18.

22. Giuliani, G.; Martelat, J.-E.; Ohnenstetter, D.; Feneyrol, J.; Fallick, A.E.; Boyce, A. Les gisements de rubis et de spinelle rouge de la ceinture met'amorphique neóoprotefozoïque mozambicaine. *Rev. Gemmol. AFG* **2015**, *192*, 11–18.

23. Garnier, V.; Giuliani, G.; Ohnenstetter, D.; Fallick, A.E.; Dubessy, J.; Banks, D.; Vinh, H.Q.; Lhomme, T.; Maluski, H.; Pecher, A. Ar-Ar and U-Pb ages of marble-hosted ruby deposits from central and southeast Asia. *Can. J. Earth Sci.* **2006**, *34*, 169–191. [CrossRef]

24. Iyer, L.A.N. The geology and gemstones of the Mogok Stone Tract, Burma. *Mem. Geol. Surv. India* **1953**, *82*, 100.

25. Searle, M.P.; Noble, S.R.; Cottle, J.M.; Waters, D.J.; Mitchell, A.H.G.; Hlaing, T.; Horstwood, M.S.A. Tectonic evolution of the Mogok metamorphic belt, Burma (Myanmar) constrained by U-Th-Pb dating of metamorphic and magmatic rocks. *Tectonics* **2007**, *26*, 1–24. [CrossRef]

26. Thu, K. The Igneous Rocks of the Mogok Stone Tract: Their Distributions, Petrography, Petrochemistry, Sequence, Geochronology and Economic Geology. Ph.D. Thesis, University of Yangon, Yangon, Myanmar, 2007; p. 139.

27. Thu, Y.K.; Win, M.M.; Enami, M.; Tsuboi, M. Ti–rich biotite in spinel and quartz–bearing paragneiss and related rocks from the Mogok metamorphic belt, central Myanmar. *J. Mineral. Petrol. Sci.* **2016**, *111*, 270–282.

28. Zaw, K.; Sutherland, L.; Yui, T.-F.; Meffre, S.; Thu, K. Vanadium-rich ruby and sapphire within Mogok gemfield, Myanmar: Implications for gem colour and genesis. *Miner. Depos.* **2015**, *50*, 25–39. [CrossRef]

29. Themelis, T. *Gems and Mines of Mogok*; A&T Pub.: Los Angeles, CA, USA, 2008; p. 352.

30. Chankhantha, C.; Amphon, R.; Rehman, H.U.; Shen, A.H. Characterisation of Pink-to-Red Spinel from Four Important Lo-calities. *J. Gemmol.* **2020**, *37*, 393–403. [CrossRef]

31. Schmetzer, K.; Berger, A. Lamellar inclusions in spinel from Morogoro area, Tanzania. *J. Gemmol.* **1992**, *23*, 93–94. [CrossRef]

32. Gübelin, E.J.; Koivula, J.I. *Photoatlas of Inclusions in Gemstones*; Opinio Verlag: Basel, Switzerland, 2005; Volume 2, pp. 662–714.

33. Chopelas, A.; Hofmeister, A.M. Vibrational spectroscopy of aluminate spinels at 1 atm and of $MgAl_2O_4$ to over 200 kbar. *Phys. Chem. Miner.* **1991**, *18*, 279–293. [CrossRef]

34. Wang, C.S.; Shen, A.H.; Liu, Y.G. Characterization of order-disorder transition in $MgAl_2O_4$: Cr^{3+} spinel using photoluminescence. *J. Lumin.* **2020**, *227*, 117552. [CrossRef]

35. Hainschwang, T.; Karampelas, S.; Fritsch, E.; Notari, F. Luminescence spectroscopy and microscopy applied to study gem materials: A case study of C centre containing diamonds. *Miner. Petrol.* **2013**, *107*, 393–413. [CrossRef]

36. Nassau, K. The origins of color in minerals. *Am. Mineral.* **1978**, *63*, 219–229.

37. Hålenius, U.; Skogby, H.; Andreozzi, G.B. Influence of cation distribution on the optical absorption spectra of Fe^{3+}-bearing spinel s.s.–hercynite crystals: Evidence for electron transitions in $^{VI}Fe^{2+}$– $^{VI}Fe^{3+}$ clusters. *Phys. Chem. Miner.* **2002**, *29*, 319–330.

38. Taran, M.N.; Koch-Müller, M.; Langer, K. Electronic absorption spectroscopy of natural (Fe^{2+}, Fe^{3+})-bearing spinels of spinel s.s.hercynite and gahnite-hercynite solid solutions at different temperatures and high-pressures. *Phys. Chem. Miner.* **2005**, *32*, 175–188. [CrossRef]

contain them [2]. The Reggio Emilia province is also rich in other minerals used in jewelry, such as datolite and prehnite, that occur in the ophiolitic lithologies scattered in the chaotic clays, and hyaline quartz, which is found in the sandstones of the upper Apennines [2]. In this work, we examined the geological context and the geochemical, mineralogical, and gemological characteristics of the black quartz to understand their genetic conditions and the causes that give the mineral its color. In particular, our work has mainly focused on the characterization of the inclusions present in these quartzes. In quartz, different types of inclusions with liquid, gaseous, or solid components are frequently found, sometimes even combined with each other [2]. The study of their composition provides important information on the genesis and physical properties, such as color, of the host minerals.

2. Geological Setting

The lithologies that include black quartz are part of the Tuscan Nappe that is the sedimentary succession deposited in the most proximal portion to the European continental margin. The Tuscan Nappe is partially covered by other nappes, like Cervarola, Modino, and Ligurian Units, and its basal sequence consists of the Triassic evaporitic Burano Formation [8–12] (Figure 1). This formation is mainly composed of the alternation of gypsum-anhydrite rocks ranging in size from meters to decameters, and dark gray dolomitic limestones; at a depth, there are also rock salt crystals, not observable on the surface [13–16] (Figure 2).

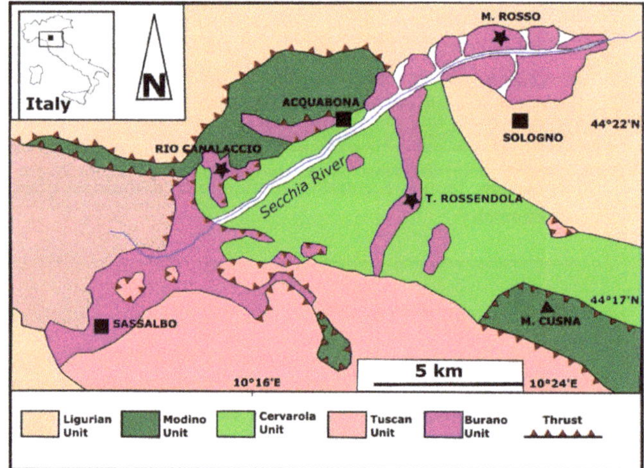

Figure 1. Simplified geological map of the upper Apennines of Reggio Emilia province with the Triassic Burano Formation; the black stars indicate the sites of the studied samples. Map redrawn from [13].

These evaporitic rocks are very soluble, and, consequently, widespread karst phenomena such as sinkholes, aquifers, and saline springs are observed along the banks of the Secchia river and its tributary Ozola [14,17–19]. Near Reggio Emilia, the Burano Formation reaches a thickness of 2200 m and has been interpreted as a transpressive system, transversal to the main tectonic lines of the northern Apennines [20,21]. During a deformation phase in the Miocene, there was a disruption in thrust slices and inclusion in younger allochthones units [22,23]; as a consequence, the outcrops along the Secchia and Ozola valleys are subjected to widespread phenomena of instability such as massive landslides (Figure 3).

Figure 2. Spectacular outcrop of the Burano Formation in Val Secchia near Sassalbo, (Massa province, Tuscany) where you can observe the alternation of lighter colored levels of gypsum-anhydrite and darker levels of dolomitic limestones. Black quartzes are found mainly in anhydrite levels.

Figure 3. Southern slope of Mt. Rosso outcropping along River Secchia Valley, with widespread phenomena of instability.

The mineralogical composition of the Burano Formation consists mainly of gypsum, quartz, sulfur, aragonite, calcite, dolomite, barite, celestine, magnesite, fluorite, and pyrite [2,16]. (Figure 4).

Figure 4. Minerals from the Burano Formation: (**a**) pyrite from Mt. Rosso (0.6 mm length); (**b**) sulfur from Rio Biola (Reggio Emilia, Italy), 5.3 mm; (**c**) fluorite from Sassalbo (Massa province), 2.1 mm; and (**d**) dolomite from Rio Canalaccio, Reggio Emilia, 3.9 mm. Photos by Enrico Bonacina.

The black quartz crystals are found exclusively in some levels of anhydrite and gypsum that never exceed two meters in thickness and emerge for a few tens of meters (Figure 5).

At the top of the reliefs where the Burano Formation outcrops, the meteoric water has dissolved the evaporitic rocks, releasing the black quartz crystals incorporated in them, and consequently the crystals are found isolated in the eluvial soils of these areas.

Figure 5. Anhydrite levels with black quartz crystals, from Mt. Rosso.

3. Materials and Methods

About 350 euhedral black quartz crystals, ranging in size from 2 to 30 mm, were sampled by M.S. or gifted by local collectors and were used for morphological study. Standard gemological analyses were performed on five gems (3.54–14.54 ct weight) cut from rough samples, to describe optical properties, specific gravity, and ultraviolet fluorescence. Density was measured using a Presidium PCS100 Sensible hydrostatic balance, and the color was evaluated with an RGB (red, green, blue) [24] color table method. The refractive index was measured with the distant vision method using a Kruss refractometer (1.45–1.80 range) and a contact liquid with an RI of 1.80. Ultraviolet fluorescence was investigated with a shortwave (254 nm) and long-wave (365 nm) UV lamp (Power 3W 20 cm distance of observation). Micro-Raman scattering measurements were conducted with a Horiba Jobin Yvon Explora Plus single monochromator spectrometer (grating of 2400 groove/mm), equipped with an Olympus BX41 microscope. The analyses were conducted on the inclusions previously observed under the microscope, to better understand their mineralogical composition. Raman spectra were recorded with 532 nm excitation. The spectrometer was calibrated to the silicon Raman peak at 520.5 cm^{-1}. The spectral resolution was ~2 cm^{-1}, and the instrumental accuracy in determining the peak positions was approximately 0.56 cm^{-1}. Raman spectra were collected in the spectral range 100–4000 cm^{-1} for 5 seconds, averaging over 40 scans accumulated. Trace elements were determined by laser-ablation inductively coupled plasma mass spectroscopy (LA–ICP–MS) at the IGG-CNR Laboratory of Pavia, combining an excimer laser (193 nm; Lambda Physik with GeoLas optics) with a triple quadrupole ICPMS (QQQ Agilent 8900). The analyses concerned the whole samples of quartz. The laser was operated at a repetition rate of 10 Hz, with a fluence of 6 J·cm^{-2} and an ablation spot size of 55 μm. The optimization of LA–ICP–MS to minimize elemental fractionation was performed by ablating NIST 612 glass and adjusting the nebulizer Ar and the carrier laser cell He gas flows to obtain the ratio of ^{232}Th and ^{238}U signals close to 1, by minimizing the ThO$^+$/Th$^+$ ratio (<1%) in order to reduce the formation of polyatomic oxides. The selected masses were acquired in MS/MS mode, and each analysis consisted of the acquisition of 1 min of background before and after about 1 min of ablation signal. Data reduction was performed with the "GLITTER" software package [21], using NIST SRM 610 glass as an external standard and ^{29}Si as an internal standard, changing the value in each analysis as from microprobe. Precision and accuracy estimated on the USGS basaltic glass standard BCR2 and NIST612 were better than 10%. Three spots per sample were measured.

4. Results

4.1. Morphology

The quartz crystals of the Burano Formation exhibit some dimensional (Figure 6) and chromatic variability: few crystals are very light grey or grey (Figure 7a), while most, particularly those examined in this work, have a black opaque color and a waxy, sometimes greasy luster (Figure 7b). Many crystals appear bi-terminated with well-developed faces (Figure 7c); they more frequently show a prismatic habit, formed by the combination of an apparently hexagonal prism (actually the combination of two trigonal prisms) with two rhombohedrons (one positive and one negative) often twinned and equally developed to simulate a hexagonal bipyramid. Sometimes the faces of the rhombohedrons are much extended and cover the faces of the prism—in this case, the crystal takes on an almost bipyramidal habit (Figures 6 and 7d). Shortened and squat crystals are also observed, and some appear disproportionate or deformed. Few samples have a twinning according to the law of Japan [25], in which two flat crystals are attached to each other with the main axes arranged at 84° 33′.

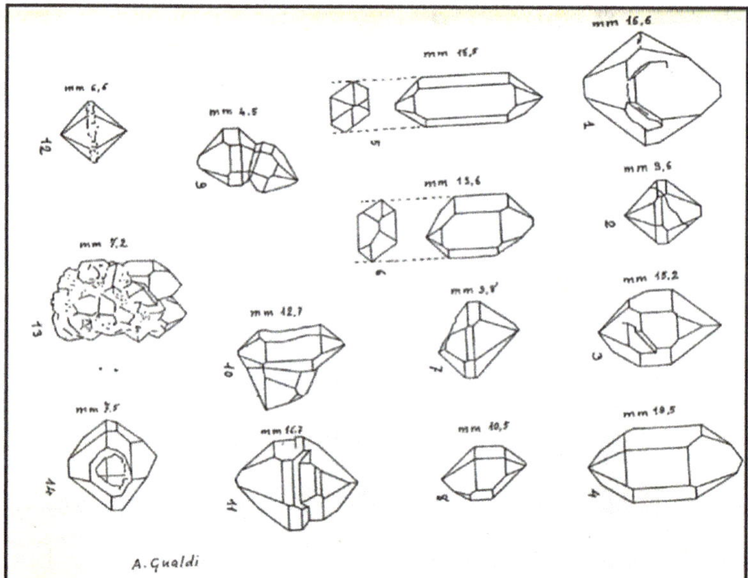

Figure 6. Morphological table of Burano Formation quartz (drawn by Alberto Gualdi).

The minimum values of length and width of 355 crystals from the Burano Formation were measured with a micrometer; the values are shown on the diagram of Figure 8 (black spots). Since the minimum length/width ratio (L/W) is directly proportional to the length of the prism, its values can be used to describe the type of habitus. The L/W ratio is a very important parameter for gem cutters who use it frequently to determine the correct patterns for cutting gemstones and also create new gem designs. The values of the L/W ratio slightly higher than 1 correspond to type III, i.e. a habitus without a prism, while the values of the L/W ratio of about 2.5 correspond to type VI, with a fairly well developed prism [26,27].

Figure 7. Quartz crystals of the Val Secchia (Monte Rosso) gypsum rock formation: (**a**) grey: length 15 mm; (**b**) black: length of the crystal on the left 30 mm; (**c**) black: length 32 mm; (**d**) black: length 30 mm. Note the rough surface due to the imprints of the white anhydrite crystals. Photos by Antonio Miglioli.

No close correlations are observed between crystal type, size, and color; but crystals longer than 1 cm with an L/W ratio of about 2 are more frequently black, while those of type III shorter than 1 cm are more frequently gray. On the contrary, a closer correlation between the color of the crystals and their location is observed, as black crystals are almost exclusively present in some anhydrite layers in the eastern outcrops of the Burano Formation. The shapes of the black quartzes of the Burano Formation are mainly prismatic-hexagonal, and the L/W ratio varies only between 1.17 and 3. For comparison purposes, the values for quartz crystals from the Alps (Val Veny, Triolet, and Ayas in Aosta Valley; Reale in Formazza Valley; and Dosso dei Cristalli in Malenco Valley) are also reported (blue spots). The quartz crystals present in the cracks of the rocks of the Alps often show a typical rhombohedral shape, very elongated and pointed. In the diagram of Figure 8 we have plotted the values of W and L of about one hundred crystals collected by M.S. in various Alpine locations; we observe that most of the crystals have an index L >> 3 (blue spots).

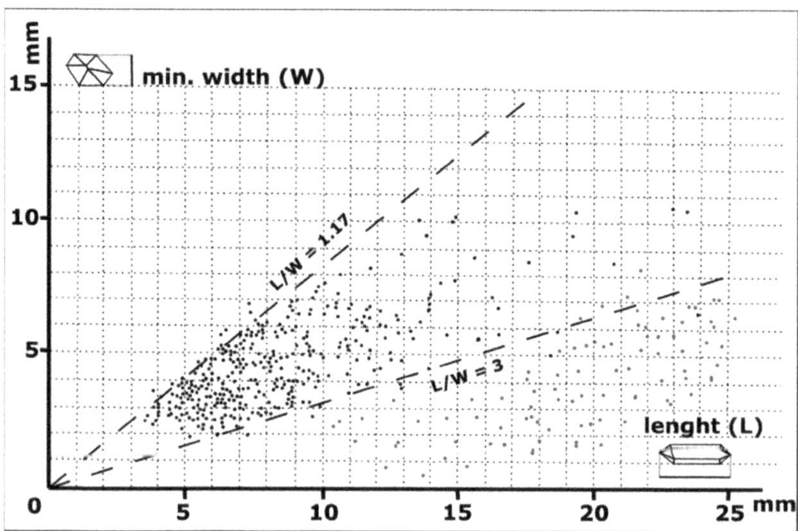

Figure 8. Diagram of length vs. minimal width on 355 crystals from the Burano Formation (black spots) and 104 hyaline quartz crystals from Alps (blue spots).

4.2. Gemological Results

Quartz crystals can be faceted even if included—in fact none of our crystals broke during cutting. Sometimes, since the crystals have a nice euhedral morphology, it is preferable to use the simple shapes already present and carry out only the polishing or regularization of any anomalous faces. The surfaces of the crystals are often pitted by the encasing rock and have high luster; the crystals are usually opaque (Figure 9).

Table 1. Gemological properties of black quartz from the Burano Formation.

Sample	Dimensions (mm)	Weight (ct)	Cut	Shape	RI	Specific Gravity	Color (RGB)
1	12 × 9 × 4	3.54	Cabochon	Pear	1.542; 1.550	2.661	Black
2	11 × 11 × 5	5.72	Cabochon	Round	1.546; 1.553	2.651	Black
3	12 × 9 × 6	5.83	Cabochon	Oval	1.544; 1.552	2.643	Black spots grey
4	19 × 15 × 9	14.54	Fancy	Hexagonal	1.543; 1.552	2.650	Black spots grey
5	14 × 12 × 6	6.11	Fancy	Rhomboid	1.542; 1.551	2.640	Black spots grey

The faceted specimens have vivid, intense, and rich colors, ranging from homogeneous black to black with grayish spots, and are of gemological interest. The luster is vitreous; the diaphaneity is opaque. The gems were cut from specimens with a form resulting from a combination of hexagonal prisms and hexagonal bipyramid (Figure 10).

The physical and optical properties of the investigated gems are reported in Table 1.

The specific gravity ranges from 2.643 to 2.661; the refractive indexes between 1.542–1.548 and 1.550–1.553 (using spot analysis and turning the sample), in agreement with the literature data for similar quartz. All the analyzed gems are inert to long and short UV.

Figure 9. Quartz round cabochon (diameter 11 mm, Table 1, sample 2) with small veins of anhydrite. Photo by Enrico Borghi.

(a) (b)

(c) (d)

Figure 10. (**a**) Oval cabochon (length 12 mm, sample 3; (**b**) pear cabochon (length 12 mm, sample 1); (**c**) fancy black quartz (length 8 mm, sample 4); and (**d**) fancy black quartz (length 14 mm, sample 5). Photo by Enrico Borghi.

4.3. Raman Results

Raman analyses on quartz and inclusions (gemstone n. 2) were carried out to confirm that the color is actually due to the presence of graphite. Since the attempts on cabochon gems were unsuccessful, we performed the analysis on a thin section of the sample; however, given that the analyzed inclusions are not superficial but completely incorporated in the quartz, we believe it is unlikely that the polishing operation has altered the quality of the analyses. The Raman analyses of quartz and relative inclusions were carried out on a thin petrographic section of gem 2 in order to confirm that the color is linked to graphite (Figure 11).

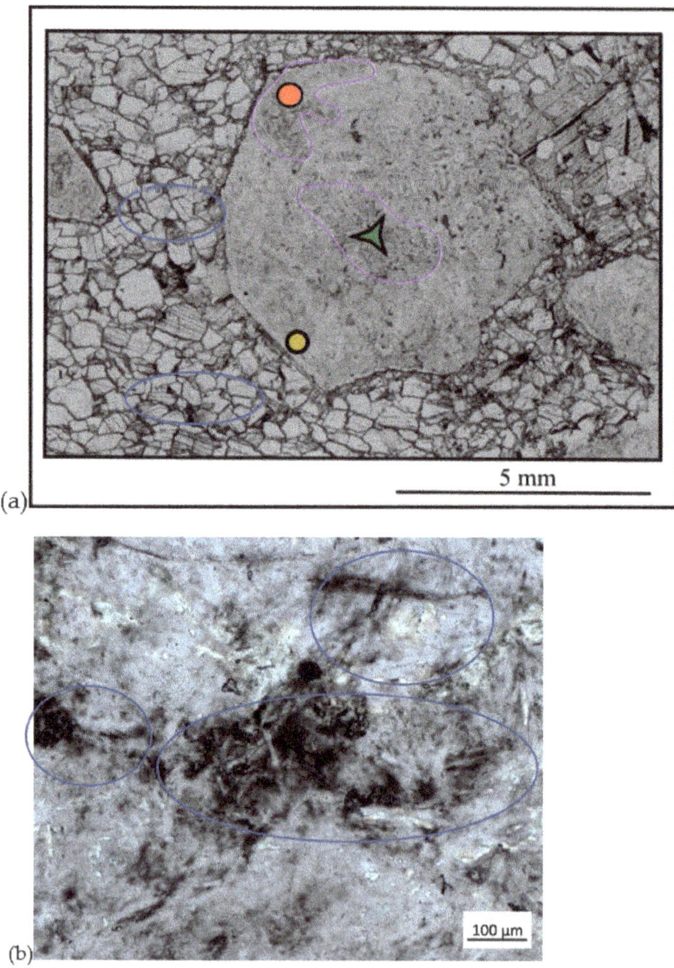

Figure 11. *Cont.*

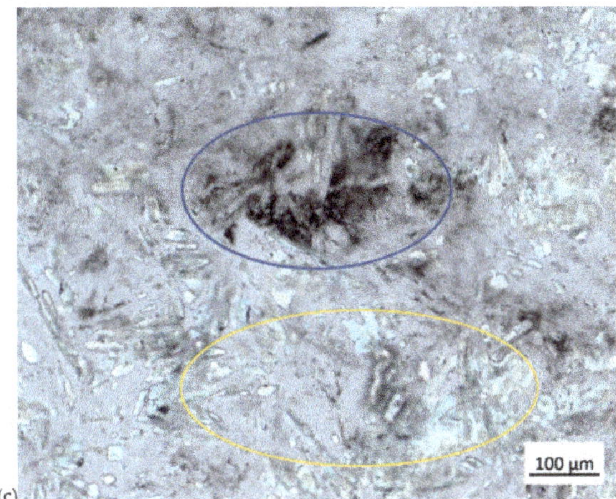

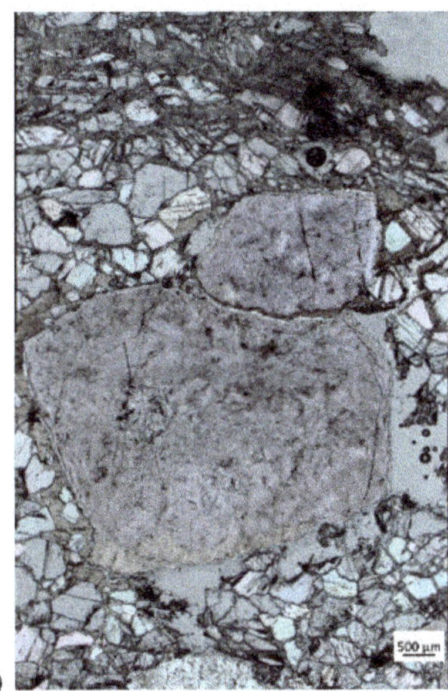

Figure 11. (**a**) petrographic thin section with disordered graphite (red circle), anhydrite (green star), and bukovskyite (orange circle) inclusions incorporated into the quartz. Inside the host rock there are also (**b**) gypsum crystals (blue ovals); (**c**) graphite (blue ovals) and anhydrite (yellow ovals) inclusions clusters (20×); and (**d**) dark spots, all of graphite clusters (2.5×).

In Figure 12 we report the Raman analysis spectra of four points (red color), which highlight the presence of anhydrite, bukovskyite, and disordered graphite, together with the host quartz, and the relative RRUFF database spectra (grey color). According to [28], the bending of Si–O–Si bond angles and twisting are localized below 300 cm^{-1} (126, 206, and 263 cm^{-1} in Figure 12c), and the bending of O–Si–O angles are localized in the region 350–500 cm^{-1} (395, 465cm^{-1} in Figure 12c). According to [29,30], in the sulfate group vibrational modes of Ca-O bonds appear in Raman spectra in the region below 400 cm^{-1}. In Figure 12 the bands at 140 and 160 cm^{-1} are due to the vibration of Ca-O bond. The bands at 416, 508 cm^{-1} and 612, 628, 680 cm^{-1} are assigned respectively to the symmetric and antisymmetric bending SO$_4$. The band at 1020 is due to the symmetric stretching SO$_4$. The peak at 1160 is due to the antisymmetric stretching SO$_4$ stretching mode. Along the edge of the quartz was recognized the bukovskyite [29], a powdery micro-crystalline aggregates of metacolloidal material, commonly associated with gypsum and kankite (Figures 11 and 12b). According to [31], the band at 814 cm^{-1} is assigned to the antisymmetric stretching of AsO4; the bands at 984 and 1010, 1131cm^{-1} are due to the symmetric and antisymmetric stretching of SO4, respectively. The band at 1652 cm^{-1} is assigned to the H$_2$O bending vibration. Raman spectroscopy also made it possible to evaluate the degree of crystallinity of the graphite inclusions [32]. The crystal lattice of graphite consists of an ordered stacking of planes within which the C atoms are organized in hexagonal rings, forming a honeycomb structure. The crystal structure of hexagonal graphite corresponding to the space group P63/mmc is realistic only for the case of an ideal single crystal. The decrease in the size of the crystalline domains leads to a decrease in the degree of order, which is greater in nanocrystalline graphite where the lattice planes are distorted and often formed by rings with an odd number of atoms. The exact interpretation of the position and shape of the graphite bands depends on several factors, such as the excitation of the laser, the orientation, and the resolution [32]. The 1581 cm^{-1} frequency band is related to a C-C vibration stretching of the individual planes and is common to all the different forms of carbon that have sp2 hybridized atoms. As the size of the crystalline domains decreases, a further broad band (not present in the monocrystalline graphite) is observed at about 1350 cm^{-1} (1340 cm^{-1} in Figure 12d), which indicates the presence of structural defects such as the deformations of the planes due to the small size of the crystal. Raman spectra of graphite show that the increase in crystallinity is accompanied by the disappearance of the band at about 1350 cm^{-1}. The peak at 1620 cm^{-1} (see Figure 12d) due to the antisymmetric translation motion is characteristic of polycrystalline graphite alone [33–36].

4.4. LA-ICP-MS Results

LA-ICP-MS analyses concerned the whole samples of five quartz; the results are reported in Table 2. We observe that the investigated quartz has low contents of trace elements; in fact, the Si–O bonds in quartz have a very stable atomic configuration so that only small amounts of other chemical elements can enter the structure of this mineral. In some cases, however, some trace elements, vicariant or interstitial, can enter the quartz structure either by replacing the Si or by entering the channels that run parallel to the c axis [1].

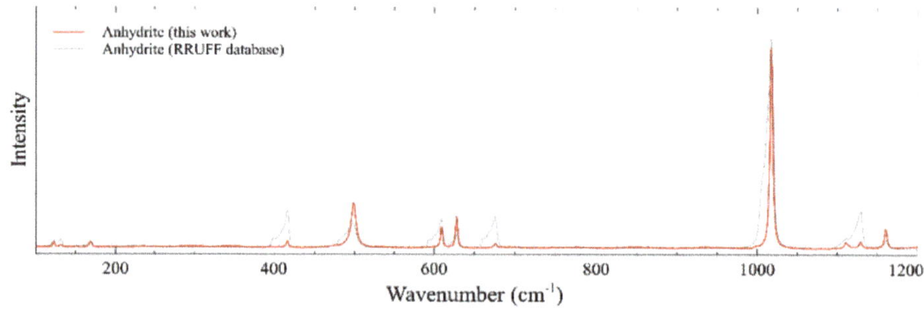

Figure 12. *Cont.*

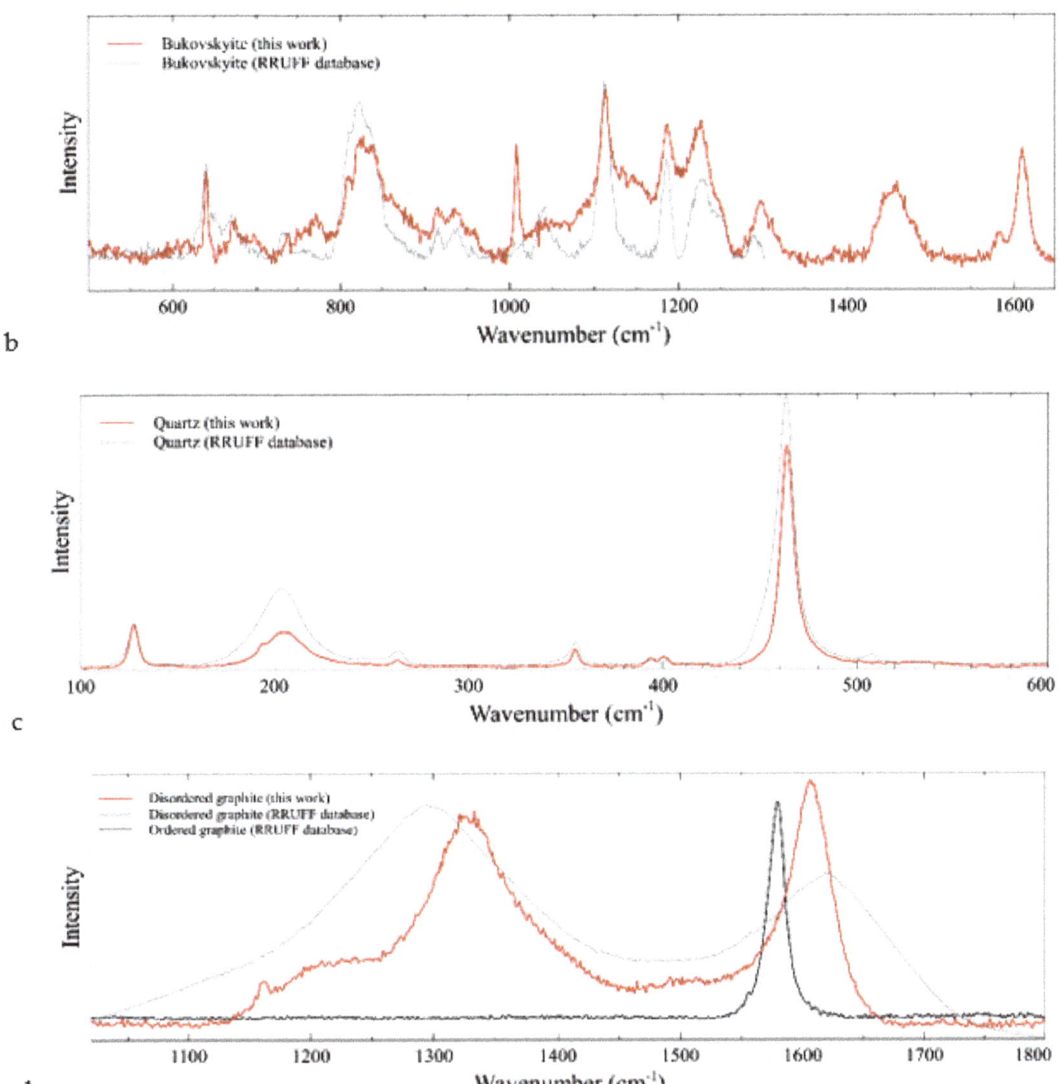

Figure 12. The Raman spectra of this work (red color) and the relative RRUFF database (grey color): (**a**) Raman spectra of anhydrite (416, 508, 612, 1129 cm^{-1}) [29,30,37]; (**b**) bukovskyite (814, 984, 1010, 1131, 1652 cm^{-1}) [29]; (**c**) quartz (128, 206, 263, 356, 465 cm^{-1}) [28,38]; and (**d**) disordered graphite (1340, 1620 cm^{-1}) [32,39] (Sample 2).

The most abundant element in the analyzed quartz is Al (194–352 ppm), followed by lower contents of Hf (26–90 ppm), B (44–50 ppm), Na (35–53 ppm), Li (16–33 ppm), Mg (4–7 ppm), Sc (6–7 ppm), and Sn (3–10 ppm). Overall we observe that these quartzes are devoid of chromophore elements such as Fe, Mn, and Ti (below detection limits). Furthermore, in accordance with the previous models in the literature, a good correlation is observed between the concentration of monovalent cations (Li + Na) and those of Al. Al is the only trace substituent of Si, while Li and Na are interstitial elements that act as charge compensators. Finally, a relevant quantity of hafnium is found, the origin of which is difficult to establish.

Table 2. Trace element composition of the quartz investigated in this work, obtained through LA-ICP-MS analysis (three spots per sample; the values in table are the averages). The elements that resulted below the detection limits (bdl) are not reported; detection limits are reported in [40].

Element	Trace Element Concentrations (ppmw) SDD (Limits of Detection)					
	Qz 1	Qz 2	Qz 3	Qz 4	Qz 5	
^{7}Li	18.92	25.6	31.69	16.48	33.47	7.52
^{11}B	44.46	45.94	44.12	45.99	50.54	2.56
^{23}Na	36.61	35.68	53.09	36.63	36.38	7.51
^{25}Mg	3.9	2.89	6.97	4.16	3.98	1.53
^{27}Al	194.49	248.27	352.12	241.14	342.18	68.56
Si (%)	99.8	99.8	99.8	99.8	99.8	
^{45}Sc	7.41	7.62	6.96	5.84	6.56	0.71
^{118}Sn	3.3	6.19	6.23	2.85	9.54	2.70
^{177}Hf	45.32	79.69	bdl	31.23	90.41	28.85

5. Discussion and Conclusions

The analyses carried out in this work allowed for the estimation of the gemological properties of black quartz, to better understand their geologic origin and the causes of their dark color. The euhedral quartz crystals include many different types of inclusions—according to our observations and those reported in [6,41], the solid ones are composed mostly of disordered graphite and anhydrite. Gypsum inclusions are absent, even in quartz crystals hosted in gypsum rich rocks and into evaporite-free residual deposits. The widespread presence of these anhydrite inclusions would indicate that the sulphate deposits, during the growth of the quartz, were mainly composed of anhydrite. However, experimental studies and observations on current evaporite deposits [42,43] have shown that anhydrite has little chance of precipitating in submarine evaporitic environments; consequently, the most common precipitate in Triassic marine waters was gypsum, not anhydrite. It can therefore be hypothesized that during the tectonic processes that affected the northern Apennine chain, the syn-depositional gypsum was dehydrated and converted into anhydrite, due to the increase in geothermal temperatures that referred to the increase in the depth of the burial. As a matter of fact, the transformation of gypsum into anhydrite due to increasing burial temperatures explains the general absence of gypsum at depths greater than 1000 m, while anhydrite is always present. This hypothesis agrees with the model of the tectonic evolution of the Burano Formation that was buried by the 2000 m-thick carbonate sequence of the Tuscan Nappe during the Cretaceous [44]. Subsequently, the exhumation and denudation of the Burano Formation produced a complete hydration of anhydrite rocks, which were converted into gypsum. The solid inclusions in the euhedral quartz crystals, therefore, provide further evidence for the geologic evolution of the Burano and correspond to the Apuan metamorphic complex of green schist facies [45]. We can therefore hypothesize that quartzes are authigenic and formed in the Oligocene–Miocene during the burial of the evaporitic formation, when gypsum converted into anhydrite. The absence of elements such as Ti, K, and Fe (bdl) and the transition elements would exclude magmatic genesis. We believe that quartz was formed from high salinity fluids, produced by an increase in pressure and temperature inside the evaporites; the assumed temperature during this process is about 260–300 °C [46–51]. The fluids would have solubilized the organic silica of the microorganisms contained in the evaporitic bodies, subsequently redepositing it as idiomorphic quartz in the surrounding anhydrites. In fact, it is known that diatoms with a siliceous shell live in waters with very variable salinities, but always at low depths, as in the photic zone. A possible confirmation of this origin of the fluids is given by the considerable presence of Hf inside the quartz crystals, as this element is present in various types of sediments, especially organic, according to the literature [52–54]. In fact, the solubility of Hf in fluids, generally modest, increases when the Hf enters complexes containing sulfate, arsenate, but above organic anions. Another source of silica could be represented by the thin layers of detrital silt, up to 50 cm thick, interspersed in the Burano Formation, which are mainly composed of quartz, muscovite, illite, chlorite, and feldspar [41]. Our analyses also allowed us to ascertain the causes of the black color of the quartz, which is not to be attributed to the chromophores, as they are present in very low quantities. [33,35,36]. We believe that the black color is not even to be attributed to the color centers, as in the case of the smoky quartz, as even after heating (280 °C/2 h, in an oxidizing atmosphere), the color of our samples remains dark, while the color centers should disappear after this treatment [1,3,55]. Additionally, smoky quartz crystals exhibit a vitreous luster, while our crystals exhibit a distinct metallic luster and opaque appearance due to the presence of graphite. Consequently, the dark color should be attributed to the disordered graphite inclusions previously described. The color of the samples becomes darker and more intense with the increase of these inclusions; therefore, these quartzes should be called black and not smoky quartz. However, it must be emphasized that, according to some articles, the polishing operations could alter the graphitic carbon, and consequently the Raman spectra may not be really representative of the composition of the inclusions [32–34]. Consequently, we believe the exact nature of these inclusions will need to be further investigated by other techniques like UV-Vis [56]. The chromatic inhomogeneity of these gems is instead to be attributed to the presence of large anhydrite inclusions.

The black quartzes of the Burano Formation represent an appreciable gemological material. Unfortunately, however, the difficulties in reaching the deposit limit the marketing of these quartzes, which are well known by Italian and European collectors. Furthermore, the locality is a protected park in the Emilia Romagna region, which, to safeguard the eco-sustainability of quartz extraction, grants permission only to goldsmiths. In Italy, a significant problem in recent decades has been the abandonment by local populations of mountain areas such as the Apennines and the Alps. In fact, for reasons of comfort of life and job opportunities, the younger generations leave the regions of the mountains to move to the cities of the plains, but this generates significant social and environmental problems; moreover, these small towns of the Apennines are considered among the most typical and beautiful elements of Italy. To stem this phenomenon and encourage young people to stay in their villages, it is necessary to create new cultural interests and job opportunities that are sustainable from an environmental point of view. We think that encouraging local artisanship of precious stones can be a valid help in this direction, as long as the extraction of the stones is done manually and in a strictly controlled manner. We also suggest creating a geopark for educational and informative purposes, which can also be combined with agrotourism activities.

Author Contributions: Conceptualization, F.C. and M.S.; methodology, F.C. and M.G.; software, M.S.; validation, all the authors; formal analysis, F.C. and M.G.; investigation and data curation, all the authors; writing—original draft preparation, F.C., M.S. and L. M.; writing—review and editing, F.C., M.S. and L.M.; visualization and supervision, all the authors. All authors have read and agreed to the published version of the manuscript.

Funding: This research received no external funding.

Acknowledgments: The authors are grateful to Antonio Langone for the LA-ICP-MS analyses, to Omar Bartoli for the thin section, and to Antonio Miglioli (Sassuolo, Mo), Enrico Borghi (Re), and Enrico Bonacina (Bg) for several images.

Conflicts of Interest: The authors declare no conflict of interest.

References

1. Caucia, F.; Ghisoli, C. *Quarzi: Definizione, Nomenclatura e Proposta di Classificazione Delle Varietà*; Museo Civico di Storia Naturale Cremona: Cremona, Italy, 2004; p. 259.
2. Scacchetti, M.; Bartoli, O.; Bersani, D.; Laurora, A.; Lugli, S.; Malferrari, D.; Valeriani, L. *Minerali della provincia di Reggio Emilia*; AMI Cremona: Cremona, Italy, 2015; p. 256.
3. Hatipoğlu, M.; Helvacı, C.; Kibar, R.; Çetin, A.; Tuncer, Y.; Can, N. Amethyst and morion quartz gemstone raw materials from Turkey: Color saturation and enhancement by gamma, neutron and beta irradiation. *Radiat. Eff. Defects Solids* **2010**, *65*, 876–888. [CrossRef]
4. Available online: https://www.geologyin.com/2020/02/is-black-quartz-natural.html (accessed on 3 August 2022).
5. Available online: https://www.mindat.org/loc-22957.html (accessed on 3 August 2022).
6. Lugli, S. Petrography of the quartz euhedral as a tool to provide indications on the geologic history of the Upper Triassic Burano Evaporites (Northern Apennines). *Mem. Soc. Geol. It.* **1994**, *48*, 61–65.
7. Lugli, S. La storia geologica dei gessi triassici della Val Secchia. *Ist. It. Spel. Mem.* **2009**, *22*, 25–36.
8. Vighi, L. Sulla serie triassica "Cavernoso-Verrucano" presso Capalbio (Orbetello—Toscana) e sulla brecciatura tettonica delle serie evaporitiche, "Roccemadri" del Cavernoso. *Boll. Della Soc. Geol. Ital.* **1958**, *78*, 221–236.
9. Ciarapica, G.; Passeri, L. Deformazioni da fluidificazione ed evoluzione diagenetica della formazione evaporitica di Burano. *Boll. Della Soc. Geol. Ital.* **1976**, *95*, 1175–1199.
10. Ciarapica, G.; Cirilli, S.; Passeri, L.; Trincianti, E.; Zaninetti, L. "Anidriti di Burano" et "Formation du Monte Cetona" (nouvelle formation), biostratigraphie de deux series types du Trias superieur dans l'Apennin septentrional. *Rev. Paleobiol.* **1987**, *6*, 341–409.
11. Martini, R.; Gandin, A.; Zaninetti, L. Sedimentology, Stratigraphy and micropaleontology of the Triassic evaporitic sequence in the subsurface of Boccheggiano and in some outcrops of southern Tuscany (Italy). *Riv. Ital. Paleontol. Stratigr.* **1989**, *95*, 3–28.
12. Regione Emilia Romagna. *Note Illustrative Della Carta Geologica D'ITALIA Alla Scala 1:50.000, Foglio 235, Pievepelago*; Selca: Firenze, Italy, 2002; p. 140.
13. Colombetti, A.; Fazzini, P. Il salgemma nella formazione dei gessi triassici di Burano (Villaminozzo, RE). *Grotte D'italia* **1986**, *4*, 209–219.
14. Forti, P.; Francavilla, F.; Prata, E.; Rabbi, E.; Chiesi, M. Hydrogeology and hydrogeochemistry of the triassic evaporite in the upper Secchia valley (Reggio Emilia, Italy) and the Poiano karst spring. *Grotte D'italia* **1986**, *4*, 267–278.
15. Lugli, S.; Parea, G.C. Halite cube casts in the Triassic sandstones from The Upper Secchia River valley (Northern Apennines, Italy): Environmental interpretation. *Accad. Naz. Sci. Lett. Arti Modena* **1996**, *15*, 341–351.
16. Lugli, S.; Morteani, G.; Blamart, D. Petrographic, REE, fluid inclusion and stable isotope study of the magnesite from the Upper Triassic Burano Evaporites (Secchia Valley, northern Apennines): Contributions from sedimentary, hydrothermal and metasomatic sources. *Miner. Depos.* **2002**, *37*, 480–494. [CrossRef]
17. Bertolani, M.; Rossi, A. La petrografia del "Tanone grande della Gaggiolina" (154 E/RE) nelle evaporiti dell'Alta Val Secchia (Reggio Emilia—Italia). *Grotte D'italia* **1986**, *4*, 79–105.
18. Chiesi, M.; Forti, P. Le sorgenti carsiche di Poiano. *Ambiente Nat. Po Degli Appennini* **1987**, *3*, 11–15.
19. Lugli, S.; Domenichini, M.; Catellani, C. Peculiar karstic features in the Upper Triassic sulphate evaporites from the Secchia Valley (Northern Apennines, Italy). In *Gypsum Karst Areas in the World, Their Protection and Tourist Development. Bologna, 26 Agosto 2003*; Istituto Italiano di Speleologia, Memoria: Bologna, Italy, 2004; Volume XVI, pp. 99–106.
20. Colombetti, A.; Zerilli, A. Prime valutazioni dello spessore dei gessi triassici mediante sondaggi elettrici verticali nella Valle del F. Secchia (Villa Minozzo- R.E.). *Mem. Soc. Geol. Ital.* **1987**, *39*, 83–90.
21. Andreozzi, M.; Casanova, S.; Chicchi, S.; Ferrari, S.; Patterlini, P.; Pesci, M.; Zanzucchi, G. Riflessioni sulle evaporiti triassiche dell'alta Val Secchia (RE). *Mem. Soc. Geol. Ital.* **1987**, *39*, 69–75.
22. Chicchi, S.; Plesi, G. Sedimentary and tectonic lineations as markers of regional deformation: An example from the Oligo-Miocene arenaceous Flysch of the Northern Apennines. *Boll. Soc. Geol. Ital.* **1991**, *110*, 601–616.
23. Chicchi, S.; Costa, E.; Lugli, S.; Molli, G.; Montanini, A.; Torelli, L.; Cavozzi, C. The Passo del Cerreto-Val Secchia evaporites and associated rocks. In Proceedings of the Field trip guide book Realmod Conference San Donato Milanese (MI), San Donato Milanese, Italy, 2–4 October 2002.
24. Available online: https://www.rapidtables.com/web/color/RGB_Color.html (accessed on 3 August 2022).
25. Laughner, J.W.; Newnham, R.E.; Cross, L.E. Mechanical Twinning in small quartz crystals. *Phys. Chem. Miner.* **1982**, *8*, 20–24. [CrossRef]
26. Grimm, W.D. Idiomorphe quarze als Leit mineralien für salinare Fazies. *Erdöl Kohle* **1962**, *15*, 880–887.
27. Tarr, W.A. Doubly terminated quartz crystals occurring in gypsum. *Am. Miner.* **1929**, *14*, 19–25.
28. Etchepare, J.; Merian, M.; Smetankine, L. Vibrational normal modes of SiO_2. I. α and β quartz. *J. Chem. Phys.* **1974**, *60*, 1873. [CrossRef]
29. Liu, Y.; Wang, A.; Freeman, J. Raman, MIR and NIR spectroscopic study of calcium sulfates: Gypsum, bassanite and anhydrite. In Proceedings of the 40th Lunar and Planetary Science Conference, The Woodlands, TX, USA, 23–26 March 2009.
30. Bagavantam, S.; Venkatara, T. The normal modes and frequencies of the sulphur molecule. In *Proceedings of the Indian Academy of Science, Section A*; Springer: Berlin/Heidelberg, Germany, 1938; pp. 101–114.

31. Loun, J.; Cejka, J.; Sejkora, J.; Plasil, J.; Novak, M.; Frost, R.; Palmer, S.; Keeffe, E.A. Raman spectroscopic study of bukovskyite Fe2(AsO4)(SO4)(OH). 7H2O, a mineral phase with a significant role in arsenic migration. *J. Raman Spectrosc.* **2011**, *42*, 1596–1600. [CrossRef]

32. Pimenta, W.M.A.; Dresselhaus, G.; Dresselhaus, M.S.; Cancado, L.G.; Jorio, A.; Saito, R. Studying disorder in graphite-based systems by Raman spectroscopy. *Phys. Chem.* **2007**, *9*, 1276–1291. [CrossRef] [PubMed]

33. Ferrari, A.C. Raman spectroscopy of graphene and graphite: Disorder, electron–phonon coupling, doping and nonadiabatic effects. *Solid State Commun.* **2007**, *143*, 47–57. [CrossRef]

34. Escribano, R.; Sloan, J.J.; Siddique, N.; Sze, N.; Dudev, T. Raman spectroscopy of carbon-containing particles. *Vib. Spectrosc.* **2001**, *26*, 179–186. [CrossRef]

35. Jehlička, J.; Urban, O.; Pokorný, J. Raman spectroscopy of carbon and solid bitumens in sedimentary and metamorphic rocks. *Spectrochim. Acta Part A Mol. Biomol. Spectrosc.* **2003**, *59*, 2341–2352. [CrossRef]

36. Potgieter-Vermaak, S.; Maledi, N.; Wagner, N.; Van Heerden, J.H.P.; Van Grieken, R.; Potgieter, J.H. Raman spectroscopy for the analysis of coal: A review. *J. Raman Spectrosc.* **2011**, *42*, 123–129. [CrossRef]

37. Available online: http://rruff.info/graphiteR050503 (accessed on 3 August 2022).

38. Available online: http://rruff.info/anhydriteR040012 (accessed on 3 August 2022).

39. Available online: http://rruff.info/quartzR04001 (accessed on 3 August 2022).

40. Miller, C.; Zanetti, A.; Thoni, M.; Konzett, J.; Klotzli, U. Mafic and silica-rich glasses in mantle xenoliths from Wau-ennamus, Lybia: Textural and geochemical evidence for peridotite melt reactions. *Lithos* **2012**, *128*, 11–26. [CrossRef]

41. Lugli, S.; (Chemical and Geological Department, University of Modena and Reggio Emilia, Italy). Personal Communication, 2012.

42. Lugli, S. Timing of post-depositional events in the Burano Formation of the Secchia valley (Upper Triassic, Northern Apennines), clues from gypsum–anhydrite transitions and carbonate metasomatism. *Sediment. Geol.* **2001**, *140*, 107–122. [CrossRef]

43. Schreiber, B.C. Arid shorelines and evaporites. In *Sedimentary Environments and Facies*; Reading, H.G., Ed.; Blackwell: Oxford, UK, 1986; pp. 128–189.

44. Shearman, D.J. Syn-depositional and late diagenetic alteration of primary gypsum to anhydrite. In *Sixth International Symposium on Salt*; Schreiber, B.C., Ed.; Salt Institute: Portland, ME, USA, 1985; Volume 1, pp. 41–55.

45. Boccaletti, M.; Decandia, F.A.; Gasperi, G.; Gelmini, R.; Lazzarotto, A.; Zanzucchi, G. *Note Illustrative Della Carta Strutturaledell'appennino Settentrionale*; CNR, Progr. Finalizzato Geodinamica; Consiglio Nazionale delle Ricerche, Progetto Finalizzato Geodinamica, Sottoprogetto: Siena, Italy, 1987; Volume 429, p. 203.

46. Murray, R.C. Origin and diagenesis of gypsum and anhydrite. *J. Sed. Geol.* **1964**, *34*, 512–523.

47. Lugli, S. Petrography and geochemistry of the Upper Triassic Burano Evaporites (Northern Apennines, Central Italy). *Plinius* **1993**, *9*, 87–92.

48. Lugli, S. The magnesite in the Upper Triassic Burano Evaporites of the Secchia River valley (Northern Apennines, Italy): Petrographic evidence of an hydrothermal-metasomatic system. *Mem. Soc. Geol. Ital.* **1996**, *48*, 669–674.

49. Lugli, S. Megabrecce solfatiche nella Formazione di Burano dell'alta val di Secchia (Trias sup., RE): Cap rock da dissoluzione di salgemma. In Proceedings of the 1° Forum Italiano di Scienze della Terra, Geoitalia, Bellaria, Rimini, Italy, 5–9 October 1997; Volume 2, pp. 36–38.

50. Carmignani, L.; Kligfield, R. Crustal extension in the northern Apennines: The transition from compression to extension in the Alpi Apuane core complex. *Tectonics* **1990**, *9*, 1275–1303. [CrossRef]

51. Mc Lennan, S.; Murray, S.R. Earth's continental crust. In *Encyclopedia of Geochemistry*; Marshall, C.P., Fairbridge, R.W., Eds.; Kluwer: Dordrecht, The Netherlands, 1999; pp. 145–151.

52. Vlasov, K.A. Geochemistry and Mineralogy of rare Elements and genetic types of their deposits. In *I Geochemistry of Rare Elements(a), II Mineralogy of Rare Elements (b)*; Israel Program for Scientific Translations: Jerusalem, Israel, 1966.

53. Kabate Pendas, A.; Szteke, B. *Hafnium [Hf, 72] In Trace Elements in Abiotic and Biotic Environments*; CRC Press: Boca Raton, FL, USA, 2015; 3p.

54. Taylor, A.; Mc Lennan, S. The geochemical evolution of the continental crust. *Rev. Geophys.* **1995**, *33*, 241–265. [CrossRef]

55. Spingardi, A. Analyses of Charge Transfer and Color Centers in Gem-Quality Minerals with Thermal Treatment. Master's Thesis, Università di Pavia, Pavia, Italy, 2006; p. 164.

56. Beysac, O.; Lazzeri, M. Application of Raman Spectroscopy to the study of graphitic carbon in the earth Sciences. *Eur. Mineral. Union Notes Mineral.* **2012**, *12*, 415–454.

Article

Agates from Mesoproterozoic Volcanics (Pasha–Ladoga Basin, NW Russia): Characteristics and Proposed Origin

Evgeniya N. Svetova * and Sergei A. Svetov

Institute of Geology, Karelian Research Centre of RAS, 185910 Petrozavodsk, Russia
* Correspondence: enkotova@rambler.ru

Abstract: Agate gemstones occurring in the Mesoproterozoic volcanic rocks of the Priozersk Formation (PrF) within the Pasha–Ladoga Basin (Fennoscandian Shield, NW Russia) were investigated to characterize the mineral and geochemical composition of the agates and provide new information concerning their origin. Optical and scanning electron microscopy, EDS microanalysis, X-ray powder diffraction, X-ray fluorescence spectrometry, Raman spectroscopy, inductively coupled plasma mass spectrometry (ICP-MS), and C-O isotope analysis were used for the study. Agate mineralization appears mostly as an infill of fissures, cavities, gas vesicles in massive and vesicular basalts, lava-breccias. The mineral composition of agates is dominated by alpha-quartz (fibrous chalcedony, microcrystalline and macrocrystalline quartz), but it also displays abundances of calcite. The characteristic red-brownish agate's coloration is caused by multiple hematite inclusions distributed in an agate matrix. The study revealed the two phases of agate formation in the PrF volcanics, which are most likely controlled by two distinctly different fluids and/or their mixture. At first, agates appeared due to post-magmatic iron-rich fluids. The late hydrothermal activity was probably triggered by intrusion of gabbro-dolerite sill and resulted in the second phase of agate formation. We suggest that the late hydrothermal fluids remobilized the iron compounds from the crust of weathering underlying the PrF volcanics, which led to additional formation of vein agates and filling of gas vesicles with hematite-rich calcite/silica matter.

Keywords: agates; Mesoproterozoic volcanics; Fennoscandia Shield; Raman spectroscopy; X-ray diffraction; SEM-EDS; C-O isotope; genesis

Citation: Svetova, E.N.; Svetov, S.A. Agates from Mesoproterozoic Volcanics (Pasha–Ladoga Basin, NW Russia): Characteristics and Proposed Origin. *Minerals* **2023**, *13*, 62. https://doi.org/10.3390/min13010062

Academic Editors: Stefanos Karampelas and Emmanuel Fritsch

Received: 12 December 2022
Revised: 24 December 2022
Accepted: 27 December 2022
Published: 30 December 2022

1. Introduction

Agates belong to the fascinating variety of gemstones that have attracted people's attention since ancient times, owing to the showiness and uniqueness of their patterns. Due to the wide variation in shape, size and colour, agates are of considerable use in stone-cutting art as a valuable decorative and ornamental material. Agates are generally defined as banded or patterned chalcedony that are mainly composed of minute crystals of α-quartz [1]. They usually contain other silica polymorphs (moganite, crystobalite, tridymite, opal-CT) and paragenetic minerals (carbonates, zeolites, clay minerals, iron oxides, etc.) [2]. Agate occurrences are distributed throughout the world and related to various volcanic and, to a lesser extent, sedimentary rocks. Agates ordinarily fill veins and cavities in volcanic rocks where the supply of silica is triggered by post-magmatic hydrothermal activity, combined with the influence of meteoric waters [3,4]. Silica accumulation in cavities and empty pore space of sedimentary rocks often takes place during sedimentation or early diagenesis, when the sediments contain enough moisture [2]. The established agate hosts range in age from 13 to 3480 Ma [5]. The oldest known agate occurrence was found in Western Australia (Warrawoona) within Archean metamorphosed rhyolitic tuffs [6]. Miocene (13 Ma) volcanic tuffs of Yucca mountain, USA, belong to the youngest known agate-bearing host [1].

Agate occurrences on the Southeast Fennoscandia Shield are mainly related with Precambrian mafic volcanic rocks. The oldest famous agate hosts are the Paleoproterozoic massive and pillow lavas of the Ludicovian Superhorizont (ca. 2100−1920 Ma) within the Onega Basin [7,8]. In the present paper, agates from the volcanic rocks of the Mesoproterozoic Pasha–Ladoga Basin (Fennoscandian Shield, NW Russia) were investigated. The occurrence of so-called Salmi agates (after the name of the nearby village) was initially discovered in 1969 during a geologic exploration to the south of Pitkäranta town and later described by Polekhovsky and Punin in 2008 [9]. The study aims to obtain new information concerning the genesis of the Salmi agates. An integrated geological, mineralogical, geochemical, and C-O isotope data provides information for the reconstruction of geological and geochemical processes leading to the agate formation in the volcanics and the interpretation of their mineralogical features.

2. Geological Setting

The Mesoproterozoic Pasha–Ladoga Basin (PLB) is located at the margin of the Archean Karelian Craton and the Paleoproterozoic Svecofenian Belt in northwest Russia (Figure 1) [10–12]. The PLB, with an area of about 70,000 km^2, is bounded by a northwest-striking faults system. The PLB was formed during the Early Mesoproterozoic rifting.

The PLB rocks with angular unconformity overlay Archean granitoids (2.70–2.66 Ga), Paleoproterozoic hornblende-biotitic gneisses, shales, amphibolites (1.97–1.95 Ga), and metaturbidites (ca. 1.88 Ga) [13], and are intruded by the Salmi anorthosite–gabbro–gapakivi-granite complex (1.55–1.53 Ga) [14,15]. PLB is separated from these Archean and Paleoproterozoic rocks by a horizon of crust of weathering. The crust of weathering with a thickness of 0.5–5 to 30 m has a wide lateral distribution. It was probably formed in a range of 1.53–1.50 Ga and represented by hematized clay-altered basement rocks [16].

The PLB comprises four Mesoproterozoic (Riphean in Russian stratigraphic scale) units (from oldest to youngest): the Priozersk, Salmi, Pasha and Pliladoga, with a total thickness of 430–1250 m and Neoproterozoic (Vendian in Russian stratigraphic scale) Yablonovka Formation with thickness 90–380 m [11] (pp. 145–147), [12]. Volcanics are recognized only in the upper parts of Priozersk, Salmi, and Pasha Formations. The studied PLB area (the northeastern side of basin) is characterized by the following sequences, respectively, [10] (pp. 74–75), [11] (pp. 145–147), [17] (pp. 49–50), [12,18] (Figure 1C):

Priozersk Formation (PrF) comprises terrigenous sedimentary sequence (with a thickness from the first to 450 m) of red-coloured arkose sandstones, mudstones, gravelstones, and conglomerates, underlying the volcanic sequence (with a total thickness of 120 m or more). Volcanics are presented by nine lava flows of basalts interbedded with horizons of tuff, tuff-breccia, tuffites, and sandstones.

Salmi Formation (SalF) with a total thickness of 120−200 m overlies erosion rock layer covering the PrF volcanics. The basement of the PrF sequence is formed by quartz-feldspar and polymictic sandstones, mudstones with thin carbonate interbeds. The sedimentary sequence is overlain by lavas of porphyritic basalt and andesibasalts interbedded with tuffs and tuffites. The thickness of volcanics is about 80 m.

Pasha Formation (PashF) has a total thickness of 150 to 500 m and is represented by coarse-grained feldspar-quartz sandstones and gravelstones with fragments of quartzites. Single lava flows and interlayers of tuffs similar to PrF volcanic rocks were found within the sedimentary sequence. The thickness of the volcanic sequence is about 20 m.

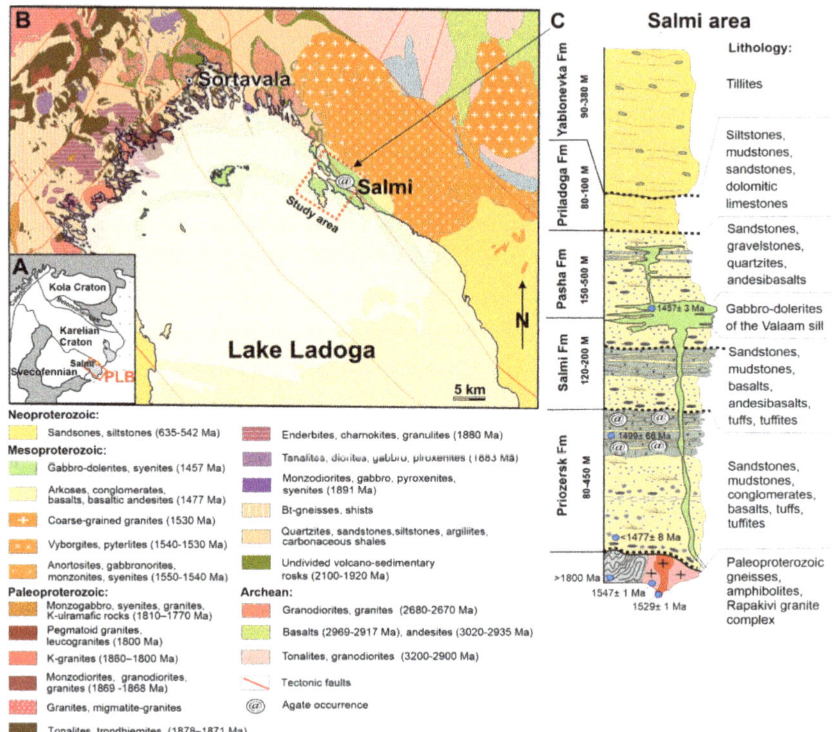

Figure 1. (**A**) Location of sampling areas. (**B**) simplified geological map of the Pasha–Ladoga Basin based on [18]. (**C**) simplified geological section of the studied area based on [12], [17] (pp. 49–50). Dotted line marks position of weathered crusts or erosion rock layers; @ represents sampled agate-bearing flows.

The erosion rock layer developed after the PashF sequence is overlain with rocks of the *Priladoga Formation (PrilF)* formed by thin-layered siltstones, mudstones and fine-medium-grained sandstones, with interlayers of dolomitic limestones in the upper part of the sequence. The total thickness of PrilF is ca. 80−100 m. The PrilF rocks with angular unconformity are overlain by the Neoproterozoic rocks of the *Yablonovka Formation (JabF)*. The JabF sequence has a thickness of 90–380 m and is represented by tillites [17] (p. 49).

The PLB volcanics (observed in PrF, SalF, PashF) age can be comprehensively estimated based on the following data. PrF overlaps the weathered crust of rapakivi granites (Salmi batholith), aged on intrusion phases from 1547 ± 1 to 1529 ± 1 Ma [14]. A depositional age of the PrF volcanics by the Sm/Nd isotopic data is 1499 ± 68 Ma [19]. The U–Pb dating of the youngest detrital zircon from the basal part of the PrF sandstone is ca. 1477 ± 8 Ma [20]. The PrF, SalF, and PashF rocks are intruded by gabbro-dolerites of the Valaam Sill aged 1457 ± 3 Ma and 1459 ± 3 Ma [21]. The data obtained indicates that the dominating part of the PLB (PrF, SalF, PashF) was formed very rapidly in the interval 1477−1457 Ma in the intraplate riftogenic regime [11] (pp. 145–147).

The volcanic-sedimentary complexes of the PLB basement have not undergone regional metamorphism, but were affected to diagenetic and local hydrothermal alterations [12]. The PLB is overlapped by the Vendian and Phanerozoic sediments [11] (pp. 145–147).

3. Agate Occurrence

The studied area is located near the Salmi village, where, in the Tulemajoki River-valley outcrop, bedrock attributed to the upper part of the PrF sequence are recorded. The volcanic sequence with a thickness about of 100 m is exposed in riverbed outcrops (on rapids and rifts). It is represented by interbedded gently dipping massive lava flows of basalts, andesibasalts, and andesites that are 1−30 m in thickness. The lava flows are alternated with slag lavas, agglomerate tuffs, tuffites, tuff sandstones, gravelstones, and small-pebble conglomerates. The thickness of these beds varies from tens of cm to 5 m.

The thick lava flows (>5 m) are differentiated (Figure 2A,B). They are formed by massive rocks with single agate nodules at the flow's base. The middle part of the flows is characterized by porphyritic structure. The number of agate nodules and veinlets increases (it can reach 40% of the volume) towards the upper parts of the lava flows. The top lava flows are formed of breccias or frothy lavas (similar to slags) and contain numerous cavities and gas vesicles partially or completely filled with agates of quartz-carbonate composition (Figure 2C,D).

Figure 2. Field outcrop photographs illustrating agates' location in the Mesoproterozoic volcanics within the Pasha–Ladoga Basin (Priozersk Formation). (**A**) the upper part of lava flow with the massive, amygdaloidal and breccia zones within the coastal cliffs on the Tulemajoki River; (**B**) vein agate in the inter-breccia lava space, (**C**) brecciated top of the lava flow; (**D**) enlarged fragment from selected area of (**C**) showing agates in the inter-clastic lavas space, (**E**) cavity infilling with Fe-enriched agate; (**F**) large gas vesicles within the massive lava filled with agate. Red arrows point to the location of agate mineralization within the lavas.

In the thin lava flows, porphyritic structures are rare; they are characterized by homogenic massive texture and top lava-breccia. The lavas exhibit a grayish-brown, rarely brownish colour. The tuff material with local gradational bedding is recognized between lava flows. The tuffs are represented by agglomerate and, to a lesser extent, by bomb varieties. The volcanic clasts with the thin chill zones range from 0.2 to 20 cm in size. The matrix is represented by chloritized pelitic material. The fragments of feldspars and granites are occasionally observed within the agglomerate tuffs [10] (pp. 74–75).

The individual zones of vesicular basalts (mandelstones) with the thickness of 5–30 cm are locally recognized within the top massive lava flows. Mineralized vesicles compose up to 80% of the rock volume, their size ranging from 2 to 20 mm (Figure 3). They have usually rounded, elongated shapes or shapes of coalesced amygdales, and they are characterized by light-green or red-brownish coloration. The amygdales have different orientation, and most of them exhibit concentric zoning.

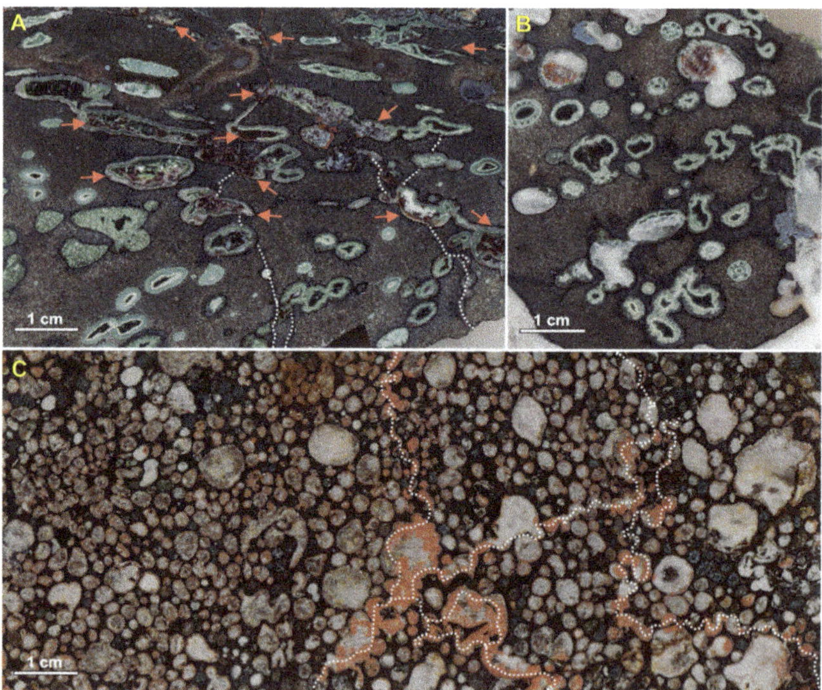

Figure 3. Photographs of the polished slabs of Mesoproterozoic vesicular basalts from the Priozersk Formation within the Pasha–Ladoga Basin. (**A,B**) Images illustrating concentrically zoned amygdales tend to be filled with chlorite. Red arrows point to deformed amygdales and fissures with iron oxides appearance. (**C**) Image showing dense accumulation of rounded and irregularly formed amygdales with red-brownish pigments. Dotted lines point to channels of Fe-bearing solution supply.

Agate mineralization is mainly exposed in the upper part of PrF lava flows. Agates are widespread as an infill of fissures, gas vesicles in lavas, and an infill of cavities in slags and lava-breccias (Figure 2B–F). Most agates are round-shaped nodules and veinlets, their size ranges from 1 to 15−20 cm. The frequency of agate occurrence is about of 3–14 per 10 m^2 of the lava flow surface (Figure 2A,C). According to the data of Polekhovsky and Punin [9], agates were also found in cores of boreholes drilled in the area of Salmi village. They fill gas vesicles in basalts, cavities in frothy lavas, as well as occur as pebbles in conglomerate and gravelstone beds of the SalF, overlying the PrF volcanic rocks.

4. Materials and Methods

The Salmi agates and parent volcanic rocks were examined using optical microscopy, scanning electron microscopy with energy-dispersive spectroscopy (SEM-EDS), X-ray fluorescence (XRF), powder X-ray diffraction (XRD), Raman spectroscopy, inductively coupled plasma mass spectrometry (ICP-MS), wet chemistry at the Institute of Geology, Karelian Research Centre, RAS (IG KRC RAS, Petrozavodsk, Russia).

Six agate-bearing volcanic rocks samples (including two samples of vesicular basalts) and two agate nodules with visually different chalcedony/calcite ratio (calcite-rich and calcite-poor samples) were selected from the broad author's collection for detailed mineralogical and geochemical investigation. Additionally, a macrocrystalline quartz sample and three crystalline calcite samples extracted from individual agate nodules were used for geochemical and C-O isotope analysis. Polished thin sections and powders were prepared for the studies.

Standard petrographic investigations were carried out using the Polam-211 optical microscope. The SEM-EDS analyses were applied to determine the chemical composition of the mineral phases comprising the agates. The experiments were performed using a VEGA II LSH (Tescan, Brno, Czech Republic) scanning electron microscope with EDS INCA Energy 350 (Oxford Instruments, Oxford, UK) on carbon-coated, polished thin section. The analyses were carried out at the following parameters: W cathode, 20 kV accelerating voltage, 20 mA beam current, 2 µm beam diameter, and counting time of 90 s. The following standards were used: calcite, albite, MgO, Al_2O_3, SiO_2, FeS_2, wollastonite, Fe, Zn, and InAs. SEM-EDS quantitative data and determination of the analysis accuracy were acquired and processed using the Microanalysis Suite Issue 12, INCA Suite version 4.01.

The concentrations of major elements in the agate-bearing volcanics were determined by wet chemistry and XRF. XRF analysis was performed using an ARL ADVANT'X-2331 (Thermo Fisher Scientific, Ecublens, Switzerland) wavelength-dispersive spectrometer with a rhodium tube, working voltage of 60 kV, working current of 50 mA, and resolution of 0.01. Preliminarily, 2 g of each powdered sample was heated in ceramic crucibles, at 1000 °C, in a muffle furnace for 30 min. The loss of ignition was determined by a change in the mass of the sample after heating. For XRF measurements, 1 g of heated sample was mixed with Li-tetraborate flux and heated in an Au-Pt crucible to 1100 °C to form a fused bead.

The concentrations of trace and rare-earth elements were determined by ICP-MS using an X Series 2 (Thermo Scientific, Bremen, Germany) mass spectrometer. The powdered samples were digested in an acid mixture following the standard procedure [22]. The accuracy of the analyses was monitored by analysing the USGS standard BHVO-2 (see Supplementary Materials, Table S2).

The concentrations of the trace element and rare earth elements were normalized to the concentrations in chondrite (C1) and primitive mantle (PM) [23]. Parameters were calculated: Nb anomaly as $Nb/Nb^* = Nb_n/(Th_n * La_n)1/2$, Sr anomaly as $Sr/Sr^* = Sr_n/(Pr_n * Nd_n)1/2$ and Eu anomaly as $Eu/Eu^* = Eu_n/(Sm_n * Gd_n)1/2$.

Powder XRD analysis was carried out using a Thermo Scientific ARL X'TRA (Thermo Fisher Scientific, Ecublens, Switzerland) diffractometer (CuKα-radiation (γ = 0.1790210 nm), voltage 40 kV, current 30 mA). Diffractograms of agate samples in the range of 5–75° 2θ were recorded at a scanning step of 0.40 2θ/min. X-ray phase analysis was performed using the program packWin XRD, ICCD (DDWiew2008). The detection limit for XRD phase identification was 3 wt%.

Raman spectroscopy was used to identify mineral phases (SiO_2, Fe-oxides, Mn-oxides) in agates. Raman spectroscopy analysis was conducted on a dispersive Nicolet Almega XR Raman spectrometer (Thermo Fisher Scientific, Waltham, MA, USA) with a green laser (532 nm, Nd-YAG (Thermo Fisher Scientific, Waltham, MA, USA). The spectra were collected at 2 cm^{-1} spectral resolution. A confocal microscope with a $50\times$ objective was used to focus an excitation laser beam on the sample and to collect a Raman signal from a 2 µm diameter area. Raman spectra were acquired in the 85–4000 cm^{-1} spectral region, with the exposition time between 30 s and 100 s for each scan, depending on the signal intensity and laser power of 2–10 mW to prevent any sample degradation.

Isotope studies of agate-associated calcite samples were carried out at the Institute of Geology, Komi Scientific Center, Uralian Branch, RAS (IG Komi SC UB RAS, Syktyvkar, Russia). The decomposition of calcite in phosphoric acid and the C-O isotopic composition measurement by flow mass spectrometry (CF-IRMS) were carried out on the Thermo Fisher Scientific analytical complex, which includes a Gas Bench II sample preparation and injection system connected to a mass spectrometer DELTA V Advantage. The $\delta^{13}C$ values are given in per mille relative to the PDB standard, $\delta^{18}O$—the SMOW standard. The international standards NBS 18 and NBS 19 were used for calibration. The analytical precision for $\delta^{13}C$ and $\delta^{18}O$ was ±0.15 ‰ (1σ).

5. Results

5.1. Host Rock Characteristic

The studied volcanics are mostly aphyric and porphyritic with phenocrysts composed of zoning plagioclase crystals with labradorite composition. Plagioclase, augite, and forsterite are the major rock-forming minerals of PrF volcanics. Dominating accessory minerals are apatite, titanite, chlorite, epidote, magnetite, titanomagnetite, ilmenite.

The geochemical differentiation of the PLB volcanic rocks is variable. Using the SiO_2 (49.24−62.01 wt%) and Na_2O + K_2O (4.42−5.93 wt%) content, volcanic rocks are classified as trachybasalt, basaltic andesites, andesites of the alkaline/subalkaline series (Table 1, Figure 4). Basalts and andesites dominate in the studied succession. In accordance with FeO_{tot} (12.61−15.76 wt%) and TiO_2 (2.13−3.87 wt%) content, volcanics belong to the Fe-Ti-rich type.

Table 1. Chemical composition of agate-bearing volcanics from the PrF, wt %.

Sample	SiO$_2$	Al$_2$O$_3$	Fe$_2$O$_3$	FeO	Na$_2$O	CaO	K$_2$O	MgO	MnO	TiO$_2$	P$_2$O$_5$	S	LOI	H$_2$O	Σ
B-1 *	46.6	13.39	9.04	4.6	4.99	3.74	0.59	6.14	0.18	3.64	1.09	0.04	4.74	0.86	99.6
B-2 *	47.5	15.03	11.75	3.45	4.95	3.02	0.74	5.15	0.09	3.54	1.22	0.03	2.72	0.72	99.6
MBG-2 *	58.72	10.72	8.33	4.3	3.86	3.15	0.43	3.86	0.15	2.5	0.92	0.02	2.45	0.55	99.94
MBG-3 *	59.98	9.9	7.6	4.6	4	2.95	0.42	4.29	0.14	2.06	0.75	0.03	2.46	0.4	99.55
MBR-5 **	43.43	5.81		9.42	BDL	17.94	2.01	3.68	0.22	1.58	0.54	-	15.32	-	99.95
MBR-6 **	42.83	6.36		10.21	BDL	16.76	2.51	3.37	0.27	1.81	0.6	-	14.98	-	99.7

Measurements were obtained using wet chemistry (*) and XRF analysis (**). Samples: B-1,2—massive basalts; MBG-2,3—vesicular basalts with light-green amygdales; MBR-5,6—vesicular basalts with red-brownish amygdales. For (**) samples, FeO values are total FeO. BDL—below XRF detection limit; (-) no data.

The massive lavas are more primitive and differ from vesicular basalts by the lower contents of SiO_2 (<50 wt%), elevated MgO (5.34−6.53 wt%), Al_2O_3 (14.24−15.58 wt%), and P_2O_5 (>1 wt%) contents.

Mandelstones with light-greenish amygdales compared to the mandelstones with red-brownish ones are enriched in Al_2O_3 (9.9−10.7 vs. 5.8−6.3 wt%) and Na_2O (3.8−4 wt% vs. below detection limit), but depleted in CaO (~3 wt % vs. ~18 wt %) and MnO (~0.15 wt % vs. ~0.3 wt %). The high content of CaO in mandelstones with red-brownish amygdales is probably caused by a local hydrothermal alteration of volcanic rocks. This observation is confirmed by the anomalously high LOI values (up to 15.32 wt %, compared to common values 2.72−4.74 wt%).

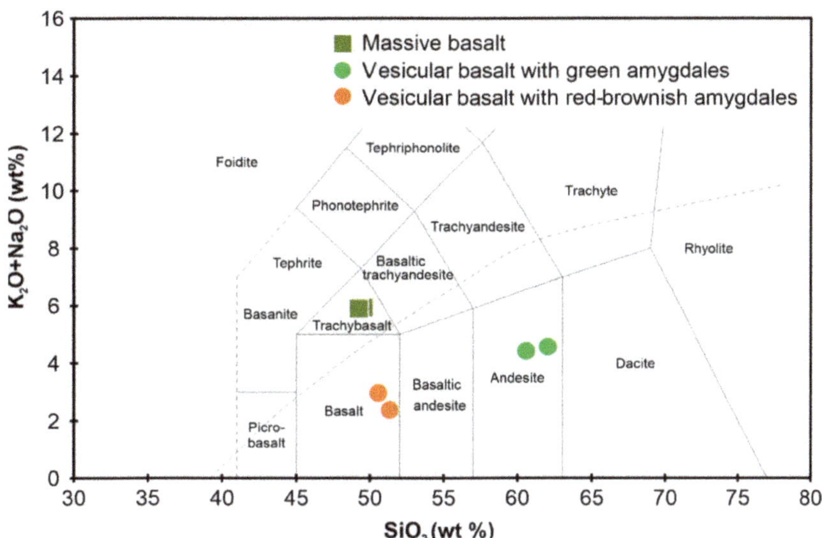

Figure 4. Total alkali silica (TAS) diagram classification (after [24]) of volcanics from the Priozersk Formation within the Pasha–Ladoga Basin.

5.2. Macro- and Microscopic Characteristics of the Agates

The Salmi agates are characterized by specific red-brownish coloration (Figure 5). The colour likely results from chromogenic Fe-oxides compounds disseminated within the agate matrix [25]. The agate nodules often exhibit concentrically zoning patterns (Figure 5A,C,E). Some of them are characterized by distinct coarse (1–2 cm) and fine (from few micrometres) banding (Figure 5A), poor-fancy agate varieties are also widespread (Figure 5B,D). Frequently agates contain large calcite and/or quartz crystals. Aggregations of chlorite inclusions were commonly observed in the contact area to the host rock (Figure 5E).

The petrographic study revealed various microtextures of agates. The silica matrix of agates is represented by micro- and macro-crystalline quartz, length-fast chalcedony (with c-axis oriented perpendicular to the fibre direction), as well as zebraic chalcedony (length-fast with a helical twisting of fibres along fibre direction) (Figure 6). These silica phases form banded zones, micro-geodes, and radial fibrous spherolites varying both in thickness and number of concentric layers (Figure 6A,C,E,G). Agates contain an appreciable quantity of calcite phases in the form of elongated or isometric crystals up to 1 cm in size, grained aggregates in the intergrowth with chalcedony, as well as fine-grained aggregates filling the fissures crosscutting the silica matrix (Figure 6C,D). Fine-dispersed Fe-oxide inclusions disseminated along the agate banding emphasize agate zoning (Figure 6B,D,F,H). It is noteworthy that the minute Fe-oxide particles are mostly concentrated within the chalcedony fibres and between microcrystalline quartz grains, whereas macrocrystalline quartz zones are depleted in it.

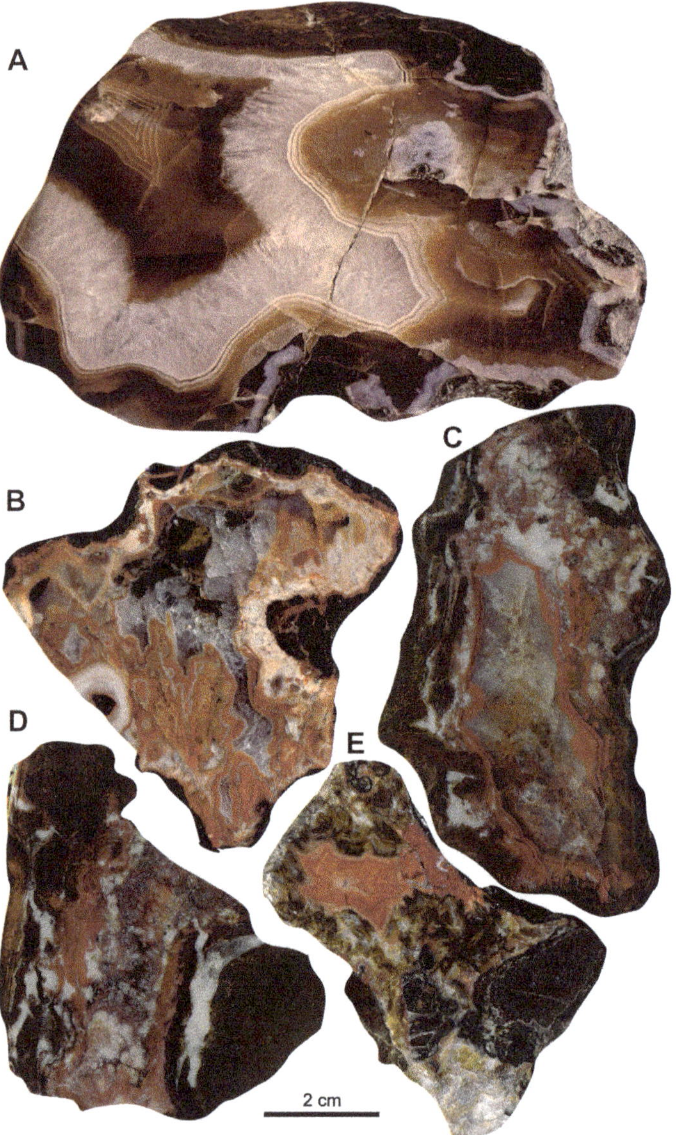

Figure 5. Photographs of agates from the Mesoproterozoic volcanics within the Pasha–Ladoga Basin (Priozersk Formation). (**A**) Monocentric agate nodule with well-developed concentric zoning and Fe-oxide pigment disseminated along individual chalcedony layers (museum expositional sample of the IG KRC RAS); (**B**) poor-fancy agate with red-brownish coloured chalcedony and calcite layers, as well as colourless quartz-rich portion; (**C**) concentrically zoned agate with red-brownish pigment distributed along the external margin of the amygdale; (**D**) fragment of the vein agate with red-brownish pigment irregularly distributed within chalcedony; (**E**) concentrically zoned agate surrounded by silica–chlorite matrix.

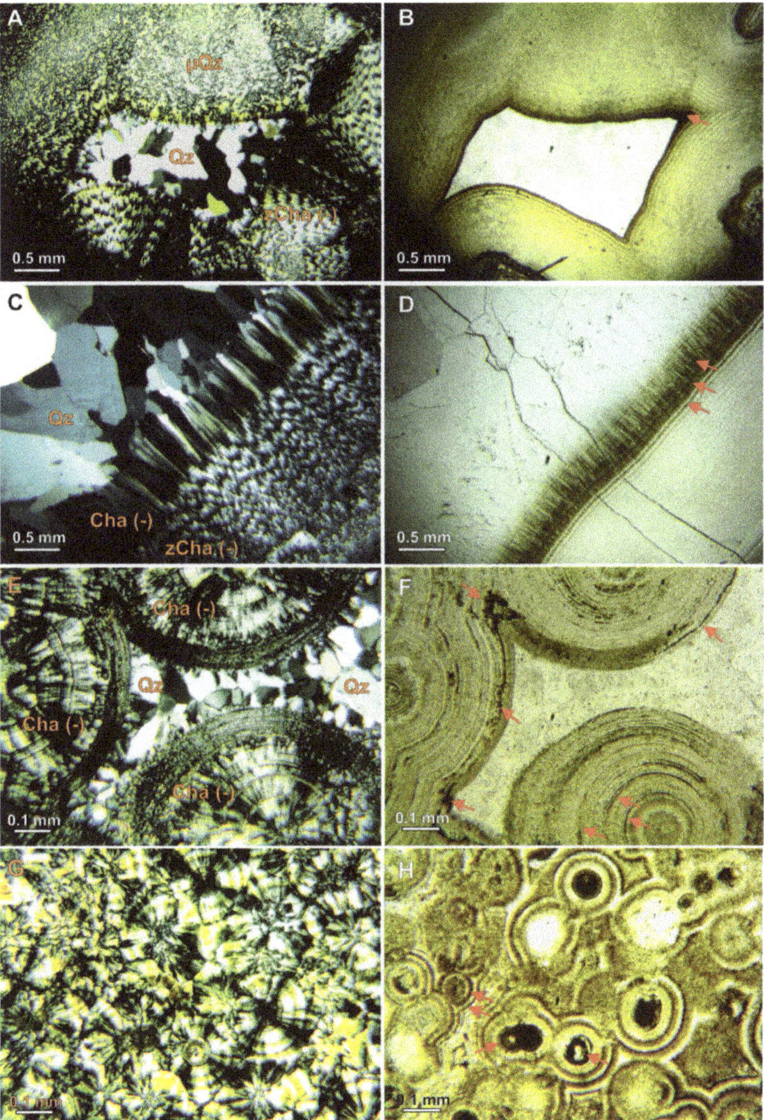

Figure 6. Microphotographs of agates with Fe-oxide pigment: (**A,B**) microgeode of macrocrystalline quartz within a matrix of microcrystalline quartz and zebraic chalcedony; (**C,D**) alternate silica phases in the internal part of the agate with microcrystalline quartz, zebraic chalcedony, length-fast chalcedony, macrocrystalline quartz and crosscutting calcite veinlet; (**E,F**) chalcedony spherulites associated with macrocrystalline quartz; (**G,H**) dense accumulation of chalcedony spherulites. (**A,C,E,G**)—polarized light, crossed nicols; (**B,D,F,H**)—plane polarized light. The micrographs (**B,D,F,H**) show that the macroscopically visible red colour is caused by the incorporation of Fe-oxide/hydroxide minerals along the chalcedony banding (see arrows). Abbreviations: Cha (-), length-fast chalcedony; zCha (-), zebraic chalcedony; Qz, macrocrystalline quartz; μQtz, microcrystalline quartz. The red arrows in figures indicate the location of chromogenic minerals along the strips.

In the studied area, at least two different types of vesicular basalts were distinguished during the macroscopic observation (Figure 3). The first one displays amygdales predominantly light-green in colour due to a high content of chlorite (Figure 3A,B). The amygdales are characterized by elongated, rounded, as well as irregular shapes. It is noteworthy that amygdales with differences in shape and texture can be located directly next to each other (Figure 3A). Additionally, tectonically deformed mineralized vesicles show a significant amount of iron oxides penetrating with the amygdale matrix, whereas intact amygdales are depleted by them (Figure 3A). The other type of vesicular basalts contains amygdales of rounded and, to a lesser extent, irregular forms with close packing (Figure 3C). The larger amygdales tend to be filled with calcite, whereas minute ones can be filled with silica, chlorite, calcite, as well as with their mixture. Most amygdales have a red-brownish tint caused by iron oxides pigmentation. The development of iron oxides is also observed along the fissures connecting the amygdales (Figure 3C).

Microscopic investigation revealed that most of the amygdales exhibiting concentric zoning are composed of successive layers of minute radial fibrous chalcedony, fine-grained, and plate-like chlorite from the margin towards the vesicle centre (Figure 7A–D). Occasionally, amygdales contain macrocrystalline quartz cores (Figure 7E,F). Another zoning type is rarer: amygdales could be filled half with coarse-crystalline calcite and half with concentrically zoned chlorite (Figure 7G). Additionally, gas vesicles could be completely filled with coarse-crystalline calcite or radial fibrous chalcedony (Figure 7H). In the latter case, dusty chlorite is scattered ordinarily between the chalcedony spherulites.

5.3. Mineral Composition

5.3.1. X-ray Powder Diffraction

The X-ray powder diffraction of two bulk agate samples was applied to determine their phase composition. Only peaks corresponding to calcite and alpha-quartz have been identified on the X-ray diffraction patterns (Figure S1). No other silica polymorphs (e.g., opal, cristobalite, tridymite, and moganite) were revealed. The calcite content in one of the examined samples was 69 wt%, while the other sample was poor in calcite.

5.3.2. Raman Spectroscopy Results

Micro-Raman spectroscopy investigation of individual silica bands of agates showed that in all cases, they are only composed of alpha-quartz recognized by the characteristic bands at 464, 353, 208, and 127 cm^{-1}. Numerous Fe-oxide inclusions appeared as needle-like, spherical, irregular grains with a size up to 200 μm distributed in agate nodules and were identified as hematite by their main Raman bands located at 408, 607, 665, 1319 cm^{-1} (Figure 8A,D) [26,27]. The grains of dark minerals surrounding the calcite in agate were evidenced as manganite (MnOOH) according to characteristic Raman bands at 386, 528, 554, 620 cm^{-1} (Figure 8B,E) [28]. Additionally, barite, which frequently occurred in the agates as prismatic crystals with a size up to 0.6 mm surrounded by calcite was identified by its characteristic Raman bands at 453, 461, 617, 647,988, 1139 cm^{-1} (Figure 8C,F) [29,30].

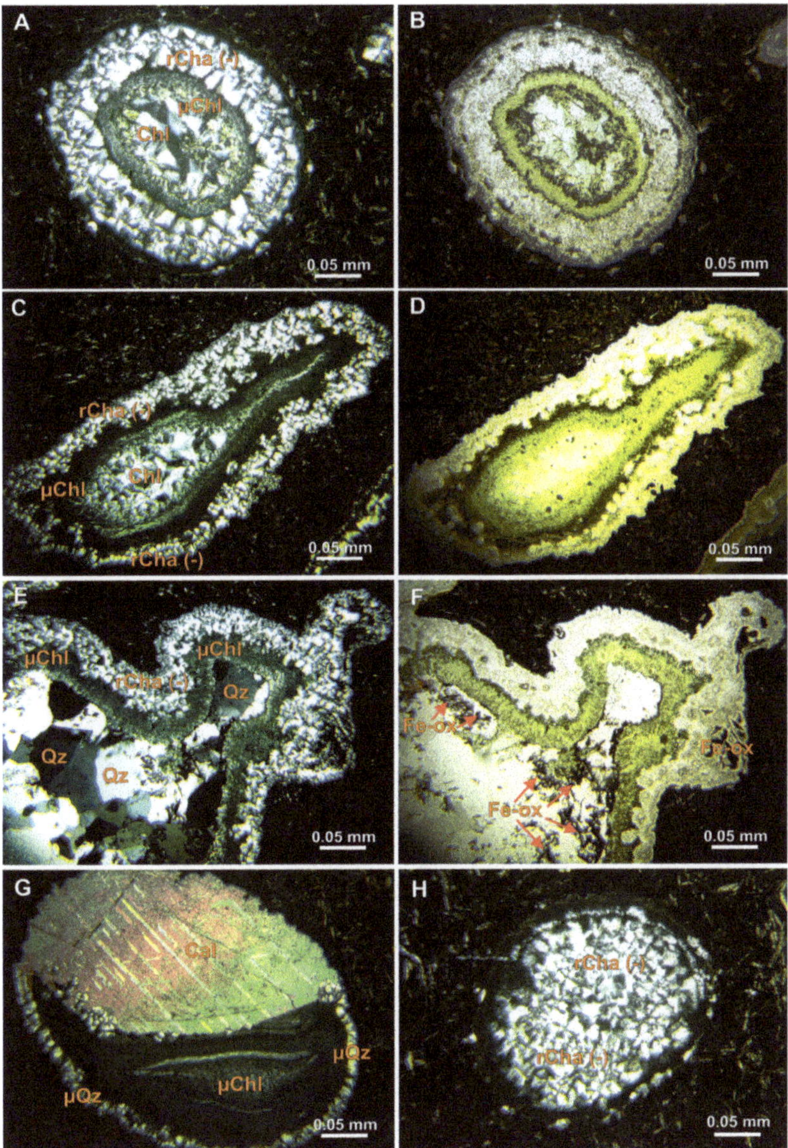

Figure 7. Microphotographs of amygdales in vesicular basalts (mandelstones) from the Priozersk Formation within the Pasha–Ladoga Basin: concentrically zoning rounded (**A,B**) and elongated (**C,D**) amygdales composed of spherulites of radial fibrous chalcedony, fine-grained and plate-like chlorite; (**E,F**) irregularly shaped amygdale consisting of radial fibrous chalcedony, fine-grained chlorite, and macrocristalline quartz core with Fe-oxides in silica matrix; (**G**) bizoning amygdale composed of calcite and concentrically zoning chlorite aggregation; (**H**) Amygdale uniformly filled with radial fibrous chalcedony. (**A,C,E,G,H**)—polarized light, crossed nicols; (**B,D,F**)—plane polarized light. Abbreviations: Cha (-), length-fast chalcedony; rCha (-), radial fibrous chalcedony; Qz, macrocrystalline quartz; µQz, microcrystalline quartz; Chl, plate-like chlorite; µCha, fine-grained chlorite.

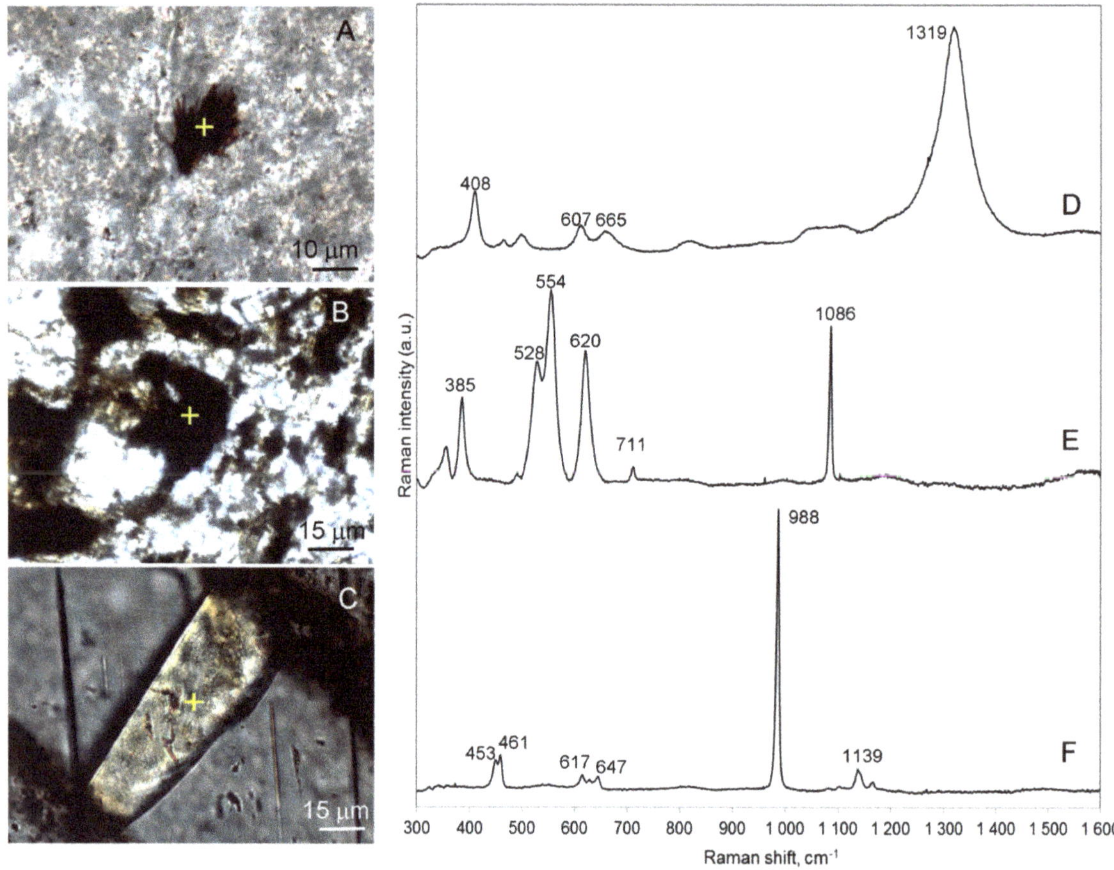

Figure 8. Micrographs (transmitted light mode) and Raman spectra of hematite (**A,D**), manganite (**B,E**), and barite (**C,F**) inclusions in the agates in the spectral range of $300-1600$ cm^{-1}. The analytical spots are marked by yellow crosses. Bands at 711, 1086 cm^{-1} on Figure (**E**) corresponds to calcite.

According to the local Raman study of mineralized gas vesicles in mandelstones, they are filled with alpha-quartz, calcite, and chlorite. The plate-like chlorite crystals predominantly fill the central parts of greenish zoned amygdales from either mandelstone types are characterized by similar Raman bands at $355-367$ cm^{-1}, $541-548$ cm^{-1}, $673-677$ cm^{-1}, and diffused bands at $3600-3657$ cm^{-1} (Figure 9). The apparent downshift in the Si-O-Si related Raman band to $673-677$ cm^{-1} (from the $681-683$ cm^{-1} of Mg-rich chlorite) [31], together with the high intensity of the band at $355-367$ cm^{-1} suggest an intermediate Mg-Fe composition of examined chlorites. In the spectral range of H_2O/OH vibrations ($3400-3700$ cm^{-1}), the Raman spectral patterns from studied chlorite samples show insignificant variations, which are most probably related to different proportions of Mg and Fe in the structure of chlorite [32].

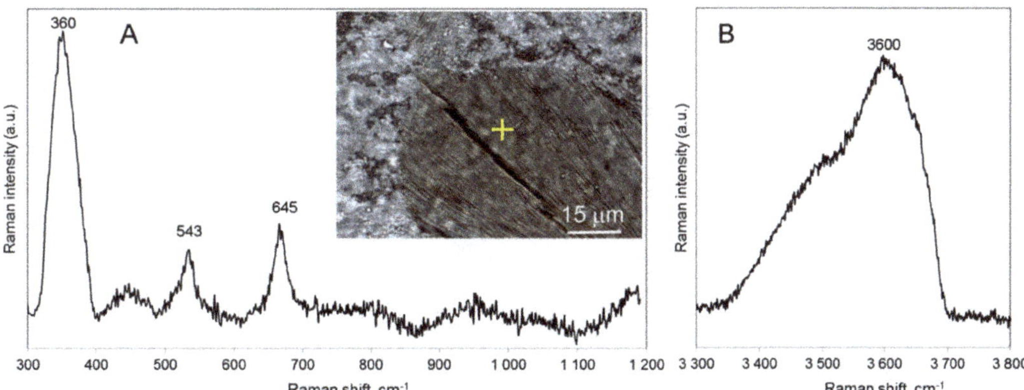

Figure 9. Micrographs and Raman spectra of plate-like chlorite in the amygdale of vesicular basalts in the spectral range of $300-1200$ cm^{-1} (**A**) and $3300-3800$ cm^{-1} (**B**). The analytical spot is marked by a yellow cross.

Red-brownish spherical, needle-like, and crystalline inclusions within the amygdales were all identified as hematite by their characteristic Raman bands. No other Fe-oxide/hydroxide phases were found.

5.4. SEM-EDS Investigation

More evidence concerning the mineral composition of the agates from PLB volcanics was obtained by SEM-EDS investigation. In general, the mineral assemblage of agate nodules and amygdales from vesicular basalts is similar. Banded agate zones are related to the difference in porosity of individual silica layers (Figure 10A). In the higher porosity area scattered needle-like crystals of Fe-oxide and minute inclusions of barite have been concentrated (Figure 10A). Inclusions of Fe-oxide composition often appear at the fissures both in the amygdales and agate nodules (Figure 10B). Fe-oxide phases in the form of large idiomorphic crystals or spherulitic aggregates randomly arranged both in nodules and amygdales are also observed (Figure 10C,D). The most examined Fe-oxide phases are Ti-rich (up to 3 wt%). Numerous barite crystals frequently form clusters embedded within the calcite matrix (Figure 10E). Pseudomorphs of calcite after silica frequently forming banded pattern were revealed in the internal parts of some agates (Figure 10B).

The chemical compositions of chlorites occurring both in agate nodules (Figure 9H) and amygdales of mandelstones (Figure 10D) are characterized by high FeO and MgO contents (Table S1). The total FeO content of chlorite hosted in nodules (21.6%−27.3%) is slightly higher than that in the amygdales (17.0%−23.4%). The MgO content is 12.9%−15.7% and 16.0%–21.6% in chlorite from nodules and amygdales, respectively. It is noteworthy that chlorite compositions of amygdales recorded for both types of mandelstones are generally similar. Alkali contents (CaO + K$_2$O + Na$_2$O) are low and do usually not exceed 1 wt%.

The EDS data from chlorite occurring both in agate nodules and amygdales of mandelstones are used to calculate the structural formulae based on the 20 oxygens (Table S1). The data indicated that chlorite composition is quite homogeneous. In the Al(IV)-Mg-Fe ternary diagram [33], chlorite markers of the amygdales are plotted in the Mg-rich chlorite domain, whereas agate nodules markers are on the limit with Mg-chlorite and Fe-chlorite domains (Figure 11A). The Si vs. Fe/(Fe + Mg) classification diagram of Hey [34] displays that examined chlorite samples mainly fall in/near the diabantite field (Figure 11B).

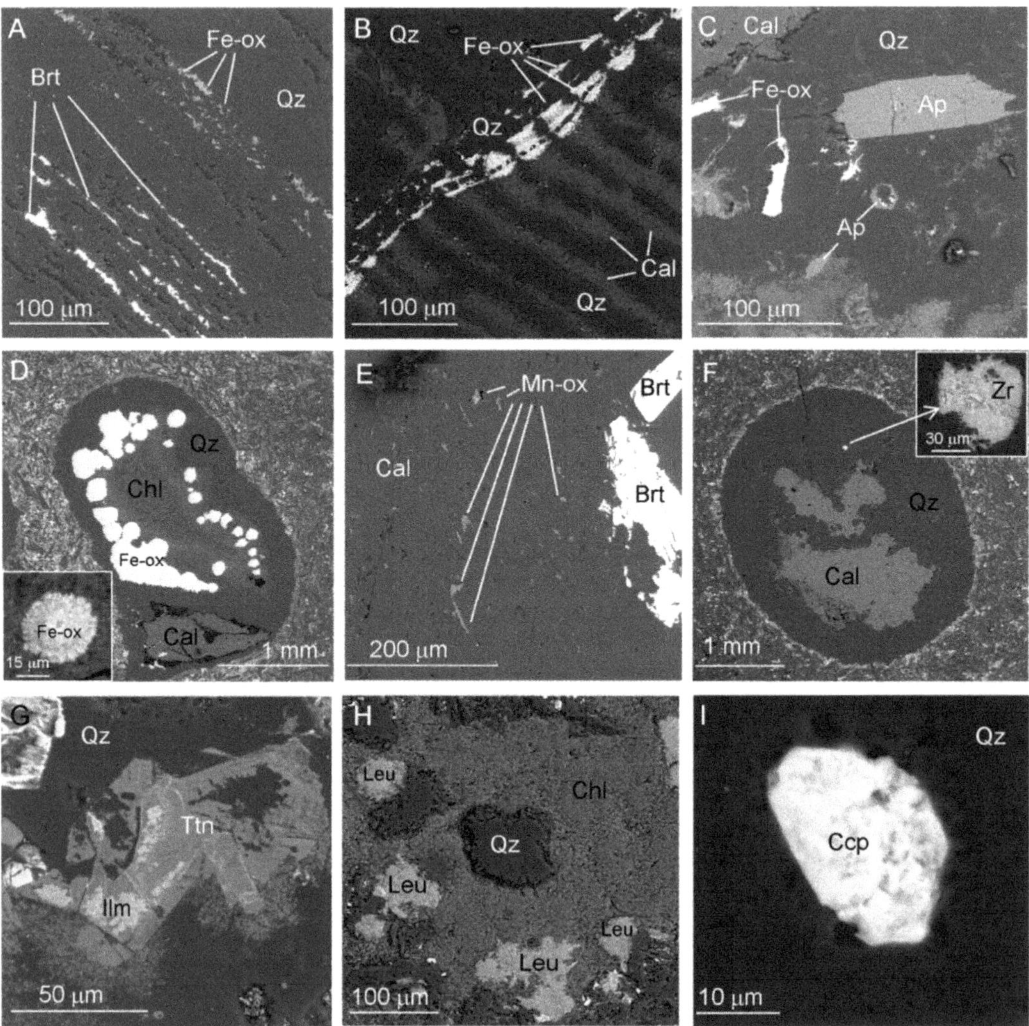

Figure 10. BSE images of microinclusions in agates: (**A**) Fe-oxides and barite (Brt) scattered along high porosity chalcedony bands in agate nodule; (**B**) a fissure filling with silica and Fe-oxides, which cross-cuts the banding agate area formed by chalcedony and calcite layers; (**C**) apatite (Ap) and Fe-oxide inclusions in silica area of agate nodule; (**D**) spherical aggregates of Fe-oxides arranged in a concentrically zoned amygdale and large calcite (Cal) crystal-filled cavity (enlarged image of spherical inclusion inserted on (**C**) from the lower left); (**E**) Mn-oxides and large barite aggregates accompanied by calcite area of agate nodule; (**F**) corroded grain of zircon (Zr) within the amygdale of quartz-calcite composition; (**G**) titanite (Ttn) and ilmenite (Ilm) aggregates surrounded by quartz (Qz) in agate nodule; (**H**) flaky leucoxene (Leu) enclosed by chlorite (Chl) in agate nodule; (**I**) grain of chalcopyrite (Ccp) in quartz (Qz) matrix within the amygdale of vesicular basalt.

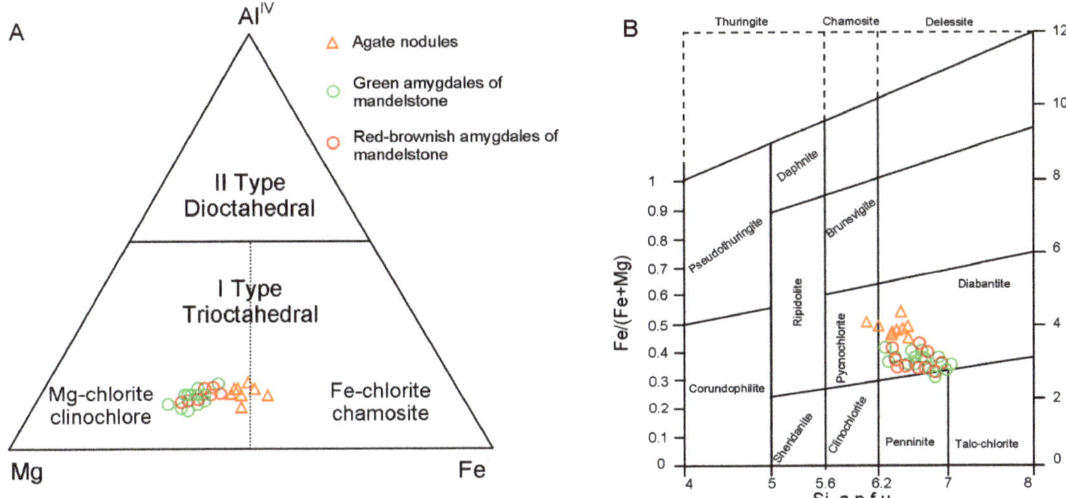

Figure 11. Chlorite classification diagrams showing the position of the investigated samples of the agates. (**A**) (Al)–Mg–Fe diagram (after [33]); (**B**) Si vs. Fe/(Fe+Mg) diagram (after [34]).

Additionally, manganese hydroxide, hydroxyapatite, titanite, leucoxene, rutile, ilmenite, muscovite were recorded as characteristic mineral inclusions for the Salmi agates, whereas sulphide minerals (pyrite, chalcopyrite) are extremely rare in them (Figure 10E,G,H,I). In a single case, a corroded grain of zircon within the amygdale of quartz-calcite composition was identified (Figure 10F).

5.5. Geochemical and Isotopic Characterization

5.5.1. Trace Elements

ICP-MS trace element analyses of Salmi agates, PrF host volcanics, as well as quartz and calcite separated from agate nodules were carried out to obtain more information about the geochemistry of agates and the mineral-forming fluids (Table S2). Massive and vesicular types of PrF volcanics were all characterized by similar chondrite-normalized rare-earth element (REE) patterns with ΣREE = 179−367 ppm, enrichment in light rare-earth elements (LREE), slight depletion in heavy rare-earth elements (HREE) ([La/Sm]n = 2.48–3.19; [Gd/Yb]n = 1.62−2.30), as well as the minor negative Eu anomaly (Eu/Eu* = 0.77−0.95) (Figure 12). Additionally, volcanic rocks are significantly enriched in Rb, Ba, Zr, Nb and F (0.16%–0.25%) and exhibit a strong negative Sr anomaly (Sr/Sr* = 0.09–0.12) (Figure 13). Elevated concentrations of Rb (37 ppm), Ba (1390 ppm), Mn (2200 ppm), Cs (0.6 ppm) and depleted concentrations of P, Sc, Ti, Cr, Co, Ni, Cu, Ga, Sr, Y, Zr, Nb, Mo, REE in mandelstones with red-brownish amygdales compared to mandelstones with green amygdales were determined.

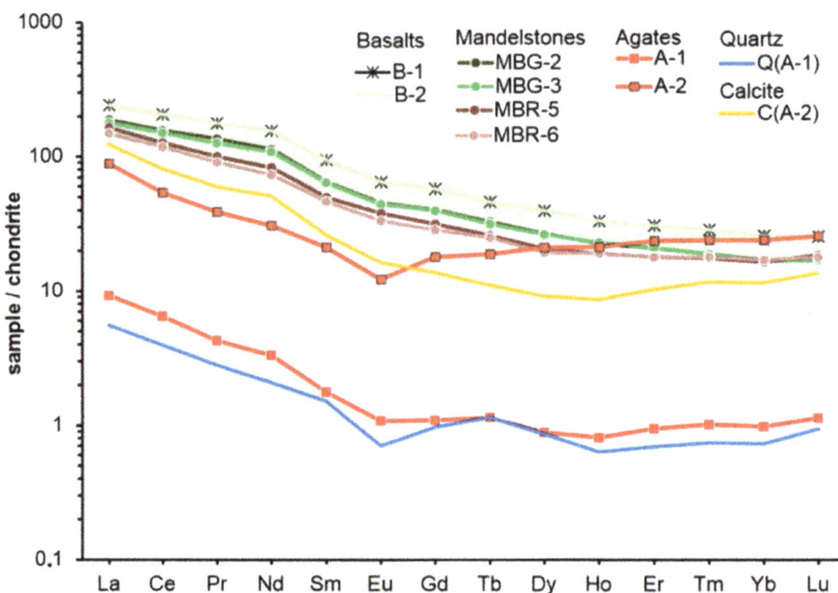

Figure 12. Chondrite-normalized REE distribution patterns for Salmi agates (including quartz and calcite samples extracted from agates) and host volcanic rocks of the Priozersk Formation. The normalization values are from [23]. Green field: the composition of the Late Permian High-Ti continental rift basalts in the Songpan–Ganzi Terrane (South China Block) [35].

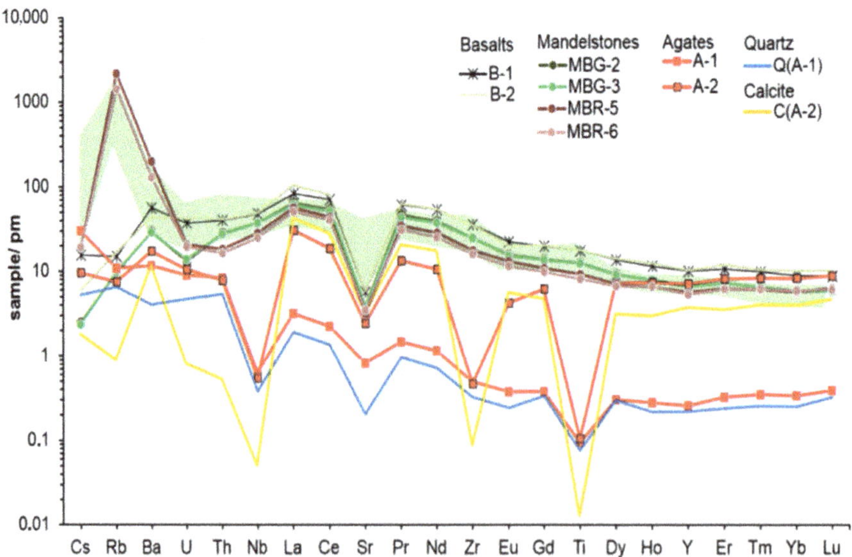

Figure 13. Primitive mantle-normalized trace element distribution patterns for Salmi agates (including quartz and calcite samples extracted from agates) and host volcanic rocks of the PLB. The normalization values are from [23]. Green field: the composition of the Late Permian High-Ti continental rift basalts in the Songpan–Ganzi Terrane (South China Block) [35].

The concentration of the most trace elements in agates is much lower than in the host volcanic rocks (Table S2). Two examined agate nodules with various quartz/calcite ratio exhibit REE distribution trends with enrichment of LREE ([La/Sm]n = 4.04–5.07), and slightly negative Eu anomaly (Eu/Eu* = 0.62–0.78) (Figure 12). However, they are characterized by difference in distribution of HREE and different levels of REE content. The calcite-rich agate has slightly enrichment of HREE, while calcite-poor agate is characterized by unfractionated HREE trend. The ΣREE are 9 and 97 ppm for calcite-poor (A-1) and calcite-rich (A-2) samples, respectively. The pattern of the REE distribution for the studied agates slightly differs from that for the host rocks. The degree of LREE enrichment of agates ([La/Sm]n = 4–5) is higher than that of volcanics ([La/Sm]n = 2–3), but the degree of HREE enrichment ([Gd/Yb]n = 0.7–1) in agates is lower than that for volcanics ([Gd/Yb]n= 1.9–2.3). Salmi agates have very low V, Cr, Co, Ni, Cu, Zn, Ga, Sr, Zr, Nb, Mo, Cd, Sn, Ba, Hf, Ta, W, Pb, Th concentrations. In contrast, elements such as Mn (max 5302), Ti (max 136 ppm), Li (max 66 ppm), Cs (max 0.96 ppm), and Tl (max 0.78 ppm) are concentrated in them. Only the contents of Rb (4−6 ppm), Sb (0.04−0.07 ppm), and Bi (0.02 ppm) are the same in agates and host rocks.

Microcrystalline quartz hosted in the agate is depleted in the most trace elements compared to the host agate, with the exception of Cr, Cu, Zn, Rb, Sn, Sb, Hf, Ta, Tl and Bi (Figure 13, Table S2). The quartz had a lower REE concentration (ΣREE = 6 ppm) compared with the agates and shows the distinct Eu anomaly (Eu/Eu* = 0.58−0.6) (Figure 12). The REE distribution pattern of quartz with enrichment in LREE is generally similar to the REE pattern for calcite-poor agate (A-1). In contract, the agate-associated calcite had a higher total REE content (ΣREE = 124) than agates and is characterized by a high concentration of Mn (3680 ppm), Ba (83 ppm), Sr (67 ppm), La (29 ppm), Ce (49 ppm), Nd (23 ppm), Y (17 ppm), Ti (16 ppm), Zn (14 ppm), P (11 ppm). The concentration of follow elements are lower: Li, Ni, Cu, Pr, Sm, Gd, Dy, Er, Yb (1–8 ppm); V, Cr, Co, Rb, Zr, Sn, Eu, Tb, Ho, Tm, Lu, Tl, Pb (0.3–0.9 ppm). Content of some trace elements are below the detection limit (0.3 ppm) of the ICP-MS method.

5.5.2. C-O Isotopic Composition

To specify the origin of carbonate material in Salmi agates, the stable isotope composition of agate-associated calcite samples was studied (Table 2). The results show that calcites have narrow δ^{13}C (–3.4‰ to –6.94‰, PDB) and δ^{18}O (24.83‰ to 25.16‰, SMOW) intervals. The correlation diagram δ^{13}C/δ^{18}O (Figure 14) displays that the δ^{13}C values fall within the interval (-2‰ to -8‰) that is typical of mantle carbonate [36]. However, the δ^{18}O values of examined calcites are significantly higher than those ones that characterize the deep carbon sources and coincide with the values (20‰ to 26‰) that are typical of sedimentary carbonates (normal marine limestone) [37].

Table 2. Carbon and oxygen isotope composition of calcite samples from Salmi agates.

Sample	δ^{13}C PDB(‰)	δ^{18}O SMOW(‰)
C-1	−6.94	24.35
C-2	−5.22	25.16
C-3	−3.46	24.83

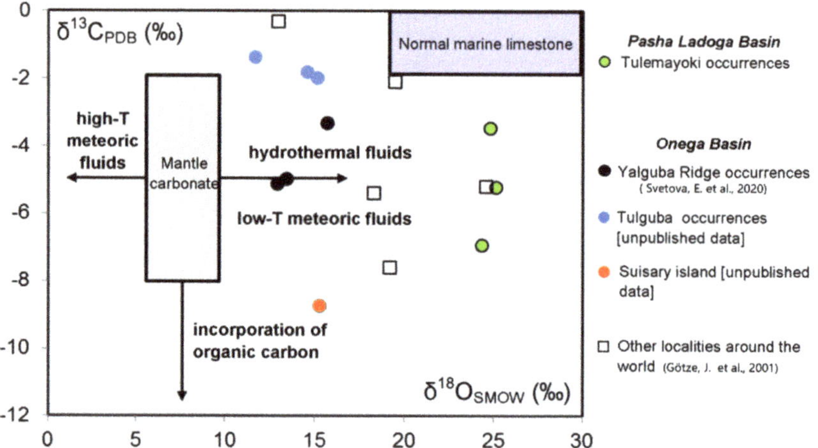

Figure 14. Plot of the oxygen-versus-carbon isotope results for agate-associated calcite samples from the Salmi occurrence (Pasha–Ladoga Basin) along with data available for agates of the Paleoproterozoic Onega Basin (NW Russia) and other localities around the world [3,7]. Mantle carbonate box and trends showing how various processes would affect the C-O isotope composition of magmatic carbonates are given according to Giuliani et al. [36], normal marine limestone [37].

6. Discussion

The obtained data by the integrated mineralogical, geochemical, and C-O isotope investigation of the Salmi agates together with the analysis of published geological materials for the studied area allow the discussion of the agate's peculiarities in more detail, speculating about coloration origin, agate provenance, and age.

6.1. Agate-Bearing Volcanic Rocks in the PLB

Mesoproterozoic volcanic rocks of the Priozersk Formation within the PLB are the youngest known agate-bearing volcanics at the Southeast Fennoscandia. Agates are mostly associated with massive varieties of basalt, andesites, as well as lava breccias and frothy lavas.

Volcanic host rocks are significantly enriched in Rb, Ba, Zr, and F, characterized by highly fractionated LREE/HREE chondrite-normalized patterns with no significant negative Eu anomaly. The chondrite-normalized REE diagram (Figure 12) exhibits patterns very similar to intracontinental rift basalts (CRB), such as Late Permian High-Ti Basalts in the Songpan–Ganzi Terrane, which mark the continental breakup of the South China Block [34] and close to Eocene-Oligocene alkaline rocks of Tibetan magmatic complex [38].

The low contents of Cr (54–61 ppm) and Ni (25–28 ppm) together with the strong negative Sr anomaly in PrF basalts (Figure 13) indicates crystallization from previously differentiated melts, which were formed at low-degree melting of a metasomatized mantle source containing amphibole and phlogopite [11] (pp. 145–147). Most likely, the primary melts were contaminated with crustal components [11] (pp.146), [19].

The comagmatic intrusive complex for PLB volcanics is the Valaam sill [11] (pp. 145–147), [17,18,21]. It forms a flat intrusive body extending over 2000 km^2 and exposed on the pre-Quaternary surface of the Valaam archipelago in the northern part of the Ladoga Lake. The Valaam sill intrudes the PrF, SalF, and PashF and is represented by subalkaline ferro-gabbro-dolerites, while more felsic rocks occur in the uppermost part of the intrusion. The U-Pb age of the intrusion is 1457 ± 3 Ma [21].

The geochemical composition of the PLB volcanics and Valaam sill rocks [11] (pp. 145–147) gives evidence of the intraplate character of magmatism (continental rifting), which took place within the Pasha–Ladoga Basin (ca. 1485−1454 Ma).

The obtained data suggest that the agates formation in the Pasha–Ladoga Basin are connected with post-magmatic processes, as well as hydrothermal fluids related to the intrusion of gabbro-dolerite sills (1459−1457 Ma).

6.2. Distribution and Morphology of Salmi Agates

Agates have been recognized in the upper part of PrF within the PLB and also observed as pebbles in conglomerate and gravelstone beds of the SalF [9], overlying the PrF volcanic rocks. Agate mineralization appears mostly as an infill of fissures, cavities, gas vesicles in massive and vesicular basalts, and lava-breccias. The agates are characterized by diverse textures and frequently exhibited concentric zoning patterns. The vein agates most often have a poor-fancy texture pattern (Figure 5D). According to Polekhovsky and Punin [9], this is determined by the quiet regime of silica deposition from the post-magmatic fluids in a connected system of open cavities. It is assumed that after filling with chalcedony and blockage of silica supply channels, a cavity passes to feeding by diffusion through the micropores of cavity walls and previously precipitated layers [24]. The diffusion of silica and its precipitation lead to oscillatory zoning. The zoning patterns in the individual cavities are varied probably because of variable permeability and rate of silica transfer (Figure 5). As for mandelstones, the vesicles were initially closed, and therefore, an individual regime of silica precipitation is established in each of them. This probably leads to a difference in the microtexture of the closely located amygdales (Figure 3). Additionally, the sporadic presence of discordant-filling vesicles (Figure 7G) indicates that the orientation of the amygdales has been changed during the agates formation under the influence of local tectonics. The abundance of fissures filled with silica, calcite and Fe-oxides in both agate nodules and mandelstones indicates that basalts were subjected to tectonic destructions and at least two phases of agate formation took place.

6.3. Mineralogy and Geochemistry of Salmi Agates

The agates exhibit diversified internal microtextures due to the individual layers composed of micro- and macro-crystalline quartz, length-fast chalcedony, and zebraic chalcedony. An insignificant moganite impurity in Salmi agates was reported by Polekhovsky and Punin according to X-ray diffraction data [9]. In the present study, X-ray diffraction analysis revealed only alpha-quartz and calcite phases in Salmi agates. No other silica phases beside alpha-quartz in the agates were also identified by the local Raman spectroscopy method. The absence of amorphous silica or other silica phases that appear during its crystallization (cristobalite, tridymite, moganite) is common for agates from relatively old volcanic host rocks that have been subjected to thermal effect [39]. Such silica phases are metastable and thus tend to transform into water-poor α-quartz.

The agates with the characteristic red coloration are widely distributed and have been described for many localities of Germany, Poland, Morocco, China, USA, Russia, etc. (e.g., [4,40–45]). It is shown that such coloration is associated with different types of iron compounds disseminated within the agate matrix.

Petrographic and Raman spectroscopy investigations revealed that the specific red-brownish colour of the Salmi agates (including amygdales in vesicular basalts) is caused by hematite inclusions. These inclusions occur in two main generations. The first one is represented by minute hematite particles (fine-grained or needle-liked) aligned along the micro-banding of agate (Figure 6B,D,F,H). Most likely, Fe-oxides and silica precipitated simultaneously from the same primary mineralizing fluids enriched in Fe. The observed agate's banding can be the result of a process of "self-purification" of silica during crystallization [44]. The second one is characterized by large spherical and irregular aggregates randomly distributed within the agate matrix (Figure 7E) or associated with fissures (e.g., Figures 3A and 10B) in which late silica fluids were transported. The abundance of two generations of hematite point to a high activity of iron both in primary silica-bearing fluids and in late hydrothermal solutions. We suggest that late hydrothermal fluids could mobilize the Fe-compounds from the crust of weathering underlying PrF volcanics, which are then accumulated in vesicles and fissures. As a result, an additional red coloration of the agates appears and hematized zones were formed in the mandelstones.

Besides the abundance of hematite in agates, elevated FeO content in chlorite also indicates a high activity of iron in the agate-forming fluid. The empirical geothermometers based on chemical composition of chlorites are widely used in different geological environments [46–48]. Since examined chlorites can be assigned to Mg-Fe variety and classified as diabantite, we applied a thermometer proposed by Kranidiotis and MacLean [47] to evaluate the chlorite formation temperatures (Table S1). This thermometer is based on the linear relation between Al(IV) in chlorite (with correction using Fe and Mg contents) and temperature. The calculation formula is T ($^\circ$C) = 106 (Al(IV)corr) + 18, where Al(IV)corr = Al(IV) + 0.7 (Fe/[Fe + Mg]). As a result, it was found that crystallization temperatures of chlorite from green amygdales (146–232 $^\circ$C, mean = 189 $^\circ$C) are slightly lower than those of chlorite from red-brownish amygdales (160–219 $^\circ$C, mean = 191 $^\circ$C). Meanwhile, the calculated results of chlorite from agate nodules (206–261 $^\circ$C, mean = 225 $^\circ$C) are unequivocally higher than those for amygdales from mandelstones. The estimated chlorite formation temperatures correspond to low-temperature hydrothermal environments. The difference in chlorite formation temperatures between agate nodules and amygdales from mandelstone is probably due to two-stage hydrothermal activities. Crystallization of chlorite apparently occurred under decrease in temperature, first in agate nodules and then in amygdales of mandelstone.

Thus, the mineral assemblage characterising the Salmi agates is predominantly represented by alpha-quartz, calcite, hematite, and chlorite, but it also contains inclusions of barite, hydroxyapatite, titanite, leucoxene, rutile, ilmenite, manganite, muscovite, as well as in rare cases, zircon, pyrite, and chalcopyrite. They were probably formed during the post-magmatic processes and/or at the hydrothermal stage.

The comparison of trace element data of agates and agate-bearing volcanic rocks are of special interest in view of the reconstruction of the hydrothermal mineral-forming process [3,49–51]. According to ICP-MS measurements, the agates are depleted in most trace elements compared to host volcanics, which is consistent with data of other investigations [3]. The general REE trends of agates and host volcanics display similarities in enrichment of LREE in regard to HREE and presence of negative Eu-anomaly, which may indicate the mobilization of some elements by circulating post-magmatic fluids from volcanic rocks. However, different fractionation trends for HREE in agate and host volcanics probably point to unrelated sources of their transporting. The common feature is also the absence of a Ce anomaly. The development of positive Ce-anomalies is associated usually with hydrothermal alteration of volcanic rocks [52]. According to petrographic and geochemical data the studied rocks were not subjected to regional alteration.

Elevated (5320 ppm) Mn concentrations (exceeding the values reported for host volcanic rocks) in calcite-rich agate possibly arise from both manganite microinclusions and structural incorporation of Mn within calcite. The quite high Ba (up to 121 ppm) and Ti (to 136 ppm) contents in agates are probably related to barite and titanite inclusions, respectively. Meanwhile, the unexpected high Ti concentrations in quartz sample (98 ppm) may provide evidence for the high temperature of agate crystallization (>200 $^\circ$C) [53], which has been also suggested on the basis of chlorite data.

The general similarity between REE distribution trends and elevated total REE contents in calcite-rich agates and extracted calcite suggest that calcite is a concentrator of REE. In contrast, the quartz together with calcite-poor agate is characterized by low REE content, which has also been reported in other studies [3,54]. Interestingly, macrocrystalline quartz from the central part of the agate nodule (which was formed on the final stage of agate crystallization) is significantly depleted in trace elements compared to host agate. The decreasing trend of trace element concentration in the banded agates from the outer zone to the core provide by local LA-ICP-MS analyses indicated the chemical purification process during crystallization [2,7,50].

The $\delta^{13}C$ and $\delta^{18}O$ values obtained for agate-associated calcite point to unrelated origin of carbon and oxygen (Figure 14). The $\delta^{13}C$ values in the range (–2 ... –8‰, PDB) probably reflect primary magmatic origin, whereas relatively high $\delta18O$ values (24 ... 25‰, SMOW) point to a sedimentary source. The close $\delta^{13}C/\delta^{18}O$ rations display agate-associated calcite samples from Phanerozoic and Precambrian volcanics of some other regions in the world (Figure 14). The majority of $\delta^{13}C/\delta^{18}O$ markers are in the transition zone between sedimentary carbonates (normal marine limestone) and magmatic carbonates with the influence of hydrothermal or low-thermal meteoric fluids. It is suggested that isotope feature of calcite can be connected with the processes of fluid mixing during formation, where carbon inherits its primary magmatic signature (volcanic H_2CO_3 as the dominant carbon species), whereas the signature of oxygen isotopes is overprinted by the secondary influence of meteoric water [2]. The close C-O isotopic characterization of agate calcite from different localities suggests that agates in volcanic rocks were formed with the active participation of post-magmatic hydrothermal fluids.

6.4. Origin of Agates

Summarizing the data on the mineralogy and geochemistry of Salmi agates from the PrF volcanics obtained in the present study and compilation of a regional geological information [10] (pp. 74–75), [11] (pp.145–147), [12], [17] (pp. 49–50] the agate mineralisation in PLB can be generalized into three stages, as illustrated in Figure 15.

Stage I involves the initial formation of PLB, including accumulation of terrigenous sedimentary rocks of PrF (after ca. 1477 Ma) and appearance of riftogenic volcanics (Figure 15A). The primary agate mineralization occurred after the formation of volcanic rocks of PrF (1485–1460 Ma) (Figure 15B). The upper lava flows were eroded simultaneously with the accumulation of terrigenous sediments of SalF, which resulted in the ingress of agate-bearing basalt fragments into the conglomerate sequence at the SalmF basement (Figure 15C). The migration of post-magmatic fluids toward sediments of SalF was blocked because of low permeability of argillaceous sandstone of SalF.

Stage II involves the formation of volcanic-sedimentary complexes of SalmF and PashF (ca. 1460 Ma) (Figure 15D). The multiphase intrusion of gabbro-dolerite Vallaam sill at 1459–1454 Ma (Figure 15E) most likely results in the formation of tectonically permeable zones at the basement of PLB rocks (Figure 15F). Fluid flows associated with late post-magmatic phases of sill were interacted with basin fluids and remobilized the iron compounds from the crust of weathering.

Stage III (1450–1400 Ma) involves the top-up of previously formed agate nodules and mineralized vesicles by hematite-rich calcite/silica matter, as well as mineralization of newly formed fissures and cavities, including gas vesicles in mandelstones (Figure 15G).

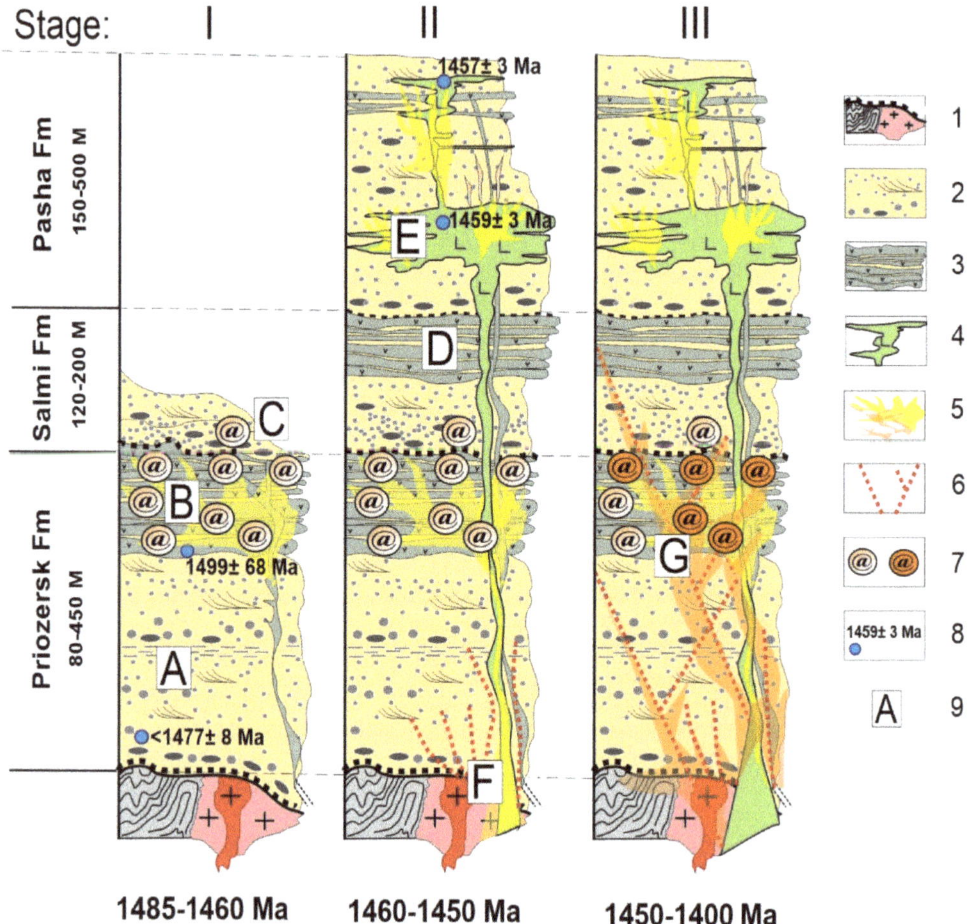

Figure 15. Schematic model of the agate formation stages in the Pasha–Ladoga Basin. Compiled using data from [10] (pp. 74–75), [11] (pp. 145–147), [12], [17] (pp. 49–50). Legend: 1, basement rocks; 2, sedimentary successions; 3, volcanic rocks; 4, Valaam sill; 5, post-magmatic and hydrothermal fluids affected zones; 6, tectonic faults, channels of fluid migration; 7, agates, agates; 8, isotope ages; 9, events associated with the agates formation. The black dotted lines indicate profiles of crust of weathering (erosion rocks areas). Stage I: formation of PLB, accumulation of sedimentary rocks of PrF, and appearance of riftogenic volcanics (A); agate formation within PrF volcanics (B); erosion of upper lava flows of PrF and accumulation of sediments of SalF (C). Stage II: formation of volcanic-sedimentary complexes of SalF and PashF (D); intrusion of gabbro-dolerite sill (E); formation of tectonically permeable zones at the PLB basement and remobilazation of iron compounds from the crust of weathering (F). Stage III: top-up of previously formed agates and mineralization of gas vesicles by hematite-rich calcite/silica matter (G).

7. Conclusions

The present contribution provides an integrated mineralogical and geochemical investigation of agate gemstones occurring in the Mesoproterozoic volcanic rocks of the PrF (ca. 1485−1460 Ma) within the Pasha–Ladoga Basin (Fennoscandian Shield, NW Russia).

The mineral composition of agates is dominated by alpha-quartz (fibrous chalcedony, microcrystalline and macrocrystalline quartz), but also displays abundances of calcite. The characteristic red-brownish agate's coloration is caused by multiple hematite inclusion distributed in an agate matrix.

The C−O isotope characteristic of agate-associated calcite points to the processes of fluid mixing during formation, where carbon inherits its primary magmatic origin, whereas the oxygen isotopes are reflected by the secondary influence of hydrothermal or meteoric waters.

The study revealed the two phases of agate formation in the PrF volcanics, which are most likely controlled by two distinctly different fluids and/or their mixture. At first, agates of quartz–calcite–chlorite composition with disseminated minute hematite inclusions appeared due to post-magmatic iron rich fluids. The late hydrothermal activity was probably triggered by the intrusion of the gabbro-dolerite Valaam sill and results in the second phase of agate formation. The hydrothermal fluids remobilized the iron oxides from the crust of weathering underlying the basement of PrF volcanics, which led to formation of vein agate, and filling of gas vesicles with hematite-rich calcite/silica matter. Fluid migration in the PLB was hindered by low permeability of argillaceous sandstone of SalF. Therefore, agates occur only locally in the PrF volcanics within the PLB sequence.

The studied agate occurrence could potentially be the source of a decorative agate gemstones, which is evidenced by single spectacular agate findings. This area is a tourist attraction, but the limited outcrop of the complex does not allow agates to be mined in large volumes.

Supplementary Materials: The following supporting information can be downloaded at: https://www.mdpi.com/article/10.3390/min13010062/s1, Table S1: Chemical composition of chlorite from the agate nodules and the amygdales of mandelstones from the PrF of the PLB (EDS microanalysis, wt%); Table S2: Trace elements composition of agate-bearing volcanics and agates from the PrF of the PLB (ppm) and ICP-MS BHVO2 standard results (ppm); Figure S1: X-ray diffractograms of the calcite-poor (A-1) and calcite-rich (A-2) agate samples.

Author Contributions: Conceptualisation, E.N.S. and S.A.S.; investigation, E.N.S. and S.A.S.; data curation, E.N.S. and S.A.S.; writing—original draft, E.N.S. and S.A.S.; visualization, E.N.S. and S.A.S.; compilation of geological published data, S.A.S. All authors have read and agreed to the published version of the manuscript.

Funding: This research was funded by state assignment to the Institute of Geology Karelian Research Centre RAS.

Data Availability Statement: All data are contained within the article and supplementary materials.

Acknowledgments: We are grateful to S.V. Burdyukh, I.S. Inina, and A.S. Paramonov (IG KRC RAS, Petrozavodsk, Russia) and to I.V. Smoleva and V.L. Andreichev (IG Komi SC, UB, RAS, Syktyvkar, Russia) for their assistance in analytical investigations. The authors thank V.I. Ivashchenko (IG KRC RAS, Petrozavodsk, Russia) for donating two agate specimens and the useful advice. We are grateful to O.B. Lavrov (IG KRC RAS, Petrozavodsk, Russia) for consultancy in the field research process. We are thankful to V.I. Ivashenko, A.A. Konyshev and Yu.N. Smirnov for assistance in field investigation in 2022. We also thank three anonymous reviewers and the Academic Editor for their constructive comments, which helped us to improve the manuscript.

Conflicts of Interest: The authors declare no conflict of interest.

References

1. Moxon, T.; Palyanova, G. Agate Genesis: A Continuing Enigma. *Minerals* **2020**, *10*, 953. [CrossRef]
2. Götze, J.; Möckel, R.; Pan, Y. Mineralogy, geochemistry and genesis of agate—A review. *Minerals* **2020**, *10*, 1037. [CrossRef]
3. Götze, J.; Tichomirowa, M.; Fuchs, H.; Pilot, J.; Sharp, Z.D. Geochemistry of agates: A trace element and stable isotope study. *Chem. Geol.* **2001**, *175*, 523–541. [CrossRef]
4. Pršek, J.; Duma'nska-Słowik, M.; Powolny, T.; Natkaniec-Nowak, L.; Toboła, T.; Zych, D.; Skrepnicka, D. Agates from Western Atlas (Morocco)—Constraints from mineralogical and microtextural characteristics. *Minerals* **2020**, *10*, 198. [CrossRef]
5. Moxon, T.; Reed, S.J.B. Agate and chalcedony from igneous and sedimentary hosts aged from 13 to 3480 Ma: A cathodoluminescence study. *Miner. Mag.* **2006**, *70*, 485–498. [CrossRef]

6. Moxon, T.; Nelson, D.R.; Zhang, M. Agate recrystallization: Evidence from samples found in Archaean and Proterozoic host rocks, Western Australia. *Aust. J. Earth Sci.* **2006**, *53*, 235–248. [CrossRef]

7. Svetova, E.; Svetov, S. Mineralogy and Geochemistry of Agates from Paleoproterozoic Volcanic Rocks of the Karelian Craton, Southeast Fennoscandia (Russia). *Minerals* **2020**, *10*, 1106. [CrossRef]

8. Svetova, E.N.; Chazhengina, S.Y.; Stepanova, A.V.; Svetov, S.A. Black Agates fromPaleoproterozoic Pillow Lavas (Onega Basin, Karelian Craton, NW Russia): Mineralogy and Proposed Origin. *Minerals* **2021**, *11*, 918. [CrossRef]

9. Polekhovsky, Y.S.; Punin, Y.O. Agate mineralization in basaltoids of the northeastern Ladoga region, South Karelia. *Geol. Ore Depos.* **2008**, *7*, 642–646. [CrossRef]

10. Khazov, R.A. Riphean complex. In *Geology of Karelia*; Sokolov, V.A., Kulikov, V.S., Stenar, M.M., Eds.; Nauka: Leningrad, Russia, 1987; 172p. (In Russian)

11. Larin, A.M. Jotnian and post-Jotnian (~1.46 Ga and younger) magmatism. In *Proterozoic Ladoga Structure (Geology, Deep Structure and Mineral Genesis)*; Sharov, N.V., Ed.; Karelian Research Centre RAS: Petrozavodsk, Russia, 2020; pp. 145–147.

12. Kuptsova, A.V.; Khudoley, A.K.; Thomas, D. Geology, alteration system, and uranium metallogenic potential of Pasha−Ladoga Basin, Russia. *Miner. Depos.* **2021**, *56*, 1345–1364. [CrossRef]

13. Glebovitsky, V.A. (Ed.) *Early Precambrian of the Baltic Shield*; Nauka: St. Petersburg, Russia, 2004. (In Russian)

14. Amelin, Y.V.; Larin, A.M.; Tucker, R.D. Chronology of multiphase emplacement of the Salmi rapakivi graniteanorthosite complex, Baltic Shield: Implications for magmatic evolution. *Contrib. Miner. Petrol.* **1997**, *127*, 353–368. [CrossRef]

15. Larin, A.M. *Rapakivi Granites and Related Rocks*; Nauka: St. Petersburg, Russia, 2011; 402p. (In Russian)

16. Velichkin, V.I.; Kushnerenko, V.K.; Tarasov, N.N.; Andreyeva, O.; Kisileva, G.D.; Krylova, T.L.; Doynikova, O.; Golubev, V.N.; Golovin, V.A. Geology and formation conditions of the Karku unconformity-type deposit in the northern Ladoga region (Russia). *Geol. Ore Dep.* **2005**, *47*, 87–112.

17. Sviridenko, L.P.; Svetov, A.P. *Gabbro-Dolerite Valaam Sill and Geodynamics of Ladoga Lake Basin*; Karelian Research Centre RAS: Petrozavodsk, Russia, 2008; 123p. (In Russian)

18. Kulikov, V.S.; Svetov, S.A.; Slabunov, A.I.; Kulikova, V.V.; Polin, A.K.; Golubev, A.I.; Gorkovets, V.Y.; Ivashchenko, V.I.; Gogolev, M.A. Geological map of Southeastern Fennoscandia (scale 1:750,000): A new approach to map compilation. *Trans. KarRC RAS* **2017**, *2*, 3–41. [CrossRef]

19. Bogdanov, Y.B.; Savatenkov, V.V.; Ivanikov, V.V.; Frank-Kamenetzkiy, D.A. Isotope age of volcanics of the Salmi Formation. In Proceedings of the II Russian Conference on Isotope Geochronology, Sankt-Petersburg, Russia, 25–27 November 2003.

20. Kuptsova, A.V.; Khudoley, A.K.; Davis, W.; Rainbird, R.H.; Kovach, V.P.; Zagornaya, N.Y. Age and provenances of sandstones from the Riphean Priozersk and Salmi formations in the eastern Pasha- Ladoga Basin (southern margin of the Baltic Shield). *Stratigr. Geol. Correl.* **2011**, *19*, 125–140. [CrossRef]

21. Rämö, O.T.; Mänttäri, I.; Vaasjoki, M.; Upton, B.G.J.; Sviridenko, L.P. Age and significance of Mesoproterozoic CFB magmatism, Ladoga Lake region, NW Russia. In Proceedings of the 2001: GSA Annual Meeting and Exposition Abstracts, Boston, MA, USA, 1–10 November 2001.

22. Svetov, S.A.; Stepanova, A.V.; Chazhengina, S.Y.; Svetova, E.N.; Rybnikova, Z.P.; Mikhailova, A.I.; Paramonov, A.S.; Utitsyna, V.L.; Ekhova, M.V.; Kolodey, B.S. Precision geochemical (ICP–MS, LA–ICP–MS) analysis of rock and mineral composition: The method and accuracy estimation in the case study of Early Precambrian mafic complexes. *Tr. KarRC RAS* **2015**, *7*, 54–73. [CrossRef]

23. Sun, S.; McDonough, W.F. Chemical and isotopic systematics of oceanic basalts: Implications for mantle composition and processes. *Geol. Soc. Spec. Publ.* **1989**, *42*, 313–345. [CrossRef]

24. Le Maitre, R.W. *Igneous Rocks: A Classification and Glossary of Terms*, 2nd ed.; Cambridge University Press: Cambridge, UK, 2002; p. 236.

25. Godovikov, A.A.; Ripinen, O.I.; Motorin, S.G. *Agates*; Nedra: Moscow, Russia, 1987; p. 368. (In Russian)

26. De Faria, D.L.A.; Venâncio, S.S.; de Oliveira, M.T. Raman microspectroscopy of some iron oxides and oxyhydroxides. *J. Raman Spectrosc.* **1997**, *28*, 873–878. [CrossRef]

27. Krolop, P.; Jantschke, A.; Gilbricht, S.; Niiranen, K.; Seifert, T. Mineralogical Imaging for Characterization of the Per Geijer Apatite Iron Ores in the Kiruna District, Northern Sweden: A Comparative Study of Mineral Liberation Analysis and Raman Imaging. *Minerals* **2019**, *9*, 544. [CrossRef]

28. Post, J.E.; McKeown, D.A.; Heaney, P.J. Raman spectroscopy study of manganese oxides: Tunnel structures. *Am. Miner.* **2020**, *105*, 1175–1190. [CrossRef]

29. Buzgar, N.; Buzatu, A.; Sanislav, I.V. The Raman study on certain sulfates. *Ann. Stiintifice Ale Univ.* **2009**, *55*, 5–23.

30. Zhou, L.; Mernagh, T.P.; Mo, B.; Wang, L.; Zhang, S.; Wang, C. Raman Study of Barite and Celestine at Various Temperatures. *Minerals* **2020**, *10*, 260. [CrossRef]

31. Wang, A.; Freeman, J.J.; Jolliff, B.L. Understanding the Raman spectral features of phyllosilicates. *J. Raman Spectrosc.* **2015**, *46*, 829–845. [CrossRef]

32. Arbiol, C.; Layne, G.D. Raman Spectroscopy Coupled with Reflectance Spectroscopy as a Tool for the Characterization of Key Hydrothermal Alteration Minerals in Epithermal Au–Ag Systems: Utility and Implications for Mineral Exploration. *Appl. Spectrosc.* **2021**, *75*, 1475–1496. [CrossRef] [PubMed]

33. Zane, A.; Weiss, Z. A procedure for classifying rock-forming chlorites based on microprobe data. *Rend. Lincei* **1998**, *9*, 51–56. [CrossRef]

34. Hey, M.H. A new review of the chlorites. *Miner. Mag. J. Miner. Soc.* **1954**, *30*, 277–292. [CrossRef]
35. Shao, Y.; Yan, D.; Qiu, L.; Mu, H.; Zhang, Y. Late Permian High-Ti and Low-Ti Basalts in the Songpan–Ganzi Terrane: Continental Breakup of the Western Margin of the South China Block. *Minerals* **2022**, *12*, 1391. [CrossRef]
36. Giuliani, A.; Phillips, D.; Kamenetsky, V.S.; Fiorentini, M.L.; Farquhar, J.; Kendrick, M.A. Stable isotope (C, O, S) compositions of volatile-rich minerals in kimberlites: A review. *Chem. Geol.* **2014**, *374–375*, 61–83. [CrossRef]
37. Valley, J.W. Stable isotope geochemistry of metamorphic rocks. *Rev. Miner. Geochem.* **1986**, *16*, 445–489.
38. Hou, Z.Q.; Griffin, W.L.; Zheng, Y.C.; Wang, T.; Guo, Z.; Hou, J.; Santosh, M.; O'Reilly, S.Y. Cenozoic lithospheric architecture and metallogenesis in Southeastern Tibet. *Earth-Sci. Rev.* **2021**, *214*, 103472. [CrossRef]
39. Moxon, T.; Carpenter, M.A. Crystallite growth kinetics in nanocrystalline quartz (agate and chalcedony). *Miner. Mag.* **2009**, *73*, 551–568. [CrossRef]
40. Götze, J.; Möckel, R.; Vennemann, T.; Müller, A. Origin and geochemistry of agates from Permian volcanicrocks of the Sub-Erzgebirge basin (Saxony, Germany). *Chem. Geol.* **2016**, *428*, 77–91. [CrossRef]
41. Powolny, T.; Dumańska-Słowik, M.; Sikorska-Jaworowska, M.; Wójcik-Bania, M. Agate mineralization in spilitized Permian volcanics from "Borówno" quarry (Lower Silesia, Poland)—Microtextural, mineralogical, and geochemical constraints. *Ore Geol. Rev.* **2019**, *114*, 103–130. [CrossRef]
42. Natkaniec-Nowak, L.; Duma´nska-Słowik, M.; Pršek, J.; Lankosz, M.; Wróbel, P.; Gaweł, A.; Kowalczyk, J.; Kocemba, J. Agates from Kerrouchen (The Atlas Mountains, Morocco). *Minerals* **2016**, *6*, 77. [CrossRef]
43. Zhang, X.; Ji, L.; He, X. Gemological characteristics and origin of the Zhanguohong agate from Beipiao, Liaoning province, China: A combined microscopic, X-ray diffraction, and Raman spectroscopic study. *Minerals* **2020**, *10*, 401. [CrossRef]
44. Götze, J.; Möckel, R.; Kempe, U.; Kapitonov, I.; Vennemann, T. Origin and characteristics of agates in sedimentary rocks from the Dryhead area, Montana/USA. *Miner. Mag.* **2009**, *73*, 673–690.
45. Goncharov, V.I.; Gorodinsky, M.E.; Pavlov, G.F.; Savva, N.E.; Fadeev, A.P.; Vartanov, V.V.; Gunchenko, E.V. *Chalcedony of North-East of the USSR*; Science: Moscow, Russia, 1987; p. 192. (In Russian)
46. Cathelineau, M. Cation site occupancy in chlorites and illites as a function of temperature. *Clay Miner.* **1988**, *23*, 471–485. [CrossRef]
47. Kranidiotis, P.; MacLean, W.H. Systematics of chlorite alteration at the Phelps Dodge massive sulfide deposit, Matagami, Quebec. *Econ. Geol.* **1987**, *82*, 1898–1911. [CrossRef]
48. Zang, W.; Fyfe, W.S. Chloritization of the hydrothermally altered bedrocks at the Igarapé Bahia gold deposit, Carajás, Brazil. *Miner. Depos.* **1995**, *30*, 30–38. [CrossRef]
49. Geptner, A.R. Hydrothermal mineralization in the Iceland rift zone (tectonic control of the formation of mineral concentrations). *Lithol. Miner. Res.* **2009**, *44*, 205–228. [CrossRef]
50. Mockel, R.; Götze, J.; Sergeev, S.A.; Kapitonov, I.N.; Adamskaya, E.V.; Goltsin, N.A.; Ennemann, T. Trace-Element Analysis by Laser Ablation Inductively Coupled Plasma Mass Spectrometry (LA-ICP-MS): A Case Study for Agates from Nowy Kościoł, Poland. *J. Sib. Fed. Univ. Eng. Technol.* **2009**, *2*, 123–138.
51. Conte, A.; Della Ventura, G.; Rondeau, B.; Romani, M.; Cestelli Guidi, M.; La, C.; Napoleoni, C.; Lucci, F. Hydrothermal genesis and growth of the banded agates from the Allumiere-Tolfa volcanic district (Latium, Italy). *Phys Chem Miner.* **2022**, *49*, 39. [CrossRef]
52. Rollinson, H.; Pease, V. *Using Geochemical Data to Understand Geological Processes*; University Printing House: Cambridge, UK, 2021; 661p.
53. Muller, A.; Herklotz, G.; Giegling, H. Chemistry of quartz related to the Zinnwald/Cinovec Sn-W-Li greisen-type deposit, Eastern Erzgebirge, Germany. *J. Geochem. Explor.* **2018**, *190*, 357–373. [CrossRef]
54. Svetova, E.N.; Svetov, S.A. Agates from Paleoproterozoic volcanic rocks of the Onega Structure, Central Karelia. *Geol. Ore Depos.* **2020**, *62*, 669–681. [CrossRef]

 *minerals*

Article

Mineralogy of Agates with Amethyst from the Tevinskoye Deposit (Northern Kamchatka, Russia)

Evgeniya N. Svetova [1], Galina A. Palyanova [2,*], Andrey A. Borovikov [2], Viktor F. Posokhov [3] and Tatyana N. Moroz [2]

[1] Institute of Geology, Karelian Research Centre of Russian Academy of Sciences, Pushkinskaya St. 11, 185910 Petrozavodsk, Russia; enkotova@rambler.ru
[2] Sobolev Institute of Geology and Mineralogy, Siberian Branch of Russian Academy of Sciences, Akademika Koptyuga Pr., 3, 630090 Novosibirsk, Russia; borovikov.57@mail.ru (A.A.B.); moroz@igm.nsc.ru (T.N.M.)
[3] Dobretsov Geological Institute, Siberian Branch of Russian Academy of Sciences, Sakh'yanovoi St. 6a, 670047 Ulan-Ude, Russia; vitaf1@yandex.ru
* Correspondence: palyan@igm.nsc.ru

Citation: Svetova, E.N.; Palyanova, G.A.; Borovikov, A.A.; Posokhov, V.F.; Moroz, T.N. Mineralogy of Agates with Amethyst from the Tevinskoye Deposit (Northern Kamchatka, Russia). *Minerals* **2023**, *13*, 1051. https://doi.org/10.3390/min13081051

Academic Editors: Stefanos Karampelas and Emmanuel Fritsch

Received: 22 July 2023
Revised: 5 August 2023
Accepted: 7 August 2023
Published: 9 August 2023

Abstract: The Tevinskoye agate deposit is located in the North of the Kamchatka peninsula (Russia) and represented by agate-bearing Eocene basaltic and andesitic rocks of the Kinkilsk complex. Agate mineralization occurs in lavas and tuffs as amygdales, geodes, lenses and veins, which are the main sources of the resupply of coastal agate placers. The present study aimed to perform a comprehensive mineralogical, geochemical, and O-isotope investigation of amethyst-bearing agates, and to evaluate data concerning the origin of mineralization and the conditions for amethyst formation. Agates exhibit spectacular textures, with variation in the sequence of silica filling of amygdales and geodes. The mineral composition of the agates is mainly represented by micro- and macro-crystalline quartz, amethyst, length-fast and zebraic chalcedony, moganite, goethite, and clinoptilolite. Carbonate forms individual bands in the outer zones of some agates. The presence of small amounts of native copper, covellite, chalcopyrite and pyrite is a feature of these agates. Copper and iron mineralization are probably typomorphic features related to the host rock composition. The measured values of crystallite size (525–560 Å) and the high moganite content (up to 50%) of agate with amethyst are evidenced by the young age (~45 Ma) of agate-hosting rocks. Agate formation temperatures (21–229 °C) were calculated from the O-isotope composition of chalcedony (+19.6 to +25.5‰), quartz (+18.1 to +22.3‰), and amethyst (+18.2 to +21.5‰). The cold-water monophase fluid inclusions revealed in amethyst crystals suggest that the mineralizing fluids have low temperatures (<100 °C) and low salinity. Magnetite grains in host rock, together with goethite inclusions identified within the amethyst crystals, point to a change in redox conditions and the presence of iron in the agate-forming fluids, which entered the quartz lattice during crystallization and influenced the formation of the violet color.

Keywords: agates with amethyst; Tevinskoye deposit (Northern Kamchatka, Russia); SEM-EDS; XRD; Raman spectroscopy; IR spectroscopy; O-isotope analysis

1. Introduction

Agates are famous, beautiful, and fascinating stones found all around the world. Amethyst is a quartz variety often used in jewelry [1,2]. There are amethyst deposits located in Southern Rhodesia (Mwakambiko), Brazil (states of Rio Grande do Sul, Minas Gerais and Bahia/Brezheno, Jacobina, Santo Se), Uruguay (departments of Takuarembo, Artigas, Salto and Paysandu), Namibia (Brandberg), India (Decan), USA (Creed and Cripple Creek), Madagascar, Sri Lanka, Mexico (in the districts of Guanajuato, Guerraro and Las Vigas), Canada (Thunder Bay), South Korea, Zambia, Greece, Italy, Czech Republic, Slovakia, Romania, Bulgaria, Russia, and other countries [3,4]. In Europe, famous amethyst localities include Idar–Oberstein, Baden–Baden and Erzgebirge (Germany), Osilo in Sardinia (Italy), Pribram (Czech Republic), Schemnitz and Kremnitz (Slovakia), Ros, Montana, Sacarâmb, Baia Sprie and Cavnik (Romania), Madjarovo (Bulgaria), and Greece (Kassiteres–Sapes, Kirki, Kornofolia, on Lesvos and Milos Islands) [3–5]. In Russia, amethyst deposits are Vatikha, Talyan, Obman, Khasavarka (Urals), Cape Korabl (the Kola Peninsula), Rusavkino (Moscow region), Karadag (Crimea), etc.

However, most of the above deposits do not belong to agate deposits. It should be noted that amethyst is not particularly common at agate deposits. The most famous deposits of agates with amethyst are Brazilian and Uruguayan deposits [6–10]. Amethysts are found in agates in Morocco (Sidi Rahal) [11], the Ijevan deposit (Tavush region, Armenia), in some agate occurrences of Russia-Nepsky (Irkutsk region), Mulina Gora (Chita region), Sededen, Olskoe plateau (Magadan region), Belorechenskoye (Northern Timan), Tulguba (Onega Basin) [12–16], etc. Amethyst is usually found in the form of crystals and their intergrowths inside agate geodes, in amygdales, and in the cracks of volcanic rocks. Despite the large number of deposits in Russia, the sites are not developed and are mothballed or abandoned. Amethyst occurrences in agates are regarded as potential deposits for possible future exploitation [17].

Amethyst forms in a wide variety of environments [3,4]. Amethyst crystallization conditions are still a matter of scientific debate. The complex process of agate formation is not yet completely understood [18–21].

The Tevinskoye deposit of agates with amethyst in coastal placers was discovered in 1982, north of Cape Tevi in the interfluve of the Mainkaptal and Eltavayam rivers that flow into the Sea of Okhotsk in the western part of the Kamchatka peninsula. Placers of agates are located on the sea beaches of the Shelikhov Bay, the width of which at low tide reaches 100 m or more. During tidal processes, the sand and gravel components are washed off the surface of the beaches and agate-bearing boulder–pebble deposits are exposed. Many collection samples of agates with amethyst from the Tevinskoye deposit (further in the text «Tevi agates») are presented in many mineralogical and geological museums of Russia (A.E. Fersman Mineralogical Museum, Moscow; Central Siberian Geological Museum, Geological Museum SNIIGGiMS, scientific and educational center «Evolution of the Earth», Novosibirsk; Scientific Museum of V&S FEB RAS, Museum of Volcanoes «Volcanarium», Petropavlovsk-Kamchatsky). The Tevinskoye agate deposit is included in the list of specially protected natural areas of the Kamchatka peninsula.

There are no data on the mineralogy and formation conditions of Tevi agates in the literature. The present study aimed to evaluate data concerning the origin of mineralization and the conditions for amethyst formation. Of particular interest was the characterization of mineral composition, the identification of silica polymorphs, the study of the isotopic composition of the fluids that formed the agates, and the estimation of the temperature at which amethyst forms in agates. The article provides new data on the mineralogical and geochemical features of Tevi agates from Kamchatka peninsula and substantiates the conditions for the genesis of amethyst in agates.

2. Geological Setting

The Tevinskoye agate deposit is located on the eastern coast of the Shelekhov Bay of the Sea of Okhotsk, 120 km northeast of the Palana village in the western part of the Kamchatka Peninsula (Figure 1A). The deposit is represented by a series of agate-bearing fields associated with the Eocene volcanic rocks of the Kinkilsk complex [22]. The volcanic rocks of the Kinkilsk complex are widespread along the Sea of Okhotsk coast from Cape Khairyuzov and the Anadyrka River mouth (Cape Kinkil) to Podkagernaya Bay and along Shelikhov Bay, as well as along the western slope of the Sredinnyi Range of the Kamchatka Isthmus. These rocks have been sufficiently studied [23–25]. The volcanics of the Kinkilsk complex inherited the distribution of the earlier (Paleocene) volcanic fields, overlapping their highly deformed older rocks with a sharp angular unconformity.

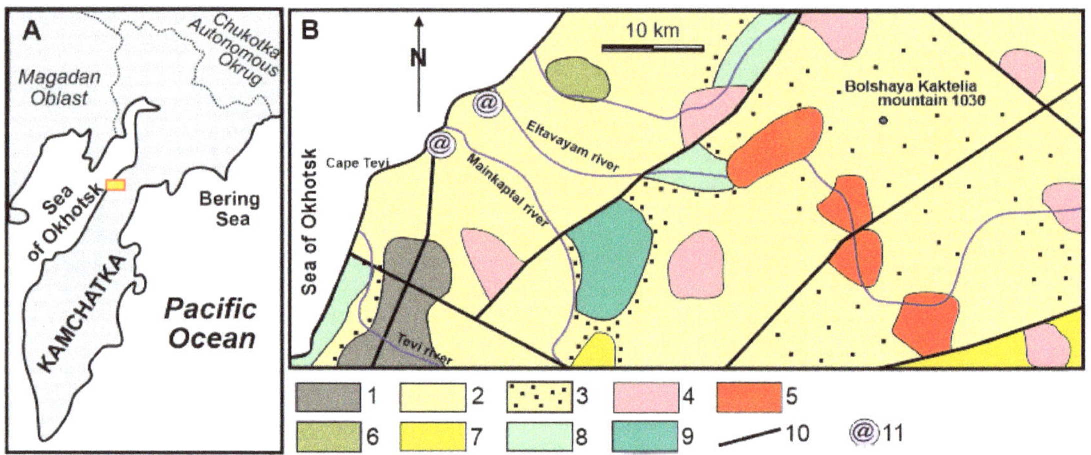

Figure 1. (**A**) Location of the studied area. (**B**) Simplified geological map of the Tevinskoe agate deposit area based on [22–25]. Legend: Quaternary glacial sediments: 1—unlayered blocky-pebble-boulder loams and sandy loams, clays with gruss, crushed stone and boulders clays. Eocene–Oligocene subvolcanics rocks: 2—andesitobasalts; 3—dacites; 4—basalts. Middle–Upper Eocene: 5—amphibole–biotite dacites and rhyolites, two-pyroxene andesites and basalts; 6—vitroclastic tuffs and tuffites with mineral and brown coals, coaly tuff-siltstones; 7—dacites, rhyolites, andesites, their tuffs, ignimbrite and tuffites. Cretaceous–Paleogene: 8—sandstones and siltstones with interbeds and lenses of coaly argillites, concretions of carbonate sandstones and marls. Upper Cretaceous: 9—argillites, sandstones, siltstones, siliceous rocks; 10—faults; 11—sampling agate placers of the Tevinskoye deposit.

2.1. Host Rocks

The volcanic rocks are represented by the subaerial lavas of differentiated basalt–andesidacite–rhyolite associations, lavabreccias, and blocky agglomerate tuffs [23]. In terms of $(Na_2O + K_2O)$–SiO_2 ratio, these rocks are moderately potassic calc–alkaline. The texture of the rocks is porphyritic, with abundant phenocrysts. Phenocrysts in basalts and andesite-basalts are represented by plagioclase-two-pyroxene paragenesis, amphibole appears in andesites, and biotite and quartz are present in dacites and rhyolites. Oxide minerals are represented by ilmenite, magnetite, and titanomagnetite. Volcanic–sedimentary sequences are locally distributed and represented by interbedded vitroclastic tuffs and tuffites, often with lenses of mineral and brown coals, and coaly tuff siltstones. The secondary alterations of the rocks are pronounced unevenly and are generally insignificant, while the areas with local hydrothermal alteration occur as quartz veinlets and lenses [23,25].

At the studied area (Cape Tevi), the most complete section of the complex is represented by volcanic rocks that unconformably overlie the Upper Cretaceous terrigenous rocks (Figure 1B) [24]. In this area, the lower part of the section is formed by amphibole–biotite dacites and rhyolites, which are replaced upward by two-pyroxene andesites and basalts. Subvolcanic rocks are widespread and presented by low-power dolerite sills [25]. The visible thickness of the section reaches up to 1500 m.

The K–Ar age of the volcanic rocks of the Kinkilsk complex is estimated as Middle–Upper Eocene (53–45 Ma) [23], while the age of volcanics from Cape Tevi is ca. 45 Ma [25]. The geodynamics of the Early Paleogene magmatism, represented in the Kamchatka region by calc-alkaline series, suggest their formation in an active continental margin setting [25].

2.2. Tevinskoye Deposit Description

The studied area is located a little north of Cape Tevi within the area of two coastal placers: Eltavayam and Mainkaptal (Figures 1B and 2A). The total length of beaches in the placers area is 6.5 km [22]. The coastal cliffs are up to 300 m high. Formed by lava flows, breccias, and tuffs of the Kinkilsk complex, the cliffs contain a large number of agate amygdales and geodes, mineralized gas vesicles and cavities (Figure 2B,C), vein chalcedony, amethyst-like quartz, and calcite. They are the main source of replenishment of coastal agate placers. Agate geodes with quartz, amethyst or quartz-carbonate filling are present in the placers of primary industrial importance. Moss agates (translucent, bluish, and bluish-gray chalcedony with black manganese dendrites) and technical agate are also found. These are colored in light gray tones, sometimes with brownish, yellowish or bluish tints. The area of the Eltavayam placer is 45,000 m^2, and reserves of technical agate are estimated at 900 kg, jewelry and ornamental agate are estimated at 315 kg, and moss agate is estimated at 180 kg. The area of the Mainkaptal placer is 61,000 m^2, the reserves of technical agate are estimated to be 1037 kg, those of jewelry and ornamental stand agate are estimated at 427 kg, and those of moss are estimated at 61 kg [22].

Figure 2. Field outcrop photographs illustrating the Tevinskoye agate deposit in the Eocene volcanic rocks within the Kinkilsk complex of Northern Kamchatka. (**A**) View of coastal outcrops of volcanics with agate placers in the bay; (**B**) rounded agate amygdule with amethyst in the amygdaloidal basalt (size 9 cm × 10 cm); (**C**) lens-shaped agates with amethyst infill cavities in the agglomerate tuff; (**D**) examples of collected agate samples with amethysts.

3. Materials and Methods

Agates with amethysts were collected during a field season in 2021. Ten representative samples of agates with amethyst, with sizes varying from 3 to 25 cm and different textures, were selected for inclusion in research. Polished and thin sections, as well as powder samples, were prepared from agate fragments for use in analytical investigation. Petrographic (optical microscopy and SEM-EDS), XRD, and Raman spectroscopy investigations of the agates were carried out at the Institute of Geology, Karelian Research Centre, RAS (IG KRC RAS, Petrozavodsk, Russia). Polished thin sections of the agates were examined using transmitted light microscopy on a Polam-211 optical microscope. SEM-EDS investigations were carried out by using a VEGA II LSH (Tescan, Brno, Czech Republic) scanning electron microscope with EDS INCA Energy 350 (Oxford Instruments, Oxford, UK) on the carbon-coated polished thin section and chips of agate samples. Analyses were performed at the following parameters: W cathode, 20 kV accelerating voltage, 20 mA beam current, 2 μm beam diameter, and counting time of 90 s. The following standards were used: calcite, albite, MgO, Al_2O_3, SiO_2, FeS_2, wollastonite, Fe, Zn, and InAs. SEM-EDS quantitative data were obtained and used in the determination of the analysis accuracy via processing performed using the Microanalysis Suite Issue 12, INCA Suite version 4.01 (Oxford Instruments, Oxford, UK).

Powder XRD analysis was carried out using a Thermo Scientific ARL X'TRA (Thermo Fisher Scientific, Ecublens, Switzerland) diffractometer (CuK-radiation, voltage 40 kV, current 30 mA). Chalcedony and amethyst crystal samples from agate were scanned for review in the range of 5–156° 2θ at a scanning step of 0.40° 2θ/min. Diffractograms in the 66–69° and 25–28° 2θ ranges were recorded at a scanning step of 0.2° 2θ/min to obtain more precise measurements of parameters of diffraction reflections. X-ray phase and structural analyses were carried out using the program pack Win XRD, ICCD (DDWiew2008). Agate samples were hand-ground to obtain grain sizes < 50 μm. The detection limit for XRD phase identification was 3 wt.%.

Raman spectroscopy was used to identify and characterize SiO_2 polymorphs at the local zones of banded agates. Raman spectroscopy analysis was conducted on a dispersive Nicolet Almega XR Raman spectrometer (Thermo Fisher Scientific, Waltham, MA, USA), using the 532 nm wavelength of Nd-YAG laser (Thermo Fisher Scientific, Waltham, MA, USA). The spectra were collected at 2 cm^{-1} spectral resolution. A confocal microscope with a $50\times$ objective lens was used to focus an excitation laser beam on the sample and to collect a Raman signal from an area with a diameter of 2 μm. Raman spectra were acquired in the 85–1200 cm^{-1} spectral region, with 30 s exposition time and laser power of 10 mW. The curve-fitting algorithm in the OMNIC software (v.8.2., Thermo Fisher Scientific, Waltham, MA, USA) was used to fit the spectra in the 400–550 cm^{-1} range using peaks of moganite at 502 cm^{-1} and quartz at 465 cm^{-1}. Raman spectral data, such as peak position and band area (i.e., integrated area), were determined by fitting the spectra with Lorentzian functions. For the calculation of moganite concentrations, we used a calibration curve provided by Götze et al. [26].

The infrared spectroscopy study was performed using a FT-801 FTIR spectrometer (NPF Simex, Novosibirsk, Russia) at the Petrozavodsk State University (Petrozavodsk, Russia). Analyses were carried out on a doubly polished wafer of agate with a thickness of about 1 mm. IR spectra were recorded from a square aperture, limiting the studied sample surface to an area of 1 cm^2. The spectra were recorded in transmission geometry at room temperature in the range of 2400–4000 cm^{-1} (0.5 cm^{-1} spectral resolution, 50 scans).

SEM-EDS, Raman spectroscopy, cryo- and thermometry were also carried out in the Analytical Centre for Multielemental and Isotope Research of the Sobolev Institute of Geology and Mineralogy, Russian Academy of Sciences (Novosibirsk, Russia). Chemical analyses of mineral phases were carried out using a MIRA LMU electron scanning microscope (Tescan Orsay Holding, Brno–Kohoutovice, Czech) with an INCA Energy 450þ X-Max energy-dispersion spectrometer (Oxford Instruments Nanoanalysis Ltd., Abingdon, UK). The operation conditions were: an accelerating voltage of 20 kV, a probe current of 1 nA and a spectrum recording time of 15 to 20 s. Raman spectra of goethite, pyrite and clinoptilolite were recorded on a Horiba Jobin Yvon LabRAM HR800 spectrometer, with a 1024-pixel liquid nitrogen-cooled charge-coupled device (LN/CCD) detector used for the excitation of 532 nm wavelengths from a frequency-doubled Nd:YAG laser operating at 1064 nm. Raman spectra were collected in a backscattering geometry using an Olympus BX41 microscope at a fixed orientation of the crystal, with the incident laser polarization perpendicular to the surface. The spectral resolution of the recorded Stokes Raman spectra was set to ~3.0 cm^{-1} at a Raman shift of 1300 cm^{-1}. This resolution was achieved by using one grating with 1800 grooves/mm, equivalent to 150 μm slits, and a pinhole. Using the microscope with an Olympus $50\times$ objective lens of WD = 0.37 mm and with a 0.75 numerical aperture for the visible spectral range produces a focal spot diameter of ~2 μm. The power of the laser radiation used was set to about 0.5 mW on the sample to avoid sample heating. The absence of sample heating was tested via comparison with the Raman spectra. The absence of sample heating was tested by comparison with the Raman spectra recorded at 7 and 0.3 mW power of the incident laser beam at the sample. The majority of the spectra were recorded using a neutral density filter, D = 1. The spectrometer was wavenumber calibrated with a silicon standard. Cryo- and thermometry methods were used to determine the temperatures of phase transitions in fluid inclusions in quartz (a THMSG-600 microthermal chamber from "Linkam" with a measurement range of $-196/+600\ ^\circ$C). The total concentrations of salts in the solutions of fluid inclusions and their belonging to one or another water–salt system were determined using the cryometry method [27–29].

The isotope composition of oxygen, as O_2, was defined by a FINNIGAN MAT 253 gas mass spectrometer at the Dobretsov Geological Institute of Siberian Branch of Russian Academy of Sciences (Ulan-Ude, Russia) using a double system of inflow in a classic variant (standard–sample). To determine $\delta^{18}O$ values, the samples were prepared using laser fluorination with the "laser ablation with oxygen extraction from silicates" mode in the presence of a BrF_5 reagent according to the method of [30]. Only pure minerals (as fragments), with a total weight of 1.5–2.5 mg, were used for oxygen isotope analysis. The calculations of $\delta^{18}O$ were performed using the international standards NBS-28 (quartz) and NBS-30 (biotite). The accuracy of the obtained data was checked via regular measurements of an inner standard GI-1 (quartz) and a laboratory standard Polaris (quartz) of IGEM RAS. The analytical precision for $\delta^{18}O$ was (1s) $\pm$ 0.2 ‰.

4. Results

4.1. Macro- and Microscopic Observation

Agate amygdales and geodes with amethyst exhibit rounded or lens-like shapes (Figure 2B,C). The mean size of the agates ranges from 3 to 12 cm, and large samples with a size of up to 30 cm occur rarely (Figure 2D). Agates mostly display monocentric concentrically zoned (Figure 3A,E,C). The outer parts of agates are characterized by a fine-banding texture and they display alternate light-gray and white-colored layers, with thickness varying from 0.5 mm to 1 cm (Figure 3A,B,G). The banded area is followed to the inside of the geode by quartz crystals. Amethyst crystals with sizes of up to 3 cm along c-axis encrusted central agate cavities as overgrowth on colorless and milky-white quartz (Figure 3A,B). The violet amethyst color commonly increases in quartz crystals towards the agate center. Sometimes amethyst can compose the individual intermediate layer between outer banded and internal crystalline agate zones (Figure 3C,E). In some agates, thick carbonate (calcite) layers form the outer zone (Figure 3F). The marginal rim of the agates at the point of contact with the host rock often comprises zeolite crystals.

Petrographic study has shown that visible silica banding of external parts of agates, corresponding to the beginning of crystallization, is represented by alternating-length fast chalcedony (with the c-axis oriented perpendicular to the fibre direction) and zebraic chalcedony (length-fast with a helical twisting of fibres along fibre direction) layers with thicknesses ranging from 50 to 300 μm (Figure 4A). The boundaries of layers correspond to the temporal gap in silica deposition and are often marked by thin interlayers of differently oriented macrocrystalline quartz (Figure 4B). Parallel-layered agate zones are usually preceded by the radiation of fibrous radial aggregates towards the agate core. The nucleation centers for these aggregates are mineral grains on the walls of host rocks (Figure 4C). Radial fibrous length-fast chalcedony and zebraic chalcedony are often associated with microcrystalline quartz (Figure 4C). The transition zone between banded chalcedony and crystalline quartz has a distinct boundary (Figure 4D). In the direction of growth of euhedral and subhedral quartz grains, oriented with long axes perpendicular to the layers of chalcedony, a gradual increase in their size is observed (Figure 4D). The crystallization sequence is completed by comb-shaped amethyst crystals. Some prismatic quartz crystals under crossed polars display feathery appearances (Figure 4D,E,F). These microtextures are characterized by subgrains, which appear as splintery or feathery patterns due to slight optical differences in maximum extinction positions [31]. The subgrains of feathery textures are usually elongated in parallel to each other in the crystal growth direction. The accumulation of fluid inclusions is commonly associated with feathery textures in the studied quartz. Iron oxide/hydroxide inclusions occasionally occur within both chalcedonic bands and quartz crystal areas.

Figure 3. Photographs of agates with amethyst from the Tevinskoye deposit (Northern Kamchatka, Russia). (**A**) Concentrically banded agate with cavities incrusted by large amethyst crystals; (**B**) polished fragment of amethyst druse overgrown on fine-banded chalcedony; (**C–E**) concentrically zoned agate amygdales with alternating bands of chalcedony, amethyst, colorless and milky-white quartz; (**F**) fragment of agate amygdule with amethyst and thick yellowish carbonate band; (**G**) cross section of agate, illustrating fine-banded chalcedony external zone with successive overgrowth of milky-white quartz and amethyst. The areas of IR spectroscopy analysis are marked with white squares (see Section 4.5).

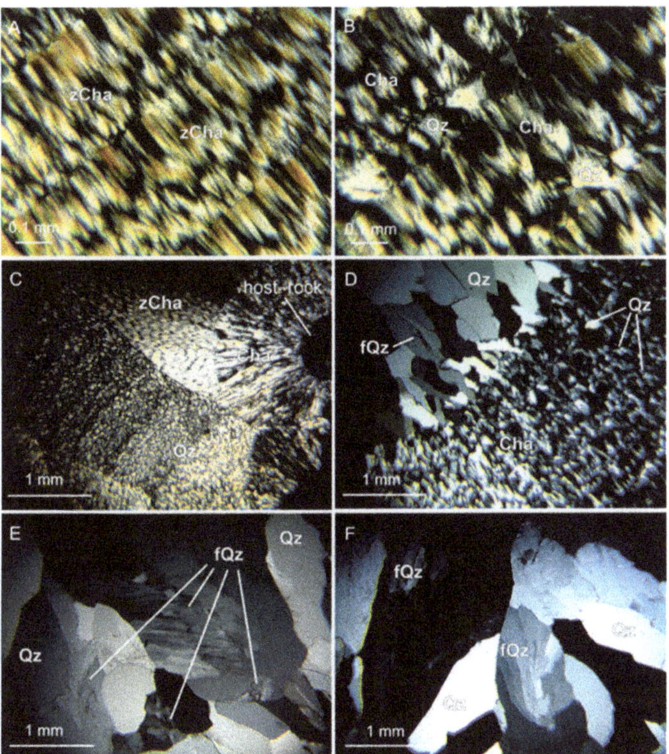

Figure 4. Micrographs of agates with amethyst from the Tevinskoye deposit in transmitted light with crossed polars showing a variety of petrographic silica textures. (**A**) Rhythmical banding composed of zebraic chalcedony layers within the outer agate zone; (**B**) alternating bands of length-fast chalcedony fibers and macrocrystalline quartz grains; (**C**) radial fibrous length-fast chalcedony accompanied by polygonal-shaped aggregates of zebraic chalcedony and microcrystalline quartz at the point of contact with host rocks; (**D**) smooth transition from banded agate to prismatic quartz crystals, which increases in size towards the geode center; (**E**,**F**) internal parts of the agates comprising euhedral prismatic quartz with feathery microtextures developed within the crystals. Abbreviations: Cha—length-fast chalcedony; zCha—zebraic chalcedony; Qz—macrocrystalline quartz; μQz—microcrystalline quartz; fQz—quartz with feathery microtexture.

4.2. SEM-EDS Investigation

SEM-EDS investigation of three typical agate samples (Figure 5A) was performed to provide more evidence concerning the micromineral composition of agates. The scanning of fine-banded agate areas revealed that individual bands are characterized by various microporosities (Figure 5D). The most frequent mineral inclusions were recognized as Mn-enriched (up to 2.5 wt.%) calcite. They are expressed in the form of both thick bands (Figure 3G) and microinclusions in quartz with a size of up to 300 μm (Figure 5E,F). Randomly distributed spherical- and paniculate-shaped aggregates of growths of needle-like goethite crystals, with sizes varying from 100 to 300 μm, were observed within the quartz and amethyst crystals (Figure 5G). Rare filamentous inclusions of native copper with sizes of 5 × 50 −150 μm (Figure 5H), sporadic chalcopyrite inclusions of 10 up to 50 μm in size, and singular covellite grains (Figure 5I,J) were recognized in the silica matrix. Ni-enriched (up to 0.4 wt.%) pyrite inclusions were observed within the calcite band at the external area of the agate (Figure 5A,E). In addition, large prismatic zeolite (clinoptilolite) crystals associated with calcite are frequently present at the point of contact with the host rock (Figure 5K). The identification of goethite and clinoptilolite was confirmed using Raman spectroscopy (see Section 4.4). We also observed brecciated textures, with fragments of host rock detached from the agate wall and cemented via agate. Magnetite, plagioclase, pyroxene, ilmenite, and amphibole were found within these fragments and in the host rock (Figure 5L).

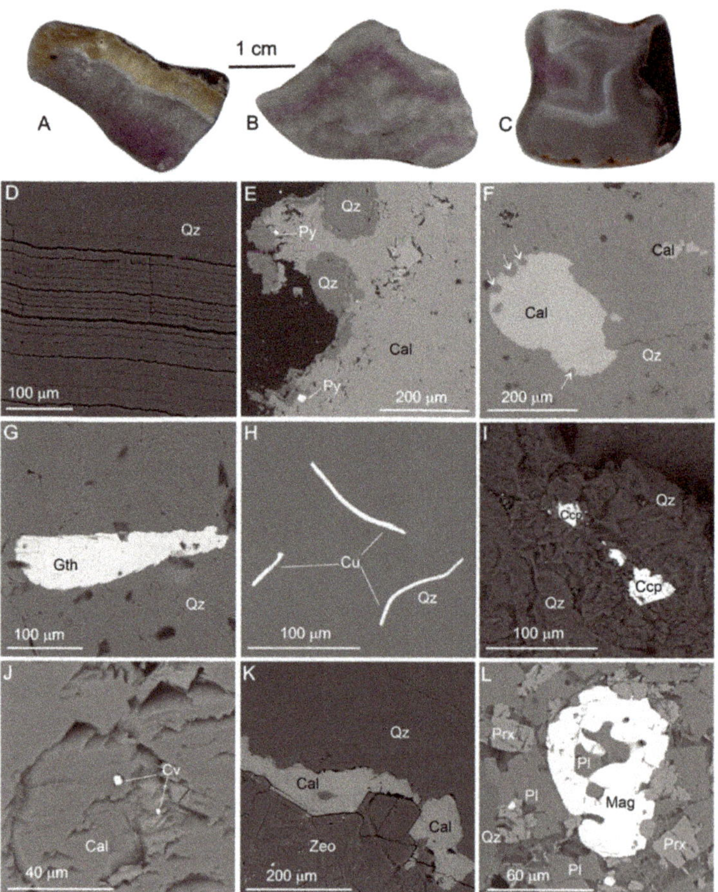

Figure 5. Photographs of the SEM-EDS-examined polished agate sections (**A–C**) and BSE images (**D–L**), illustrating mineral assemblages of agates with amethyst from the Tevinskoye deposit. (**D**) variations in porosity between chalcedony and macrocrystalline quartz (Qz) in outer agate zone; (**E**) Ni-rich pyrite (Py) inclusions accompanied by calcite (Cal) layer at the point of contact with host rock; (**F**) Mn-rich calcite aggregate, which is partially corroded by silica (arrows); (**G**) paniculate-shaped growth of needle-like goethite (Gth) crystals; (**H**) filamentous native copper (Cu) inclusion in quartz matrix; (**I**) grains of chalcopyrite (Ccp) surrounded by quartz; (**J**) grains of covellite (Cv) in calcite matrix; (**K**) prismatic crystals zeolite/clinoptilolite (Zeo) associated with calcite and quartz at the point of contact with host rock; and (**L**) main rock-forming minerals of the agate host volcanics: pyroxene (Prx), plagioclase (Pl), magnetite (Mag), and quartz. Images (**D,I,K**) correspond to agate sample identified by the X-ray diffraction (Section 4.3); images (**E,F**) correspond to sample from photo (**A**); image (**G**) corresponds to sample from photo (**B**); and images (**H,J,L**) correspond to sample from photo (**C**).

Additionally, SEM investigation of the surface morphology of the agate chips revealed the presence of different internal microstructures. Textures caused by well-shaped quartz crystals with sizes up to 150 μm (Figure 6A), as well as textures with an unclear granular surface (Figure 6B), are the most typical of the examined agate. The absence of a petrographically revealing fibrous texture of chalcedony is unexpected.

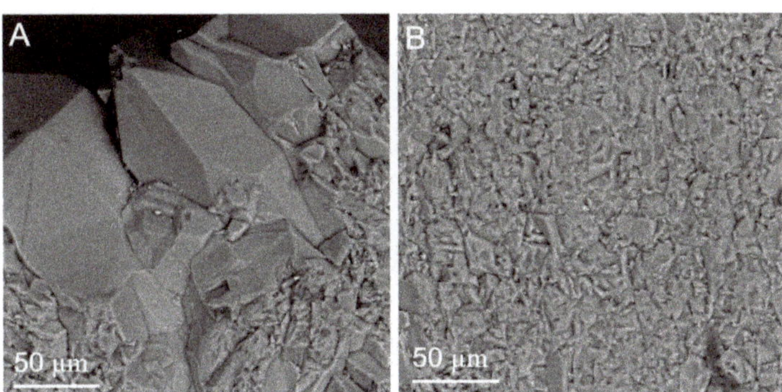

Figure 6. BSE images illustrating surface morphology of agate chip. (**A**) Transition zone of the banded chalcedony to well-shaped quartz crystals; (**B**) texture caused by granular surface of the outermost agate zone.

4.3. X-ray Diffraction

X-ray powder diffraction was applied to determine the structural parameters of banded chalcedony and amethyst crystal zones of agate sample (Figure 7A). Only peaks corresponding to alpha-quartz have been identified in the X-ray diffraction patterns in both cases (Figure 7B,C). Their differences are a slight broadening and a change in the intensity of some reflections (Figure 8).

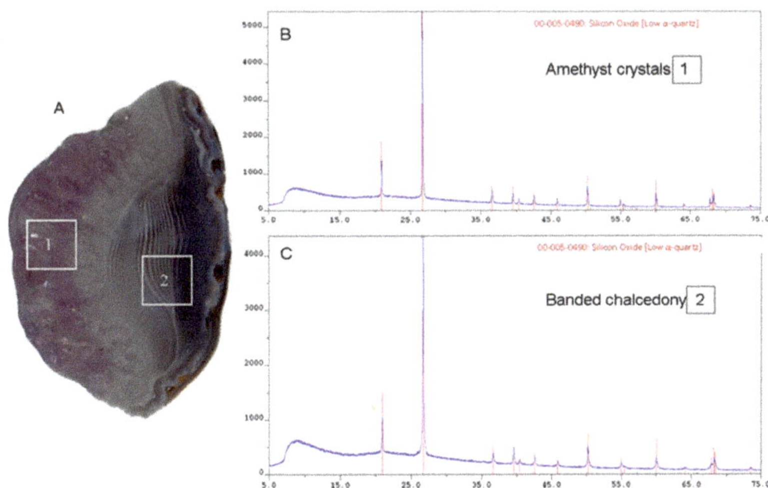

Figure 7. Photograph of agate with amethyst (**A**) and diffractograms corresponding to the amethyst crystals (**B**) and banded chalcedony (**C**) areas in the range of 5–75 °2θ. The samples areas selected for XRD analysis are marked squares on (**A**). The red lines on (**B,C**) point to peaks corresponding to alpha-quartz.

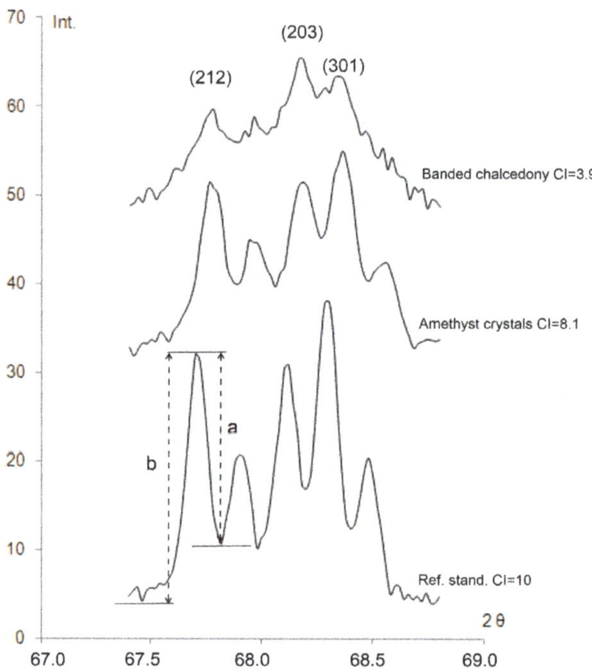

Figure 8. Selected sections of X-ray diffractograms with the 2θ 68° quintuplet peak of the banded chalcedony and amethyst crystals from agate sample (Figure 7A) with added reference-standard rock crystal (Ref. std., Subpolar Urals). *a* and *b* parameters are measured for crystallinity index (CI) calculation.

The relative intensities of reflection on the X-ray diffraction pattern are used for the calculation of a quartz crystallinity index value (CI). The CI of examined samples is defined, following the specifications of Murata and Norman [32], based on the peak resolution in the (212) reflection at 2θ 68° as CI = 10Fa/b, where the scaling factor F is 1.2 (Figure 8). A quartz crystal from the hydrothermal metamorphogenic veins of the Subpolar Urals (Russia) [33] served as a reference standard, with CI = 10. The diffractogram of amethyst crystals is characterized by well-developed 2θ reflections (212), (203), (301) and higher CI values (8.1) as compared to banded chalcedony, with a CI of 3.9 (Figure 8, Table 1).

Table 1. Unit cell parameters, crystallite size (Cs) and crystallinity index (CI) of crystalline quartz and banded chalcedony areas.

Sample	a ± Δa, Å	c ± Δc, Å	V, Å³	Cs, Å	CI
Amethyst crystals area	4.9138 ± 0.0001	5.4054 ± 0.0002	113.03	525	8.1
Banded chalcedony area	4.9140 ± 0.0001	5.4059 ± 0.0001	113.05	560	3.9
Quartz crystal, ref. std.	4.9133 ± 0.0001	5.4052 ± 0.0001	113.00	935	10

The mean crystallite size (Cs) of the samples was determined using the Scherrer equation: Cs = Kλ/(βcosθ). The shape factor K was taken as 0.9, λ was the wavelength of Cu-Kα1 radiation (1.540562 Å), and β was the full width at the half maximum (FWHM) of the peak (101) at 2θ ≈ 26.6°. Cs (101) was taken to be representative of the average crystallite diameter. The mean crystallite sizes in examined amethyst and banded chalcedony samples were 525 and 560 Å, respectively (Table 1).

4.4. Raman Spectroscopy

Raman micro-spectroscopy was used to identify silica polymorphs across the agate sample. The Raman spectra for macrocrystalline quartz from the thin interlayer within the chalcedony banded area (Figure 9A,C, point 1) show characteristic alpha-quartz bands at 465, 353, 208, and 127 cm^{-1}. It is noteworthy that the Raman spectra of fibrous chalcedony area display the occurrence of an additional, less intense band at 502 cm^{-1}, which is attributed to the presence of the unstable monoclinic SiO$_2$ modification moganite (Figure 9A,C, point 2). The 502 cm^{-1} band is also observed in the Raman spectra of microcrystalline quartz zones at the point of contact with the host rock (Figure 9B,C, point 3). Using the algorithm proposed by Götze et al. [26], based on the measurement of the intensity ratio of moganite and α-quartz bands I$_{(502)}$/I$_{(465)}$, the local moganite content in moganite-rich area was calculated. The maximum ratios I$_{(502)}$/I$_{(465)}$ were found in fibrous chalcedony layers, varying between 3.9%–19% (Table S1). According to the calibration curve [26], these values correspond to moganite content of 20%–50%. For microcrystalline quartz area, the intensity ratio I$_{(502)}$/I$_{(465)}$ ranges from 3.4 to 6.6%, corresponding to a moganite content of 15%–22% (Table S1). According to our data, moganite is absent in the areas composed of macrocrystalline quartz, including prismatic colorless and amethyst crystals. No moganite has been found within the quartz with feathery textures (Figure 4E,F).

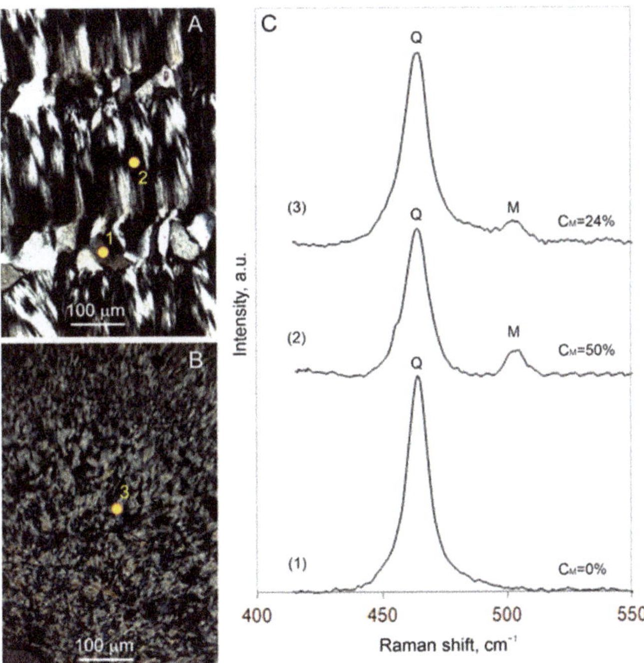

Figure 9. Micrographs (transmitted light mode with cross polars) and Raman spectra of local analyses of banded area composed of fibrous chalcedony and micrograined quartz (**A**), and microcrystalline quartz area (**B**) at the boundary with host rock. Raman spectra (1–3) on (**C**) correspond to analytical spots, which are marked by yellow circles on the photos. Spectrum (1) shows the main characteristic, i.e., symmetric stretching-bending vibrations of α-quartz (Q) at 465 cm^{-1}, whereas spectra (2,3) show the peaks of α-quartz and moganite (Q) at 502 cm^{-1}. CM is moganite content.

Local Raman spectroscopy analyses confirmed the presence of clinoptilolite phases in Tevi agates, which have been identified using their main Raman broad bands located at 412 and 478 cm^{-1} [34] (Figure 10C). Randomly distributed growths of needle-like crystals in agates (Figure 10A) were identified as goethite by their characteristic Raman bands at 166, 205, 245, 300, 387, 418, 479, 551 and 683 cm^{-1} (Figure 10C). Mineral inclusions occurring as clusters within the silica matrix were evidenced as pyrite by the presence of its marker bands at 344, 379 and 431 cm^{-1} [35] (Figure 10B,C). The small band at 325 cm^{-1} can arise from the marcasite admixture [35].

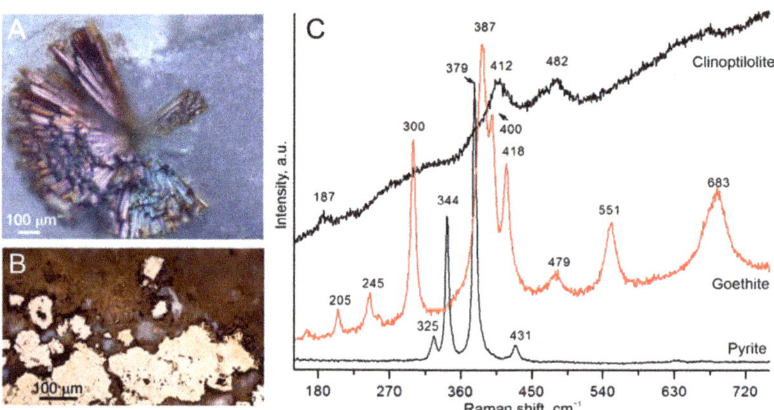

Figure 10. Micrographs (reflected light mode) and Raman spectra of mineral inclusions in Tevi agates with amethyst: growth of needle-like goethite crystals (**A**,**C**); pyrite (**B**,**C**); clinoptilolite (**C**).

4.5. IR Spectroscopy

FT-IR spectroscopic investigation was applied to analyze water speciation of the two areas (banded and amethyst) of the Tevi agate sample, as shown in Figure 3G. The IR spectra on the principal water-stretching region (2800–3800 cm^{-1}) in both cases are represented by an asymmetric broad absorption band near 3400 cm^{-1} and a shoulder near 3260 cm^{-1} (Figure 11). These bands are attributed to symmetric and asymmetric O–H stretching vibrations in water molecules. Molecular water in quartz and chalcedony can be contained in micropores and fluid inclusions [36,37]. The intensity of the 3400 cm^{-1} band is proportional to the content of molecular water. The IR spectrum of the amethyst area shows a much higher intensity of the water band in comparison with the banded area. Additionally, both spectra show a relatively sharp band at 3585 cm^{-1} on the background of a broad band. The assignment of this band is considered in the Discussion section.

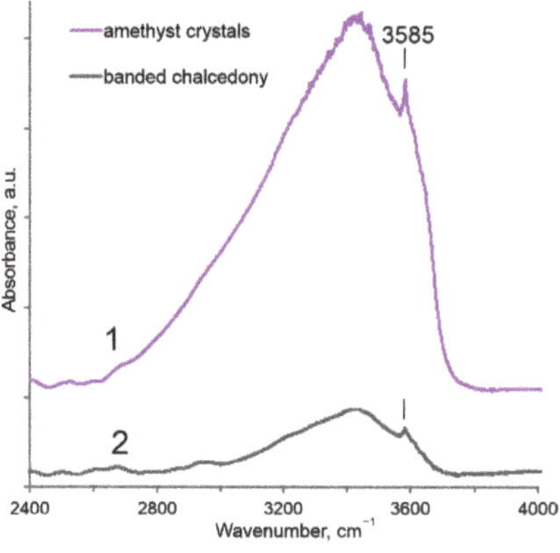

Figure 11. IR spectra corresponding to amethyst crystals (1) and banded chalcedony (2) areas of the agate sample shown in Figure 3G.

4.6. Fluid Inclusions

Fluid inclusions in doubly polished thin sections of amethyst crystals, taken from selected Tevi agate samples (Figure 3A,B,G), were analyzed. Predominantly cold-water monophase fluid inclusions, 5–20 μm in size, were revealed in all samples (Figure 12). They are isometric or elongated in shape, with crystallographic elements. It is assumed that such monophase inclusions are captured at temperatures below 100 °C from hydrothermal fluids of low salinity (not more than 1–3 wt.% NaCl-equivalent) [6,7,38]. Similar fluid monophase inclusions were found in not only colorless quartz and amethyst, but also in calcite, barite and gypsum in Brazilian agates with amethyst [38].

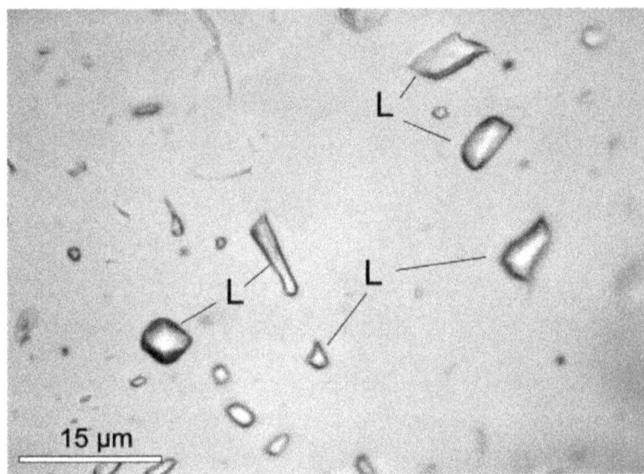

Figure 12. Monophase fluid inclusions in the amethyst crystal from Tevi agate (Figure 3G). L—liquid phase.

4.7. Oxygen Isotope Composition

To specify the equilibrium formation temperature of Tevi agates, oxygen isotope analyses were performed on the profiles of three concentrically banded agate samples with different band sequences (Figure 13). In sample#6, the silica sequence is represented by fibrous chalcedony, amethyst, and quartz. In sample#2, it is composed of fibrous chalcedony, milky-white quartz, amethyst, and colorless quartz. Conversely, in sample#12, it consists of banded chalcedony as well as colorless quartz and is completed with amethyst. In general, the measured values $\delta^{18}O$ in silica bands range from +18.1 to +25.5 ‰ (Table 2).

Table 2. Oxygen isotope composition and calculated temperatures of isotope equilibration (agate formation) of different silica phases in agate samples.

Analysis Points	Bands Characterization	$\delta^{18}O_{SMOW}$ (‰)	T_{met} (°C)	T_{oc} (°C)	T_{mag} (°C)
#6_1	fibrous chalcedony (rim)	25.5	21	69	130
#6_2	amethyst	21.3	39	97	179
#6_3	quartz (core)	18.1	54	124	229
#2_1	fibrous chalcedony (rim)	21.7	36	94	172
#2_2	quartz	21.4	38	96	177
#2_3	amethyst	21.5	37	95	175
#2_4	quartz (core)	22.3	34	90	165
#12_1	chalcedony (rim)	21.1	39	99	181
#12_2	banded chalcedony	19.6	46	110	202
#12_3	amethyst	18.9	49	116	214
#12_4	quartz (core)	18.2	53	122	226

Note. The temperatures are calculated according to [39] for fractionation with meteoric water (−10%), oceanic water (0%) and magmatic water (+8%).

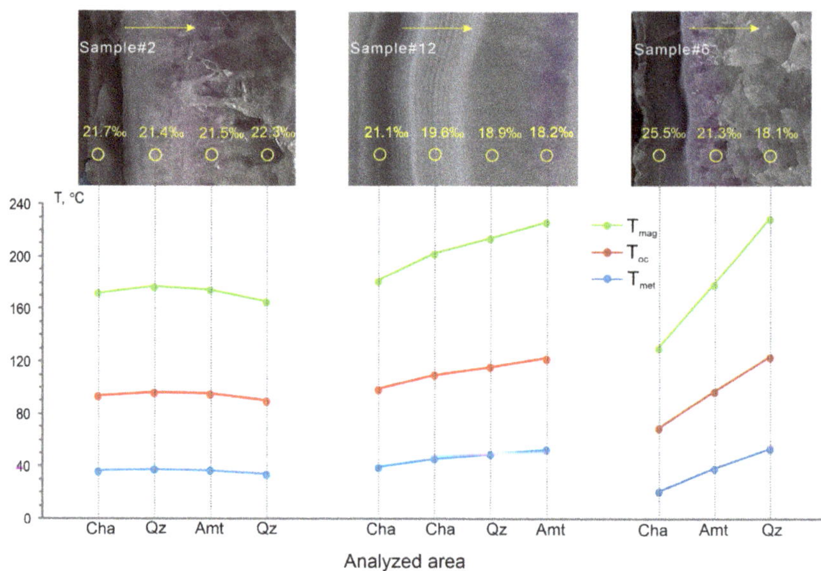

Figure 13. Equilibrium crystallization temperature of individual bands of agate samples with various bands sequences. The yellow arrows point to the growth direction of the agates, open cycles are analyzed areas. Cha is chalcedony, Amt is amethyst, and Qz is quartz. Calculated temperatures for fractionation with meteoric water (T_{met}), oceanic water (T_{oc}) and magmatic water (T_{mag}).

The results of oxygen isotope analyses in the profiles of sample#6 and #12 reveal heterogeneities in the oxygen isotope composition for individual mineral zones (Table 2, Figure 13). A gradual decrease in the heavy isotope O^{18} in the direction of agate growth from the outer chalcedony band to the center is observed. It is noteworthy that sample#2, with a different sequence of silica phases, is characterized by an almost-homogeneous isotopic composition along the profile, with a slight increase in the presence of O^{18} isotope in the central part. This portion of the sample is composed of colorless quartz (Table 2, Figure 13).

To reconstruct the conditions of agate formation and to explain the isotopic compositions of the Tevi agates, it is necessary to know the formation temperature, and/or the O-isotope composition of the mineral-forming fluid [39]. However, no precise data about the formation temperatures and O-isotope composition of the fluids have been obtained from the fluid inclusion study of Tevi agates. Therefore, the formation temperature of each mineral zone of agates was estimated using three different sources: meteoric water ($-10‰$), oceanic water ($0‰$), and magmatic water ($+8‰$). The equation of Matsuhisa et al. [39] was used to calculate the isotope fractionation temperature: $\delta^{18}O = 1000 \ln \alpha_{QW} = 3.34 (10^6 \text{ T}^{-2}) - 3.31$, where α_{QW} is the O-isotope fractionation factor between quartz and water, T is the temperature in Kelvin. The calculations showed the formation temperatures of agates from 21 to 229 °C (chalcedony 21–202 °C, quartz 34–229 °C, amethyst 37–214 °C) (Table 2).

5. Discussion

Agates with amethyst have been recognized in the Eocene volcanic rocks of the Kinkilsk complex of Northern Kamchatka (Russia). They are represented by basalt–andesite–dacite lavas, tuffs, breccias and agglomerates. These agates usually occur as infills of gas vesicles in massive lava flows and cavities in agglomerate tuffs. Additionally, they are also often observed as rounded fragments in coastal placers of the bay.

5.1. Micro-Texture of Agate and Amethyst

The Tevi agates exhibit diversified internal microtextures due to the individual bands composed of micro- and macro-crystalline quartz, length-fast chalcedony, and zebraic chalcedony. The feathery textures revealed in comb-shaped quartz crystals within the agates are typical of many hydrothermal quartz veins. The origin of these textures could be explained by (1) the epitaxial overgrowth of small quartz crystals on large existing quartz crystals [31] and (2) re-crystallization from water-rich microcrystalline silica polymorphs (moganite and/or chalcedony) [40–42]. In the present study, X-ray powder diffraction analysis only revealed alpha quartz in both amethyst crystals and banded chalcedony areas. However, local Raman spectroscopy analyses showed an admixture of moganite within the outer agate zones. The distribution of moganite within individual silica bands is not homogenous. Elevated moganite content was identified within the length-fast chalcedony (up to 50%) and microcrystalline quartz zones (up to 22%), whereas moganite was not found in macrocrystalline quartz areas. The variations in moganite content in individual bands are likely accompanied by different porosity, which can then be revealed under a microscope (Figure 5D). The absence of moganite reflections on the X-ray diffractograms of banded chalcedony area may be explained by the presence of moganite nanocrystals or moganite nano-range lamellae that are not large enough (in the sense of coherently scattering lattice domains) to be detected via X-ray diffractometry, but which are detected using vibrational spectroscopy [26]. Thus, the variation in textural characteristics, including zoned distribution of fibrous chalcedony, moganite, micro- and macro-crystalline quartz, provides evidence for multi-stage deposition of silica due to a fluctuation in physico-chemical conditions during agate formation.

The relationship between agate-bearing rock age and corresponding moganite concentration was described by Moxon and co-authors [43–45]. The moganite content in the agates sharply decreases with the increasing age of host rocks from 13 to 60 Ma, and then remains approximately constant up to 410 Ma. In the ancient agates, moganite occurs in trace amounts or is not detected. The high moganite concentration in the Tevi agates is evidently determined by the young age (~45 Ma) of host rocks.

A correlation between the main crystallite size in agates and age of the host rocks was provided by T. Moxon [46]. The measured values of crystallite size (525–560 Å) for examined agates with amethyst from Eocene volcanics are significantly lower than those previously obtained for agates from Paleoproterozoic (2100–1920 Ma) [16] and Mesoproterozoic (1485–1460 Ma) [47] volcanics of the Fennoscandian Shield (700–1050 Å and 638–751 Å, respectively), just lower values (960–1120 Å) were obtained for agates of Proterozoic and Archean metamorphosed hosts from western Australia [48] (Figure 14). The crystallite size for the Tevi agates is consistent with the values (536–567 Å) obtained for agates in hosts aged from 1100 to 412 Ma [43]. It should be noted that the changes in moganite content and quartz crystallite size with respect to host age is due to the release of structural silanol water [20].

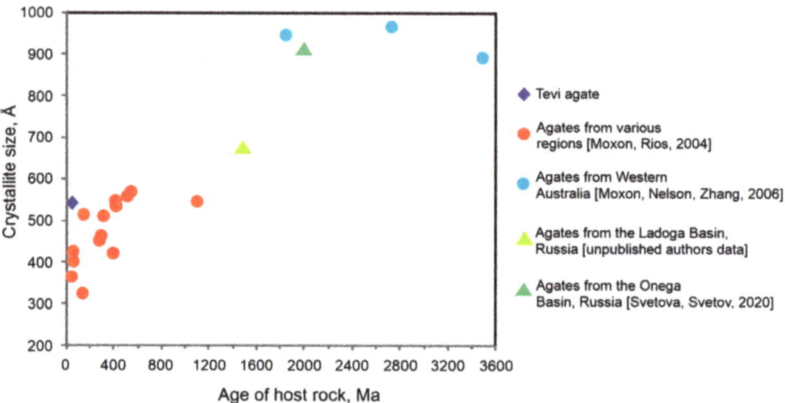

Figure 14. Crystallite growth in agate as a function of age. Data (circles) for agate hosts 3480–38 Ma from different regions of the world, with added data (triangles) for Paleoproterizoic (Onega Basin) and Mesoproterozoic (Ladoga Basin) agate hosts of Fennoscandian Shield (Russia) and for examined Tevi agate (rhomb). Markers are the mean crystallite size for each region [16,43,48].

5.2. Water in Quartz and Chalcedony

According to the IR spectroscopy, the signal of molecular water in the amethyst area of agate is significantly higher than that in the banded one, which is probably due to the abundance of fluid inclusions within the amethyst crystals. Besides molecular water, quartz and chalcedony usually contain some other OH defects. For example, Al-OH and Li-OH complexes arise from the substitution of $Si^{4+} \rightarrow Al^{3+}$ and $Si^{4+} \rightarrow Li^{3+}$ in SiO_4 tetrahedra, where H^+ acted as a compensator of the missing positive charge [36]. Therefore Al-related (3431, 3379 cm^{-1}) and Li-related (3480 cm^{-1}) OH^- bands may be overlapped by the broad band of molecular water in the IR spectrum of amethyst and contribute to total signal intensity. It should be noted that IR spectra were recorded using a square aperture, limiting the studied sample surface to an area of 1 cm^2 so that the IR spectra display the summary water characteristics of the analyzed areas. The distribution of molecular water in the banded area of agates is not homogenous. The spatial water distribution across the growing sequence (presented by the alternation of moganite-rich and moganite-poor chalcedony layers, and macrocrystalline quartz) of the agate sample from the Allumiere-Tolfa volcanic district (Latium, Italy) was studied by Conte and his colleagues using coupled Raman and FT-IR micro-spectroscopy [49]. In particular, these authors revealed an anticorrelation between the moganite content and water signal. Moreover, the IR spectrum, recorded from the macrocrystalline quartz layer, showed a lower intensity of water signal in comparison with chalcedonic layers, contrasting with our data. The 3585 cm^{-1} infrared absorption band is occasionally observed in natural and synthetic rock crystals, including amethysts [50]. It was also recognized in the IR spectra of agates from young volcanic hosts, but has not been found in the IR spectra of agates in volcanics aged from 1100 to 89 Ma [44,48,49]. Up to now, the assignment of this band has been quite vague. One of the proposed explanations for this band is a «hydrogarnet [4H]Si defect ($Si^{4+} \rightarrow 4H^+$)» [51,52]. The presence of the band at 3585 cm^{-1} (Figure 11) is also related to the stretching vibrations of OH groups in the Si-OH silanol groups, which appear when the SiO_2 bond is broken [53]. Stünitz and co-authors [54] experimentally proved that the appearance of the 3585 cm^{-1} band in the IR spectrum of a deformed natural milky quartz crystal is a consequence of the redistribution of fluid inclusions. These authors showed that primary fluid inclusions up to 100 μm in size, occurring after experimental deformation (recrystallization) of quartz crystals at 900 and 1000 °C, were transformed into structurally bound OH defects inside dislocations. These showed a clear band at 3585 cm^{-1} in the IR spectra. The 3585 cm^{-1} band in the IR spectra of examined agate with amethyst can be associated with the recrystallization of quartz observed in thin sections as feathery microtextures within some quartz crystals (Figure 4E,F). Moreover, the data provided by Conte with colleagues [49] point to a coevolution of the molecular water and Si-OH silanol components in chalcedony and quartz of studied agates.

5.3. Mineral Assemblage of Amethyst-Bearing Agate Occurrences

Most agates consist of more than one silica phase [19]. The mineral composition of Tevi agates with amethyst is dominated by micro- and macro-crystalline alpha quartz, length-fast and zebraic chalcedony, and moganite. Besides the different silica phases, other minerals are formed with agates. Carbonates and zeolites are typical minerals in volcanic agates. Mn-rich calcite and clinoptilolite are present in the outer zone of some agates at the point of contact of chalcedony with the host rock (Figure 5A,E,J,K). Iron oxide (magnetite) is observed in the host rock and hydroxide (goethite) is present in the amethyst zone (Figures 5G,L and 10A). Native copper, covellite, chalcopyrite and Ni-enriched pyrite form rare micro-inclusions that are found in the agates and reflect the host rock composition. Copper minerals and, especially, native copper are rare mineral phases in agates [55]. They have been identified in some deposits (the Ayaguz agate deposit (East Kazakhstan); the Wolverine Mine, Wolverine, Houghton County, Michigan (USA); Rudno near Krzeszowice in Lower Silesia (Poland) and others). The simultaneous presence of native copper, Cu sulfide (covellite, CuS), and Cu–Fe sulfide (chalcopyrite, $CuFeS_2$) was revealed in Tevi agates. A similar association native copper with Cu- and Cu–Fe-sulfides and Cu-oxide (cuprite, CuO) was reported for amethyst-free agates from the Avacha Bay Eastern Kamchatka (Russia) [55–57]. Native copper together with Cu- and Cu–Fe-sulphides associated with hematite and pyrite have been observed in the amethyst obtained from the Thunder Bay mine (Canada) [58]. Copper minerals are absent in agate geodes derived from the Amethysta do Sul (Brazil) and Ijevan (Armenia) deposits, which are similar in both morphology and texture to examined agates and to agate geodes from Artigas and Los Catalanes (Uruguay) and like amethysts from the Boudi deposit (Morocco) (Table 3) [7,10,11,38,59,60]. The copper mineralization is probably a typomorphic feature of copper-rich agates that can be related to the composition of the volcanic host rocks. A distinctive characteristic of Amethysta do Sul Brazilian agate geodes compared to other amethyst-containing agates is the abundance of minerals of the sulfate class (barite, anhydrite, gypsum) [38,61]. A common feature of many amethyst-bearing agates is the abundance of Fe-oxides and/or Fe-hydroxides, indicating the significant role of iron in the amethyst-forming fluid (Table 3).

Table 3. Summary of mineral assemblages of some amethyst and amethyst-bearing agate occurrences.

Amethyst Occurrences	Host Rocks	Age	Mineral Assemblage	References
Tevinskoye (Russia)	Basalts, andesites, dacites	Eocene	Chalcedony, quartz, amethyst, native copper, covellite, chalcopyrite, Ni-pyrite, goethite, calcite, and clinoptilolite.	our data
Amethysta do Sul (Brazil)	Basalts, andesites	Early Cretaceous	Chalcedony, quartz, amethyst, celadonite, pyrite, goethite, anhydrite, calcite, gypsum, barite, and fluorite.	[38,61]
Arttigas (Uruguay)	- " -	- " -	Chalcedony, quartz, amethyst, celadonite, goethite, and calcite.	[7]
Los Catalanes (Uruguay)	- " -	- " -	Chalcedony, quartz, amethyst celadonite, calcite, fluorite, pyrite–in geodes; zeolites–in amygdales.	[10]
Thander Bay and Blue Point (Canada, Ontario)	Silicified mudstone breccia	Proterozoic	Quartz, amethyst, barite, calcite, rutile, native copper, chalcopyrite, galena, marcasite, pyrite, sphalerite, hematite, goethite, kaolinite, and smectite.	[58,62]
Boudi (Morocco)	Siltstone, sandstone	Lower Cambrian	Quartz, amethyst, moganite, hematite, calcite-covered amethyst crystals.	[11,60]
Ijevan (Armenia)	Tuffs, tuffsandstones	Late Cretaceous	Chalcedony, quartz, amethyst, calcite, goethite, heulandite. Mordenite and chlorite are rare.	[59]

5.4. T,P,X-Conditions of Amethyst Formation in Agate Geodes in Altered Volcanic Rocks

There are several studies that have investigated fluid inclusions in amethysts from agate geodes in altered volcanic rocks [6,7,10,38]. For example, Morteani et al. [7] determined that the amethyst-bearing geodes in the flood basalts of the Arapey formation at Artigas (Uruguay) were formed as protogeodes by bubbles of CO_2-rich basalt. These bubbles were derived from the artesian water of the Guaraní aquifer and were present in the footwall of these basalts. The temperature of amethyst formation is estimated from fluid inclusion data to be between 50 and 120 °C. The crystallization of late calcite typically occurs on top of the amethyst crystals at a much lower temperature and from cold surficial water of about 20 °C. Commin-Fischer et al. [6] reported that the amethyst-bearing geodes from Serra Geral Formation basalts (Brazil) were formed at a temperature lower than 100 °C, probably lower than 50 °C, and with fluid salinity as high as 3 wt.% NaCl-eq. Gilg et al. [38] determined that amethyst and colorless quartz in amethyst-bearing geodes in Ametista do Sul (Brazil) were formed at temperatures between 40 and 80 °C, while different calcite generations and late gypsum precipitated at temperatures below 45 °C. They revealed that the variation in homogenization temperatures and in $\delta^{18}O$ values of amethyst showed evidence of repeated pulses of ascending hydrothermal fluids of up to 80–90 °C. According to the results of these studies amethyst of the Brazilian and Uruguayan deposits crystallized at temperatures from 50 to 130 °C, the salinity of mineral-forming fluids varied from 0.3 to 5.6 wt.% NaCl eq. The main salt component of the fluid was NaCl. The pressure upon filling the geode with silica was estimated to range from 0.21 to 0.29 kbar (Table 4).

Table 4. T,P,X-conditions of amethyst formation in agate geodes occurring within the altered volcanic rocks and in hydrothermal veins.

Deposits	Amethyst Occurs	T °C	Salinity, wt.% NaCl eq.	P (кbar)	References
Artigas (Uruguay)	geodes	50–120	- " -	- " -	[7]
Serra Geral (Brazil)	geodes	<100	0.9–5.6	- " -	[6]
Ametista do Sul (Brazil)	geodes	95–130	5.3–0.3	0.21–0.29	[38]
Artigas, Uruguay	geodes, breccia	100- " -150	- " -	1.2–5.5	[9]
Boudi (Morocco)	veins	191–445	5.71–13.94	0.64–1.31	[63]
Silver Hill (Greece)	- " -	189–205	0.9–2.1	- " -	[17]
Kassiteres (Greece)	- " -	211–275	0.5–3.4	- " -	- " -
Megala Therma (Greece)	- " -	219–246	3.1–4.8	- " -	- " -
Chondro Vouno (Greece)	- " -	204–221	5.3–8.0	- " -	- " -
Kalogries (Greece)	- " -	139–209	3.4–5.6	- " -	- " -
Eonyang (South Korea)	miarolitic cavities in the aplite	156–333	34–4	1–1.5	[64]
Cape Korabl (Russia)	stockworks, veins, breccias	260–20	- " -	- " -	[65]

Note. - " - no data.

The estimated crystallization temperatures of amethyst in hydrothermal veins are much higher and range from 139 to 420 °C, while the salinity of mineral-forming fluids is higher, up to 34 wt.% NaCl eq. [9,17,63–65]. The fluids, in addition to NaCl, often contain KCl, Fe, Mn, SO_4^{2-}, CO_2, CH_4, and N_2. The pressure of mineral formation varies from 0.2 to 5.5 kbar (Table 4).

Amethyst was synthesized in hydrothermal environments that differ from natural conditions in terms of their chemical composition. Aqueous solutions of ammonium fluoride (salinity 15 wt.%) were mainly used, as well as solutions of K_2CO_3 with the addition of Fe, Mn, Li, and Co impurities to improve the color of the amethyst. The appearance of amethyst coloration of synthetic crystals occurred when exposed to ionizing radiation. The T and P ranges of synthesis were limited by temperatures from 150 to 500 °C at pressures from 0.13 to 1.2 kbar [66–68] (Table 4).

The variations in the O-isotope composition of individual bands (chalcedony, amethyst, quartz) of concentrically banded Tevi agate (Table 2, Figure 13) can be associated with the long geological period of their formation, during which the temperature and isotopic composition of the environment can change. It should also be taken into account that the isotopic composition of the volcanic hosts significantly differs from that of agates. The case when quartz has lower $\delta^{18}O$ values compared to chalcedony (samples #6 and #12 in Figure 13) suggests that it either formed at the same temperature as chalcedony in the presence of a fluid with a lower $^{18}O/^{16}O$ ratio compared to chalcedony, or crystallized at higher temperatures but in the presence of the same fluid and, therefore, at a different time. Similar O-isotope ratios between chalcedony and associated quartz have been observed in some studies [69–71]. Götze and his colleagues [70] suggested that variation in the O-isotope composition within single agate samples can be explained either by the kinetic effects during isotope fractionation (e.g., the agate formation from a solution via an amorphous interstage) or by mixing process of meteoric and magmatic fluids.

5.5. Genesis of Amethyst in Agates

The observed mineral assemblage of examined agates suggests a preferred formation under low-temperature conditions. The appearance of zeolites and calcite can be a result of alteration processes of the volcanic host rocks and emphasizes that, besides enormous amounts of SiO_2, Al, Fe, Ca, Na, and K are also released during these processes [19]. The presence of clinoptilolite in association with quartz (or chalcedony) characterizes a low-temperature (~90–130° C) stage of zeolite facies metamorphism of agate hosts [72] (p. 19). The hydrothermal fluids that formed quartz druses were apparently enriched in iron. At the beginning of the crystallization of colorless crystals, iron was pushed into the residual solution. As they grew, the concentration of iron gradually increased and eventually led to the formation of amethyst with inclusions of goethite [73]. Simultaneously with amethyst, in the heads of colorless quartz crystals, needle-like aggregates of goethite crystallized on the surfaces of its faces (Figure 5G). Hematite and (or) goethite/hydrogoethite are permanent inclusions in amethyst of agates (Table 3). Carbonates (calcite, siderite, and ankerite), iron sulfides (pyrite, marcasite), sulfates (barite, anhydride, and gypsum), and fluorite are found in agates with amethyst at some deposits (Table 3). Judging by the mineral associations (Table 3), the composition of liquid inclusions in amethyst crystals (Table 4), as well as the results of temperature estimation based on the obtained data on O-isotope composition (Table 2), hydrothermal solutions were low-temperature, low-concentrated, oxidized, and essentially siliceous with a high iron content. In the anionic part of the solutions, chloride ions, and to a lesser extent bicarbonate and sulfate ions, played a significant role.

Amethyst comes in an attractive violet color with a reddish or bluish tint. This was originally interpreted as being a result of the presence of Fe^{3+}, which turns into Fe^{4+} after irradiation and substitutes for Si^{4+} in a deformed tetrahedral position [74]. However, recent studies questioned this result and, by using modern analytical techniques, like EPR, Mössbauer and synchrotron X-ray absorption spectroscopy, attributed the coloration of amethyst to the presence of Fe^{3+} [75–78]. It is generally accepted that amethyst is formed by irradiation, in which Fe^{3+} loses electrons and forms a new Fe^{4+} color center responsible for its violet color [74]. Amethyst requires oxidizing conditions to incorporate Fe^{3+}, which may be the result of mixing of oxidized meteoric and/or seawater with ascending hydrothermal fluids [79]. Natural radiation can be probably explained by the moderate concentrations of uranium and thorium in the surrounding volcanic wall rocks. The contents of U (1.4–2.2 ppm) and Th (3–5.39 ppm) in the volcanics of Cape Tevi [25] are higher than their Clarke values in basalts—0.6 and 2.2 ppm, respectively [80]. Shallow underwater conditions most likely predominated in studied areas, adding sea and meteoric water to the hydrothermal fluids.

6. Conclusions

The present contribution provides the first detailed mineralogical investigation of amethyst-bearing agates from the Tevinskoye deposit associated with the Middle–Upper Eocene volcanics within the Kinkilsk complex of Northern Kamchatka (Russia). The study aimed to characterize the agate with amethyst and obtain data for the reconstruction of the conditions for their genesis. The field observation showed that agate mineralization mainly occurs in lavas and tuffs as amygdales, geodes, lenses and veins, which are the main source of resupply of coastal agate placers.

The examined agates exhibit diversified internal microtextures due to the individual bands composed of micro- and macro-crystalline quartz, length-fast chalcedony, and zebraic chalcedony. Elevated moganite content was revealed in some length-fast chalcedony bands (up to 50%) and microcrystalline quartz zones (up to 22%), forming the outer agate rim. Conversely, in macrocrystalline quartz areas, including the amethyst crystals, it was not found.

The agates with amethyst from the Tevinskoye deposit are similar in mineralogy to the agates from Cretaceous volcanic rocks of Ijevan (Armenia) and Ametista do Sul deposits (Brazil) and some other occurrences. Their mineral assemblages are presented by chalcedony, quartz, amethyst, goethite, calcite, and zeolites. The presence of minor amounts of native copper, covellite, chalcopyrite, as well as Mn-enriched calcite and Ni-enriched pyrite is a feature of the Tevi agates. The copper and iron mineralization is probably a typomorphic feature related to the composition of volcanic host rocks.

Ametyst in Tevi agates could be formed at a low temperature (<100 °C) from low-salinity fluids under an oxidizing environment. The presence of goethite inclusions in amethyst points to a high iron concentration in the agate-forming fluid.

Supplementary Materials: The following supporting information can be downloaded at: https://www.mdpi.com/article/10.3390/min13081051/s1, Table S1. Fitting parameters for the Raman spectra of moganite-rich layers from banded agate sample in the 400–550 cm^{-1} range.

Author Contributions: Conceptualization, data curation, writing, E.N.S. and G.A.P.; field investigation, G.A.P.; analytical investigation, E.N.S., G.A.P., A.A.B., V.F.P. and T.N.M.; visualization, E.N.S. and G.A.P.; supervision, G.A.P. All authors have read and agreed to the published version of the manuscript.

Funding: This research was funded by state assignment to the Sobolev Institute of Geology and Mineralogy, Siberian Branch of RAS and Institute of Geology, Karelian Research Centre of RAS.

Data Availability Statement: All data are contained within the article and Supplementary Materials.

Acknowledgments: We thank G. Rychagova and I. Rychagov, participants in field investigations at the Tevinskoye agate deposits. Our thanks also go to Inina I.S. (IG KRC RAS, Petrozavodsk, Russia) and Pikulev V.B. (Petrozavodsk State University, Petrozavodsk, Russia) for their assistance with X-ray diffraction and FT-IR spectroscopy data collection. We are grateful to N.S. Karmanov and M.V. Khlestov (The Analytical Center for Multi-elemental and Isotope Research in the IGM SB RAS (Novosibirsk, Russia) for electron microprobe data. We would like to deeply acknowledge three anonymous reviewers for their careful revisions and for their comments, which helped us to improve the manuscript.

Conflicts of Interest: The authors declare no conflict of interest.

References

1. Götze, J. Chemistry, textures and physical properties of quartz—Geological interpretation and technical application. *Mineral. Mag.* **2009**, *73*, 645–671. [CrossRef]
2. Götze, J.; Möckel, R. *Quartz: Deposits, Mineralogy and Analytics*; Springer: Berlin/Heidelberg, Germany, 2012.
3. *Guidelines for the Search and Prospective Assessment of Deposits of Colored Stones (Jewelry, Ornamental, Decorative Facing)*; v.1. Amethyst; Mingeo USSR: Moscow, Russia, 1974; p. 43. (In Russian)
4. Lieber, W. *Amethyst: Geschichte, Eigenschaften, Fundorte*; Christian Weise Verlag: München, Germany, 1994; 188p.
5. Klemme, S.; Berndt, J.; Mavrogonatos, C.; Flemetakis, S.; Baziotis, I.; Voudouris, P.; Xydous, S. On the Color and Genesis of Prase (Green Quartz) and Amethyst from the Island of Serifos, Cyclades, Greece. *Minerals* **2018**, *8*, 487. [CrossRef]

6. Commin-Fischer, A.; Berger, G.; Polvé, M.; Dubois, M.; Sardini, P.; Beaufort, D.; Formoso, M.L.L. Petrography and chemistry of SiO_2 filling phases in the amethyst geodes from the Serra Geral Formation deposit, Rio Grande do Sul, Brazil. *J. S. Am. Earth Sci.* **2010**, *29*, 751–760. [CrossRef]

7. Morteani, G.; Kostitsyn, Y.; Preinfalk, C.; Gilg, H.A. The genesis of the amethyst geodes at Artigas (Uruguay) and the paleohydrology of the Guarani aquifer: Structural, geochemical, oxygen, carbon, strontium isotope and fluid inclusion study. *Int. J. Earth Sci.* **2010**, *99*, 927–947. [CrossRef]

8. Hartmann, L.A.; Duarte, L.C.; Massonne, H.J.; Michelin, C.; Rosenstengel, L.M.; Bergmann, M.; Theye, T.; Pertille, J.; Arena, K.R.; Duarte, S.K.; et al. Sequential opening and filling of cavities forming vesicles, amygdales and giant amethyst geodes in lavas from the southern Paraná volcanic province, Brazil and Uruguay. *Int. Geol. Rev.* **2012**, *54*, 1–14. [CrossRef]

9. Duarte, L.C.; Hartmann, L.A.; Vasconcelos, M.A.Z.; Medeiros, J.T.N.; Theye, T. Epigenetic formation of amethyst-bearing geodes from Los Catalanes gemological district, Artigas, Uruguay, southern Paraná Magmatic Province. *J. Volcanol. Geotherm. Res.* **2009**, *184*, 427–436. [CrossRef]

10. Duarte, L.C.; Hartmann, L.A.; Ronchi, L.H.; Berner, Z.; Theye, T.; Massonne, H.J. Stable isotope and mineralogical investigation of the genesis of amethyst geodes in the Los Catalanes gemological district, Uruguay, southernmost Paraná volcanic province. *Miner. Depos.* **2011**, *46*, 239–255. [CrossRef]

11. Pršek, J.; Dumańska-Słowik, M.; Powolny, T.; Natkaniec-Nowak, L.; Toboła, T.; Zych, D.; Skrepnicka, D. Agates from Western Atlas (Morocco)—Constraints from Mineralogical and Microtextural Characteristics. *Minerals* **2020**, *10*, 198. [CrossRef]

12. Godovikov, A.A.; Ripinen, O.I.; Motorin, S.G. *Agates*; Nedra: Moscow, Russia, 1987; p. 368. (In Russian)

13. Goncharov, V.I.; Gorodinsky, M.E.; Pavlov, G.F.; Savva, N.E.; Fadeev, A.P.; Vartanov, V.V.; Gunchenko, E.V. *Chalcedony of North.-East. of the USSR*; Science: Moscow, Russia, 1987; p. 192. (In Russian)

14. Rozhkova, V.V. Morphology and properties of the North Timan agates. *Proc. Inst. Geol. Komi Phil. Acad. Sci. USSR* **1984**, *46*, 70–77. (In Russian)

15. Sedov, B.M. *Upper Olsky Agates*; Okhotnik, Russia, 2019; 244p.

16. Svetova, E.; Svetov, S. Mineralogy and Geochemistry of Agates from Paleoproterozoic Volcanic Rocks of the Karelian Craton, Southeast Fennoscandia (Russia). *Minerals* **2020**, *10*, 1106. [CrossRef]

17. Voudouris, P.; Melfos, V.; Mavrogonatos, C.; Tarantola, A.; Götze, J.; Alfieris, D.; Maneta, V.; Psimis, I. Amethyst Occurrences in Tertiary Volcanic Rocks of Greece: Mineralogical, Fluid Inclusion and Oxygen Isotope Constraints on Their Genesis. *Minerals* **2018**, *8*, 324. [CrossRef]

18. Kigai, I.N. The genesis of agates and amethyst geodes. *Can. Miner.* **2019**, *57*, 867–883. [CrossRef]

19. Götze, J.; Möckel, R.; Pan, Y. Mineralogy, Geochemistry and Genesis of Agate—A Review. *Minerals* **2020**, *10*, 1037. [CrossRef]

20. Moxon, T.; Palyanova, G. Agate Genesis: A Continuing Enigma. *Minerals* **2020**, *10*, 953. [CrossRef]

21. Palyanova, G. Editorial for Special Issue "Agates: Types, Mineralogy, Deposits, Host Rocks, Ages and Genesis". *Minerals* **2021**, *11*, 1035. [CrossRef]

22. Slyadnev, B.I.; Borovtsov, A.K.; Burmakov, Y.A.; Sidorenko, V.I.; Sapozhnikova, L.P.; Tararin, I.A.; Badredinov, Z.G.; Surikov, S.N.; Sidorov, M.D.; Sidorov, E.G.; et al. State Geological Map of the Russian Federation. Scale 1: 1,000,000 (Third Generation). Series Koryaksko-Kurilskaya. Sheet O-57—Palana. Explanatory Letter. St. Petersburg: VSEGEI Cartographic Factory, 2013. 296p. (In Russian). Available online: https://www.vsegei.ru/ru/info/pub_ggk1000-3/Koryaksko-Kurilskaya/o-57.php (accessed on 10 January 2023).

23. Fedorov, P.I.; Kovalenko, D.V.; Bayanova, T.B.; Serov, P.A. Early Cenozoic Magmatism in the Continental Margin of Kamchatka. *Petrology* **2008**, *16*, 261–278. [CrossRef]

24. Fedorov, P.I.; Kovalenko, D.V.; Ageeva, O.A. Western Kamchatka–Koryak Continental-Margin Volcanogenic Belt: Age, Composition, and Sources. *Geochem. Int.* **2011**, *49*, 768–792. [CrossRef]

25. Fedorov, P.I.; Kovalenko, D.V.; Perepelov, A.B.; Dril, S.I. Composition of sources of the Kinkil complex of western Kamchatka according to isotope-geochemical data. *Bull. KRAUNC Earth Sci.* **2019**, *1*, 54–72. [CrossRef]

26. Götze, J.; Nasdala, L.; Kleeberg, R.; Wenzel, M. Occurrence and distribution of "moganite" in agate/chalcedony: A combined micro-Raman, Rietveld, and cathodoluminescence study. *Contrib. Mineral. Petrol.* **1998**, *133*, 96–105. [CrossRef]

27. Borisenko, A.S. Analysis of the Salt Composition of solutions of gas-liquid inclusions in minerals by cryometry. In *The Use of Methods of Thermobarogeochemistry in the Search and Study of Ore Deposits*; Laverov, N.P., Ed.; Nedra: Moscow, Russia, 1982; pp. 37–47. (In Russian)

28. Roedder, E. Fluid inclusions. *Rev. Mineral.* **1984**, *12*, 79–108.

29. Bodnar, R.J.; Vityk, M.O. Interpretation of microthermometric data for $NaCl–H_2O$ fluid inclusions. In *Fluid Inclusions in Minerals: Methods and Applications*; De Vivo, B., Frezzotti, M.L., Eds.; Virginia Polytechnic Inst State Univ: Blacksburg, VA, USA, 1994; pp. 117–131.

30. Sharp, Z.D. A laser-based microanalytical method for the in-situ determination of oxygen isotope ratios of silicates and oxides. *Geochim. Cosmochim. Acta* **1990**, *54*, 1353–1357. [CrossRef]

31. Dong, G.; Morrison, G.; Jaireth, S. Quartz textures in epithermal veins, Queensland—Classification, origin, and implication. *Econ. Geol.* **1995**, *90*, 1841–1856. [CrossRef]

32. Murata, J.; Norman, M.B. An index of crystallinity for quartz. *Am. J. Sci.* **1976**, *276*, 1120–1130. [CrossRef]

33. Kuznetsov, S.K.; Svetova, E.N.; Shanina, S.N.; Filippov, V.N. Minor Elements in Quartz from Hydrothermal-Metamorphic Veins in the Nether Polar Ural Province. *Geochemistry* **2012**, *50*, 911–925. [CrossRef]
34. Tsai, Y.-L.; Huang, E.; Li, Y.-H.; Hung, H.-T.; Jiang, J.-H.; Liu, T.-C.; Chen, H.-F. Raman Spectroscopic Characteristics of Zeolite Group Minerals. *Minerals* **2021**, *11*, 167. [CrossRef]
35. White, S.N. Laser Raman spectroscopy as a technique for identification of seafloor hydrothermal and cold seep minerals. *Chem. Geol.* **2009**, *259*, 240–252. [CrossRef]
36. Kats, A. Hydrogen in Alpha-quartz. *Philips Res. Rep.* **1962**, *17*, 201–279.
37. Aines, R.D.; Rossman, G.R. Water in minerals? A peak in the infrared. *Geophys. Res.* **1984**, *89*, 4059–4071. [CrossRef]
38. Gilg, H.A.; Krüger, Y.; Taubald, H.; van den Kerkhof, A.M.; Frenz, M.; Morteani, M. Mineralisation of amethyst-bearing geodes in Ametista do Sul (Brazil) from low-temperature sedimentary brines: Evidence from monophase liquid inclusions and stable isotopes. *Miner. Depos.* **2014**, *49*, 861–877. [CrossRef]
39. Matsuhisa, Y.; Goldsmith, J.R.; Clayton, R.N. Oxygen isotopic fractionation in the system quarz-albite–anorthite-water. *Geochim. Cosmochim. Acta* **1979**, *43*, 1131–1140. [CrossRef]
40. Sander, M.V.; Black, J.E. Crystallization and recrystallization of growth-zoned vein quartz crystals from epithermal systems; implications for fluid inclusion studies. *Econ. Geol.* **1988**, *83*, 1052–1060. [CrossRef]
41. Marinova, I.; Ganev, V.; Titorenkova, R. Colloidal origin of colloform-banded textures in the Paleogene low-sulfidation Khan Krum gold deposit, SE Bulgaria. *Miner. Depos.* **2014**, *49*, 49–74. [CrossRef]
42. Yilmaz, T.I.; Duschl, F.; Di Genova, D. Feathery and network-like filamentous textures as indicators for the re-crystallization of quartz from a metastable silica precursor at the Rusey Fault Zone, Cornwall, UK. *Solid Earth* **2016**, *7*, 1509–1519. [CrossRef]
43. Moxon, T.; Rios, S. Moganite and water content as a function of age in agate: An XRD and thermogravimetric study. *Eur. J. Mineral.* **2004**, *4*, 693–706. [CrossRef]
44. Moxon, T.; Reed, S.J.B.; Zhang, M. Metamorphic effects on agate found near the Shap granite, Cumbria: As demonstrated by petrography, X-ray diffraction spectroscopic methods. *Miner. Mag.* **2007**, *71*, 461–476. [CrossRef]
45. Moxon, T.; Carpenter, M.A. Crystallite growth kinetics in nanocrystalline quartz (agate and chalcedony). *Miner. Mag.* **2009**, *73*, 551–568. [CrossRef]
46. Moxon, T. Agates: A study of ageing. *Eur. J. Mineral.* **2002**, *14*, 1109–1118. [CrossRef]
47. Svetova, E.N.; Svetov, S.A. Agates from Mesoproterozoic Volcanics (Pasha–Ladoga Basin, NW Russia): Characteristics and Proposed Origin. *Minerals* **2023**, *13*, 62. [CrossRef]
48. Moxon, T.; Nelson, D.R.; Zhang, M. Agate recrystallization: Evidence from samples found in Archaean and Proterozoic host rocks, Western Australia. *Aust. J. Earth Sci.* **2006**, *53*, 235–248. [CrossRef]
49. Conte, A.; Della Ventura, G.; Rondeau, B.; Romani, M.; Guidi, M.C.; La, C.; Napoleoni, C.; Lucci, F. Hydrothermal genesis and growth of the banded agates from the Allumiere-Tolfa volcanic district (Latium, Italy). *Phys. Chem. Miner.* **2022**, *49*, 39. [CrossRef]
50. Li, J.; Zheng, Y.; Liu, X.; Li, G.; Yu, X.; Wang, Y.; Li, H.; Liu, H.; Shan, G.; Li, T.; et al. Thermal Process of Rock Crystal: Cause of Infrared Absorption Band at 3585 cm^{-1}. *Crystals* **2021**, *11*, 1083. [CrossRef]
51. Aines, R.D.; Kirby, S.H.; Rossman, G.R. Hydrogen speciation in synthetic quartz. *Phys. Chem. Miner.* **1984**, *11*, 204–212. [CrossRef]
52. Stalder, R.; Konzett, J. OH-defects in quartz in the system quartz–albite–water and granite–water between 5 and 25 kbar. *Phys. Chem. Miner.* **2012**, *39*, 817–827. [CrossRef]
53. Fukuda, J. Water in Rocks and Minerals—Species, Distributions, and Temperature Dependences. In *Infrared Spectroscopy—Materials Science, Engineering and Technology*; Tohoku University: Sendai, Japan, 2012. [CrossRef]
54. Stünitz, H.; Thust, A.; Heilbronner, R.; Behrens, H.; Kilian, R.; Tarantola, A.; Fitz Gerald, J.D. Water redistribution in experimentally deformed natural milky quartz single crystals-Implications for H_2O-weakening processes. *J. Geophys. Res. Solid Earth* **2017**, *122*, 866–894. [CrossRef]
55. Palyanova, G.; Sidorov, E.; Borovikov, A.; Seryotkin, Y. Copper-Containing Agates of the Avacha Bay (Eastern Kamchatka, Russia). *Minerals* **2020**, *10*, 1124. [CrossRef]
56. Sidorov, E.G.; Kutyev, F.T.; Anikin, P.P. Native Copper Agates of the Kuril-Kamchatka Province. In *Native Metals in Postmagmatic Formations*; Yakutsk Publishing House: Yakutsk, Russia, 1985; pp. 72–73. (In Russian)
57. Saveliev, D.P. A scattering of agates at Cape Vertikalny, Eastern Kamchatka. *Bull. Kamchatka Reg. Assoc. Educ. Sci. Cent. Ser. Earth Sci.* **2020**, *47*, 3. [CrossRef]
58. McArthur, J.R.; Jennings, E.A.; Kissin, S.A.; Sherlock, R.L. Stable-isotope, fluid-inclusion, and mineralogical studies relating to the genesis of amethyst, Thunder Bay Amethyst Mine, Ontario. *Can. J. Earth Sci.* **2011**, *30*, 1955–1969. [CrossRef]
59. Khakimov, A.K. Some features of the mineralogy and genesis of agate bodies in the Ijevan region of Armenia. *News High. Educ. Inst. Geol. Explor.* **1965**, *7*, 45–56. (In Russian)
60. Troilo, F.; Harfi, A.; El Bittarello, E.; Costa, E. Amethyst from Boudi, Morocco. *Gems Gemmol.* **2015**, *51*, 32–40. [CrossRef]
61. Gilg, H.A.; Morteani, G.; Kostitsyn, Y.; Preinfalk, C.; Gatter, I.; Strieder, A.J. Genesis of amethyst geodes in basaltic rocks of the Serra Geral Formation (Ametista do Sul, Rio Grande do Sul, Brazil): A fluid inclusion, REE, oxygen, carbon, and Sr isotope study on basalt, quartz, and calcite. *Miner. Depos.* **2003**, *38*, 1009–1025. [CrossRef]
62. Kile, D.E. Mineralogy of the Amethyst Mines in the Thunder Bay Area, Thunder Bay, Ontario, Canada. *Rocks Miner.* **2019**, *94*, 306–343. [CrossRef]

63. Dumańska-Słowik, M.; Toboła, T.; Jarmołowicz-Szulc, K.; Naglik, B.; Dyląg, J.; Szczerba, J. Inclusion study of hourglass amethyst from Boudi (Morocco) by Raman microspectroscopy and microthermometric measurements. *Spectrochim. Acta Part A Mol. Biomol. Spectrosc.* **2017**, *187*, 156–162. [CrossRef] [PubMed]
64. Yang, K.H.; Yu, S.H.; Lee, J.D. A fluid inclusion study of an amethyst deposit in the Cretaceous Kyongsang Basin, South Korea. *Mineral. Mag.* **2001**, *65*, 477–487. [CrossRef]
65. Frishman, N.I. Stage and formation conditions of amethyst in "Cape Korabl" deposit. *Proc. Fersman Sci. Sess. GI KSC RAS.* **2015**, *12*, 339–341.
66. Balitsky, V.S.; Machina, I.B.; Mar'in, A.A.; Shigley, J.E.; Rossman, G.R.; Lu, T. Industrial growth, morphology and some properties of bi-colored amethyst-citrine quartz (ametrine). *J. Cryst. Growth* **2000**, *212*, 255–260. [CrossRef]
67. Balitsky, V.S.; Balitskaya, L.V.; Balitsky, D.V.; Semenchenko, S.P.; Humenny, M.V. A Method for Growing Quartz or Amethyst Crystals or Amethyst Druse by the Hydrothermal Method of Temperature Difference in Aqueous Solutions of Ammonium Fluoride. RU 2707771, 29 November 2019. (In Russian).
68. Zebrev, Y.N.; Chernaya, T.N.; Konovalov, N.I.; Novoselov, V.P.; Oleinik, N.A.; Abdrafikov, S.N.; Popolitov, V.I.; Pisarevsky, Y.V. Method for Obtaining Synthetic Amethyst. RU 2040596 C1, 25 July 1995. (In Russian).
69. Harris, C. Oxygen-isotope zonation of agates from Karoo volcanics of the Skeleton Coast, Namibia. *Am. Mineral.* **1989**, *74*, 476–481.
70. Götze, J.; Tichomirowa, M.; Fuchs, H.; Pilot, J.; Sharp, Z.D. Geochemistry of agates: A trace element and stable isotope study. *Chem. Geol.* **2001**, *175*, 523–541. [CrossRef]
71. Shen, M.; Lu, Z.; He, X. Mineralogical and Geochemical Characteristics of Banded Agates from Placer Deposits: Implications for Agate Genesis. *ACS Omega* **2022**, *7*, 23858–23864. [CrossRef]
72. Spiridonov, E.M.; Ladygin, V.M.; Yanakieva, D.Y.; Frolova, J.V.; Semikolennykh, E.S. Agates in Metavolcanics. Bulletin of the Russian Federal Property Fund. 2014, 71p. (In Russian). Available online: https://www.rfbr.ru/rffi/ru/bulletin/o_1923809#8 (accessed on 10 January 2023).
73. Merino, E.; Wang, Y.; Deloule, E. Genesis of agates in flood basalts; twisting of chalcedony fibers and trace-element geochemistry. *Am. J. Sci.* **1995**, *295*, 1156–1176. [CrossRef]
74. Rossman, G.R. Colored varieties of the silica minerals. *Rev. Mineral.* **1994**, *29*, 433–467.
75. Czaja, M.; Kądziołka-Gaweł, M.; Konefał, A.; Sitko, R.; Teper, E.; Mazurak, Z.; Sachanbińskiet, M. The Mossbauer spectra of prasiolite and amethyst crystals from Poland. *Phys. Chem. Miner.* **2017**, *44*, 365–375. [CrossRef]
76. Dedushenko, S.K.; Makhina, I.B.; Mar'in, A.A.; Mukhanov, V.A.; Perfiliev, Y.U.D. What oxidation state of iron determines the amethyst colour? *Hyperfine Interact.* **2004**, *156*, 417–422. [CrossRef]
77. Di Benedetto, F.; Innocenti, M.; Tesi, S.; Romanelli, M.; D'Acapito, F.; Fornaciai, G.; Montegrossi, G.; Pardi, L.A. A Fe K-edge XAS study of amethyst. *Phys. Chem. Miner.* **2010**, *37*, 283–289. [CrossRef]
78. SivaRamaiah, G.; Lin, J.; Pan, Y. Electron paramagnetic resonance spectroscopy of Fe^{3+} ions in amethyst: Thermodynamic potentials and magnetic susceptibility. *Phys. Chem. Miner.* **2011**, *38*, 159–167. [CrossRef]
79. Fournier, R.O. The behaviour of silica in hydrothermal solutions. Geology and geochemistry of epithermal systems. *Rev. Econ. Geol.* **1985**, *2*, 45–61.
80. Taylor, S.R. Abundance of chemical elements in the continental crust: A new table. *Geochim. Cosmochim. Acta* **1964**, *28*, 1273–1285. [CrossRef]

MDPI
St. Alban-Anlage 66
4052 Basel
Switzerland
www.mdpi.com

Minerals Editorial Office
E-mail: minerals@mdpi.com
www.mdpi.com/journal/minerals